2nd International Conference on Efficient Building Design 2016

Materials and HVAC Equipment Technologies

Beirut, Lebanon
22 - 23 September 2016

ISBN: 978-1-5108-3708-9

Printed from e-media with permission by:

Curran Associates, Inc.
57 Morehouse Lane
Red Hook, NY 12571

Some format issues inherent in the e-media version may also appear in this print version.

Printed by Curran Associates, Inc. (2017)

ISBN: 978-1-5108-3708-9
ASHRAE ISBN: 978-1-939200-49-5 (USB Flash Drive)

For permission requests, please contact American Society of Heating, Refrigerating
and Air-Conditioning Engineers (ASHRAE) at the address below.

American Society of Heating, Refrigerating and Air-Conditioning Engineers (ASHRAE)
1791 Tullie Circle
N.E. Atlanta, GA 30329
USA

Phone: (404) 636-8400
Fax: (404) 321-5478

ashrae@ashrae.org

Additional copies of this publication are available from:

Curran Associates, Inc.
57 Morehouse Lane
Red Hook, NY 12571 USA
Phone: 845-758-0400
Fax: 845-758-2633
Email: curran@proceedings.com
Web: www.proceedings.com

TABLE OF CONTENTS

2nd International Conference
Efficient Building Design:
Materials and HVAC Equipment
Technologies

Beirut, Lebanon
September 22-23, 2016
ashrae.org/beirut2016

General Information

The Second International ASHRAE Conference and Showcase on Efficient Building Design organizes seven technical sessions with more than 34 presentations by local and international speakers over two days. The Conference presents the latest research and development to improve building design and state-of-the-art technologies in building material and HVAC Equipment Technologies. The sessions cover the following topics:

1. Alternative Solutions for Energy savings in Buildings I
2. Energy Conservation Strategies
3. Energy Efficiency and Climate
4. Alternative Solutions for Energy savings in Buildings II
5. Indoor Air Quality
6. Alternative Solutions for Energy savings in Buildings III
7. Modeling, Simulations, and Standards

In addition to the above, the conference hosts a number of presentations on new concepts and advances in HVAC design and industry.

Selective best ideas in building efficiency demonstrated by major HVAC industry companies will take part at an exclusive showcase held on September 22 & 23, 2016 at Olayan School of Business and Irani Oxy Engineering Complex , AUB campus, Beirut, Lebanon.

Organizers

- **ASHRAE**
  ASHRAE founded in 1894, is a global society advancing human well-being through sustainable technology for the built environment. The Society and its members focus on building systems, energy efficiency, indoor air quality, refrigeration and sustainability within the industry. Through research, standards writing, publishing and continuing education, ASHRAE shapes tomorrow's built environment today. ASHRAE was formed as the American Society of Heating, Refrigerating and Air-Conditioning Engineers by the merger in 1959 of American Society of Heating and Air-Conditioning Engineers (ASHAE) founded in 1894 and The American Society of Refrigerating Engineers (ASRE) founded in 1904.

- **Munib and Angela Masri Institute of Energy and Natural Resources (MI) at AUB**
  MI provides a vehicle for promoting research and advanced study in the petroleum, water, and energy disciplines, as well as a focal point for collaborative research among scientists, engineers, and professionals in Lebanon and in the region at large. The institute serves as an interfaculty coordinating unit in AUB and a catalyst for advanced research in the sciences and engineering for the management and conservation of natural resources and energy.

- **Department of Mechanical Engineering, AUB**
  The undergraduate program in Mechanical Engineering seeks to empower students to pursue successful careers and to create a learning environment in which they can develop their creative and critical thinking, their ability to grow into lifelong learners in the light of ever-increasing challenges of modern technology, and their commitment to the ethical and professional responsibilities required in their calling at the global level while focusing on the needs of Lebanon and the region.

- **The Lebanese ASHRAE Chapter**
  Lebanese ASHRAE Chapter is representing ASHRAE in Lebanon and serves the local industries in all matters related to heating, refrigerating and air-conditioning as well as associated activities.

Show case Sponsors

Selective best ideas in building efficiency demonstrated by major HVAC industry companies will take part at an exclusive exhibit that will be held on September 22 & 23, 2016 at Irani Oxy Engineering Complex, AUB campus, Beirut, Lebanon. The following companies and organizations are sponsoring the conference and contributing to the exhibition:

- **Gold Sponsors**
  - Camfil Middle East FZCO, www.camfil.com
  - Climatec, http://www.climatec.com/
  - DAIKIN Air Conditioning , www.daikin.com
  - Dar Al-Handasah (Shair & partners), www.dargroup.com
  - Iklimlendirme Sanayi Ihracatcilari Birligi, ISIB, www.ik.isib.gov.tr
  - Johnson Controls, http://www.johnsoncontrols.com
  - Order of Engineers, http://www.oea.org.lb/Arabic/Default.aspx?pageid=7
- **Silver sponsor**
  - Khatib and Alami, www.khatibalami.com/home.htm

Conference Sponsors and Showcase

The Executive Organizing Committee thanks the following sponsors for their support of the 2016 International Conference on Efficient Building Design:

camfil
CLEAN AIR SOLUTIONS

DAIKIN
AIR CONDITIONING

dar al-handasah
shair and partners

İSİB
İKLİMLENDİRME SANAYİİ
İHRACATÇILARI BİRLİĞİ

YORK®

آل سالم جونسون كنترولز
Al Salem Johnson Controls

Johnson Controls

خطيب و علمي

شركة الإتحاد الهندسي

Khatib & Alami

Consolidated Engineering Company

Organizing Committee

- Nesreen Ghaddar, Conference Chair, American University of Beirut
- Bassam Elassaad, ASHRAE sub-region Chair for Europe
- Saleh El Zein, Lebanese ASHRAE Chapter
- Walid Chakroun, Vice President ASHRAE
- Fadl Moukalled, American University of Beirut
- Kamel Ghali, American University of Beirut
- Samir Traboulsi , American University of Beirut, Faculty of Engineering & Architecture, WAAAUB Council & Alumni Chapters Committee member

Scientific Committee

- Nesreen Ghaddar, American University of Beirut, Chair
- Samir Traboulsi, American University of Beirut, CEC Representative
- Fadel Moukalled, American University of Beirut
- Kamel Ghali, American University of Beirut
- William Bahnfleth, the Pennsylvania State University
- Mohamad Hosni, Kansas State University
- Walid Chakroun, Kuwait University
- Michel El-Hayek, Notre Dame University
- Bassem Bakhache, Lebanese University

Student Members

(Pending)

Support Staff

- Sandrine Assaad, American University of Beirut
- Abbas Toufeili, American University of Beirut
- Hind Hajjar, American University of Beirut
- Omar Kheir, American University of Beirut

Keynotes and their Biographies

Keynote Talk 1
Thursday September 22, 2016, 10:15 AM - 11:15 AM

HFC phase-down under the Montreal Protocol – a unique opportunity to increase the energy efficiency of air-conditioning

Mr. David Underwood: ASHRAE Presidential Member
Session Chair: Dr. Kamel Ghali, AUB

Room: Maamari Auditorium, OLAYAN SCHOOL OF BUSINESS

David Underwood, a Professor of Engineering, Fellow ASHRAE, Life Member, CPMP, and ASHRAE's president for 2015-2016, chaired the Society's Board of Directors and Executive Committee. Underwood was awarded a B.S. in Civil Engineering from the University of Manitoba. In 1975, Underwood founded Isotherm Engineering Ltd., an HVAC&R design-build enterprise, which continues to be operational. He also served on the Canadian Standards Association B52 committee on Mechanical Refrigeration, the advisory committee that assisted the Ontario government in developing refrigeration regulations for Ontario. He co-authored a trade refrigeration safety manual and served on a trade advisory committee for the refrigeration trade in Ontario. As a member of a Canadian Federal committee he participated in writing the Building Science and Plumbing section of the 2015 edition of the Model National Building Code. He is an ASHRAE Distinguished Lecturer and a member of Technical Committee 7.2, HVAC&R Construction & Design Build Technologies, and of the Building Performance Alliance Ad Hoc Committee. Mr. Underwood is an ASHRAE Fellow and has received an Exceptional Service Award, a Distinguished Service Award and the William J. Collins Research Promotion Award.

Keynote Talk 2
Thursday September 22, 2016, 14:30 - 15:30

Hybrid and Integrated Power Production for the Transition towards Clean Energy

Dr. Ahmed Ghoniem; PhD Ronald C. Crane (1972) Professor; Director, Center for Energy and Propulsion Research; Director, Reacting Gas Dynamics Laboratory, MIT
Session Chair: Dr. Nesreen Ghaddar, AUB
Session Co-Chair: Dr. Issam Lakkis, AUB

Room: Maamari Auditorium, OLAYAN SCHOOL OF BUSINESS

Ahmed F. Ghoniem is the Ronald C. Crane Professor of Mechanical Engineering, Director of the Center for Energy and Propulsion Research and the Reacting Gas Dynamics Laboratory. He received his B.Sc. and M.Sc. degree from Cairo University, and Ph.D. at the University of California, Berkeley. His research covers computational engineering, clean energy with focus on CO2 capture technologies, renewable energy and alternative fuels. He made fundamental contributions to multiscale simulations, thermochemistry and combustion, energy systems and materials. He has supervised more than 100 M.Sc., Ph.D. and post-doctoral students, published more than 350 refereed articles in leading journals and conferences, lectured extensively around the World, and consulted for the aerospace, automotive and energy industry. He served as head of the thermo-fluid science division and energy science and engineering. He is Fellow of ASME and associate fellow of AIAA. He received several awards including the KAUST Investigator Award, the ASME James Harry Potter Award in Thermodynamics, and the AIAA Propellant and Combustion Award.

Keynote 3
September 23, 2016, 8:30 - 9:30

Microscale Heat and Mass Exchange to Enable Compact Vapor Compression and Absorption Heat Pumps

Dr. Srinivas Garimella Ph.D., Hightower Chair in Engineering, Director of the Sustainable Thermal Systems Laboratory, Georgia Institute of Technology, Georgia, USA
Session Chair: Dr. Fadl Moukalled, AUB

Room: Maamari Auditorium, OLAYAN SCHOOL OF BUSINESS

Dr. Srinivas Garimella received his M. S. and Ph.D. degrees from The Ohio State University. He has held prior positions as Research Scientist at Battelle Memorial Institute, Senior Engineer at General Motors Corp., and Associate Professor at Western Michigan University and Iowa State University. He is a Fellow of the American Society of Mechanical Engineers and Editor of the International Journal of Air-conditioning and Refrigeration, Associate Editor of the ASME Journal of Energy Resources Technology, Associate Editor of the ASHRAE HVAC&R Research Journal and Chair of the ASHRAE Technical Committee on Absorption and Heat Operated Machines. He has mentored over 70 graduate students, with his research resulting in over 225 archival journals and conference publications, 7 patents, a textbook on Heat Transfer and Fluid Flow in Mini-channels and Micro-channels, and a book on Condensation Heat Transfer. He is the recipient of the NSF CAREER Award, the ASHRAE New Investigator Award, the SAE Ralph E. Teetor Educational Award for Engineering Educators, was the Iowa State University Miller Faculty Fellow, and Woodruff Faculty Fellow at Georgia Tech. He received the ASME Award for Outstanding Research Contributions in the Field of Two-Phase Flow and Condensation in Micro-channels, the Thomas French Distinguished Educator Achievement Award from The Ohio State University, and the Zeigler Outstanding Educator Award at Georgia Tech.

Keynote 4
September 23, 2016, 9:45- 10:45

Indoor Environment - Health Comfort and Productivity

Dr. Bjarne W. Olesen, ASHRAE President Elect, Professor, Intl. Centre for Indoor Environment and Energy, Technical University of Denmark, Lyngby, Denmark
Session Chair: Dr. Nesreen Ghaddar, AUB

Room: Maamari Auditorium, OLAYAN SCHOOL OF BUSINESS

Dr. Bjarne Olesen received his M.S. in civil engineering, and PhD in Laboratory of Heating and Air Conditioning from the Technical University of Denmark. He served as a research scientist at the Laboratory of Heating and Air Conditioning, part time affiliate product manager at Brüel & Kjaer, senior research scientist at the College of Architecture, Virginia Tech, head of research & development at UPONOR-VELTA GmbH KG & Co., Norderstedt, Germany, full professor and director of Indoor Environment & Energy at the Technical University of Denmark. Dr. Olesen has been awarded the Ralph Nevins Award, Distinguish Service Award, Fellow Award and Exceptional Service Award from ASHRAE, the Medal of Honor from the German Engineering Society and was recognized as an International Honorary Member of SHASE, the Society of Heating, Air-Conditioning and Sanitary Engineers of Japan, President Elect of ASHRAE and Honorary member of AICARR (Italian Society for HVAC). Bjarne W. Olesen is active in several ASHRAE-CEN-ISO-DIN standard committees regarding indoor environment and energy performance of buildings and HVAC systems. He has published more than 350 papers and more than 60 in peer reviewed journals.

Keynote 5
September 23, 2016, 14: 30- 15:30

HFC phase-down under the Montreal Protocol – a unique opportunity to increase the energy efficiency of air-conditioning

Cornelius Rhein, Policy Officer for the European Commission, DG Climate Action, Unit A.2: Climate Finance, Mainstreaming, Montreal Protocol

Session Chair: Dr. Samir Trabousli, AUB

Room: Maamari Auditorium, OLAYAN SCHOOL OF BUSINESS

Dr. Cornelius Rhein studied law in Germany and works for the European Commission since 2000. Since 2007 Cornelius works on ozone-depleting substances and fluorinated greenhouse gases and participates in the meetings of the Montreal Protocol. He was involved in the drafting and negotiation of the EU Regulation on fluorinated greenhouse gases and follows now the negotiations on a HFC phase-down under the Montreal Protocol.

Moderator of ME Regional Chair Meeting

Session Chair: **Mohammad Hosni**, PhD, Professor and the Frankenhoff Chair in Engineering, Director, University Engineering Alliance , Department of Mechanical and Nuclear Engineering Kansas State University

Session Co- Chair: Dr. Andreas Christoforou, Kuwait University

Dr. Mohammad Hosni is a Professor of Mechanical and Director of the University Engineering Alliance at Kansas State University. He joined the faculty in mechanical engineering at Kansas State University in 1991 and in 1993 assumed a leadership role as the director of the Institute for Environmental Research (IER), a multidisciplinary research center at Kansas State University, where he was responsible for development of research programs in human thermal comfort and indoor air quality. He served as the director of IER until 2001 when he was named department head of mechanical and nuclear engineering (MNE). After serving eight years as the MNE department head, he was appointed as the Director of the University Engineering Alliance in 2009.

Dr. Hosni served as the ASME Vice President for Education and currently is serving as a member of the Council on Standards and Certification of ASME. He is also active at ASHRAE, serving on the Publishing and Education Council. Dr. Hosni is the ABET Executive Committee member and Chair Training Committee.

Acoustic Solutions to Noise and Vibration Problems in Buildings with Mechanical Plant Rooms Adjacent to Occupied Spaces

Mete Oguc

Okan Sever
Member ASHRAE

Eren Kalafat
Member ASHRAE

ABSTRACT HEADING

An increase in population and limited land availability in modern cities means that concrete and steel frame high-rise buildings are being constructed more than ever. These buildings are technologically advanced, offering occupants comfortable environments, which is achieved by the use of building services installations. Mechanical rooms form part of these services and with multi-storey buildings they are inevitably located on intermediate floors, close to occupied areas. Noise and vibration generated from equipment and other installations in these rooms is emitted to surrounding spaces in the form of airborne and structure borne noise. Installing the correct type of vibration isolation can eliminate structure borne noise from machinery that is transmitted as impact sound and reduce the vibration to acceptable limits. To overcome the air borne noise, floating systems can be implemented, which are based on creating an air gap with a resilient surface. These floating systems are very effective in eliminating air borne noise by acting as barriers, which can be installed on walls, floors and ceilings in any room that requires acoustical treatment. Since MEP equipment and installations differ in size and shape and have their own operating characteristics, acoustical treatment and vibration isolation requires a technical and specialized approach. To achieve higher performance levels each piece of equipment will need individual attention besides investigating the whole installation. A general strategy together with key points for implementing acoustic insulation for mechanical spaces are evaluated in this paper.

INTRODUCTION

The most effective noise control measure is to locate plant rooms as far away as possible from noise-sensitive areas. This is not the case for high-rise multistory buildings in which the plant rooms are typically located on intermediate floors, close to the occupied areas they serve. In such a case, appropriate constructive layers should be selected for walls, ceilings, and floors once the amount of noise is assessed within the plant room. It is required that engineers and architects work in an interdisciplinary manner since it is essential to restrict equipment in the plant room to protect them against seismic loads whereas staying within the projects architectural design criteria, insulating the plant room noise, and isolating the equipment vibration. Also, economic costs of noise pollution results in devaluation in house prices, productivity losses from health related impacts. Therefore, general noise and vibration control strategies which are based on international codes and practical calculations are explained below.

EQUIPMENT NOISE LEVELS

Meaningful sound level measurements require a comprehensive measurement plan. This requires the selection of proper instrumentation and measurement procedures which addresses the source of the noise problem and helps out to draw a solution map. Unfortunately, it is usually possible to have such measurements when noise complaints arise. However, noise and vibration control design should begin during a project's schematic design phase and continue throughout the entire design process. In response to demands by engineers and architects for accurate sound data to be used during the design phase, industry standards have been developed for manufacturers to declare the acoustical performance of various types of HVAC equipment. Although most of the test standards require performance data in octave bands, manufacturers usually give a single value to represent the total sound power level of their products' noise. Furthermore, it is generally very difficult to predict sound power level generated by equipment or machinery in a plant room. The possible noise-generating mechanisms vary in noise characteristics, and magnitude of the noise generated also depends upon the enviroenment presented to the noise source. Therefore, current noise-prediction schemes are largely empirical. Those emprical relationships for much of the equipment that are used in plant rooms can be found in literature. To name a few, procedures for estimating the noise radiated by refrigeration machines and chillers, air compressors and vacuum pumps, cooling towers, pumps, boilers, fans, transformers and generators are available to designers (Bies et.al 2009, Ver et al. 2006).

SOUND INSULATION PREFORMANCE OF SEPARATING STRUCTURES

Once the location of the plant room is set and the preliminary equipment selection is completed, the acoustic insulation for the plant room shall be designed according to the equipment type, expected noise levels and adjacent occupied spaces' purpose of use. Sound generated in a room can be transferred to neighboring rooms via a number of transmission paths, such as through walls, floors, building frameworks and interconnecting ducts. Noise can either be evaluated as airborne or impact noise. Evaluating the impact sound insulation performance is often discarded in practice since building services that are found in plant rooms do not cause impact noise. The working machinary causes vibrations, not impact noise unless it is a metal stamping press or similar as can be found in manufacturing facilities.

Airborne Sound Insulation

The simplest case of sound transmission is a plane sound wave normally incident on a uniform homogeneous, isotropic, flat plate. Based on the normal-incidence mass law, airborne sound transmission loss, *R*, for such cases can be calculated from:

$$R_S \cong 10\log_{10}\left[1+\left(\frac{q\,\omega}{2\,\rho_0 c_0}\right)^2\right] \tag{1}$$

where m is the mass per unit area of the separating structure, ω is the radian frequency, ${}_0\rho$ and c_0 are the density of the gas and speed of sound inside this gas which is assumed to be the same on both sides of the separating structure.

Since the highest sound transmission loss obtainable by a single partition is limited by the mass law, use of multilayered lighter partitions are inevitable. Though, the calculation of double leaf partitions are a bit more complicated, Goesele has proposed a simplified method to predict the sound transmission loss of double partitions (Goesele 1980). According to this method, for a double partition with a gap distance l the formula is:

$$R_D \cong R_I + R_{II} + 20\log_{10}\left(\frac{4\pi f \rho_0 c_0}{s}\right) \quad (2)$$

where R_I is the measured sound transmission loss for the first partition, R_{II} is the measured sound transmission loss for the second partition, f is the frequency and s is the dynamic stiffness per unit area of the gap which can be calculated from:

$$s = \begin{cases} \frac{\rho_0 c_0^2}{d} & for\ f < f_l = \frac{c_0}{2\pi d} \\ 2\pi f \rho_0 c_0 & for\ f > f_l \end{cases} \quad (3)$$

Predicting models presented above are useful for the calculation of sound reduction performance of simple structures. However, advanced models such as finite element model, transfer matrix model and improved impedance model are usually preferred for the calculation of sound insulation performance of complex multilayered structures. (Mak et al. 2015; Kurra 2012).

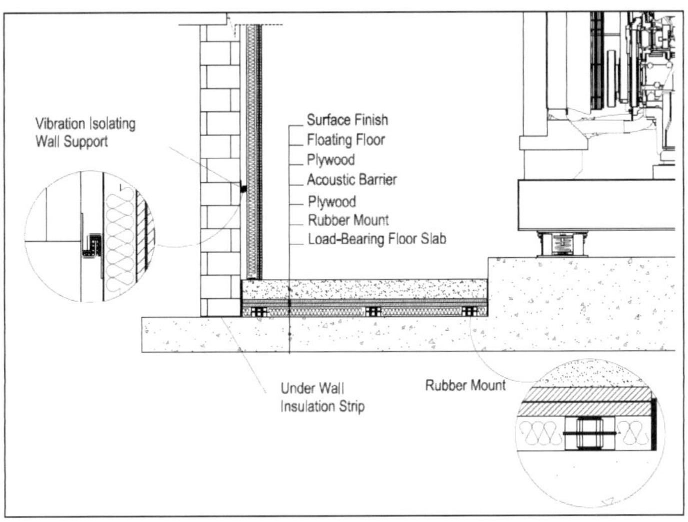

Figure 1 Floating floor system built around a housekeeping pad to insulate airborne noise generated inside a plant room. Flanking sound transmission paths are avoided by the use of vibration isolating wall supports, insulation strips and rubber mounts.

Since it is required to properly support heavy equipment such as chillers and generators to account for additional loads such as seismic loads, housekeeping pads that are rigidly attached to the buildings structure are usually utilized as shown in Figure 1. This application causes another problem to arise. When evaluating the airborne sound insulation performance of such a system within a plant room which is adjacent to a noise sensitive space, the contribution of airborne sound insulation performance of rigidly attached housekeeping pads are questioned since they create a short cut of airborne noise transmission even if a floating floor system is built around the housekeeping pad. This system which consists of a floating floor surrounding a housekeeping pad which is rigidly attached to the load bearing slab beneath the floating floor can be treated as a composit structure. It is possible to calculate sound transmission loss from this composite system from:

$$R_C = -10\log_{10}\left(\frac{S_{FF}\times 10^{-R_{FF}/10} + S_{HP}\times 10^{-R_{HP}/10}}{S_{FF} + S_{HP}}\right) \tag{4}$$

where S_{FF} is the surface area of the floating floor, S_{HP} is the surface area of the housekeeping pad, R_{FF} is the sound transmission loss of floating floor and R_{HP} is the sound transmission loss of housekeeping pad.

Impact Sound Insulation

Impact isolation quantifies the ability of a construction to prevent transmission of impact noise which can be measured using a standard tapping machine. Whether a floor is excited by impact sound or by an airborne sound field in the source room, it will in both cases radiate sound into the receiving room. The distinction between airborne sound and impact sound insulation performance of a floating floor can be made by making both airborne sound insulation performance measurement for R, and impact sound insulation performance L_n. The flanking paths can be immediately identifined by measuring the acceleration level on the wall surfaces in the source room and receiving rooms during acoustical and impact excitation. The floating floors in plant rooms are generally not designed as part of a vibration isolation or impact noise insulation scheme (ASHRAE 2015). Vibration isolation products should be used when the foundation or base of a vibrating machine is to be protected against large unbalanced forces or impulsive forces.

Vibration Isolation

The perception of vibration involves a large range of sensory systems. When people are asked about they often describe the associated noise and visual sensations alongside the physical sensation of feeling vibration. Vibration is often accompanied by noise so that the residents are usually uncertain of the differences between noise and vibration (Whittle et al. 2015). Therefore, it is significant to examine and solve the vibration problems step by step.

First step is to determine the source, in this case any equipment with a rotating element. Second step is to find out the path which the way vibration is transferred. Almost all building structural and installation components, such as beams, columns, walls, etc. transmits vibration. Third step is to define the receiver and the criteria for the disturbancy. The vibration isolation requires the use of relatively resilient elements to reduce the vibrotary forces or motions that are transmitted from one structure or mechanical component to another to acceptable limits.

The procedure to reduce the undesirable effects of vibration is called vibration isolation. Vibration isolation aims to reduce the dynamic response of the system by the insertion of a vibration isolator between the vibrating equipment and the source of vibration. Vibration isolation efficiency I of vibration isolators that are installed to a vibration source equiment can be accounted by:

$$I = 1 - T \tag{5}$$

where T is transmissibility which indicates the fraction of the disturbing motion or force that is transmitted. So that the isolation efficiency indicates the fraction by which the disturbance is less than the excitation (Rao 2004). Many aspects of vibration isolation can be understood from analysis of an ideal, linear, one-dimensional, spring-mass system. For such a single-degree-of-freedom model, the transmissilibilty of the system for a disturbing frequency f_d is:

$$T = \left|\frac{1}{1-(f_d/f_n)^2}\right| \tag{6}$$

where f_n is the natural frequency for a system with an isolator stiffness k which can be calculated from:

$$f_n = \frac{1}{2\pi}\sqrt{\frac{k}{m}} \quad (7)$$

The structure or foundation to which the isolator is connected cannot be assumed rigid when an equipment is mounted on a floating floor system. Eventhough this sort of a system is not ideal from a noise control perspective, it is requested by architects and investors for its ease of application. For such cases, the impedance of the supporting structure can no longer be ignored. Transmissibility of this system should be calculated from:

$$T = \frac{1}{\left[1-\left(\frac{f_d}{f_{n1}}\right)^2\right] - \frac{1}{\left(\frac{f_{n1}}{f_d}\right)^2 \frac{k_2}{k_1} - \frac{m_2}{m_1}}} \quad (8)$$

The two-degree-of-freedom system can model a system where an equipment with mass m_1 is mounted to an isolator with a stiffness k_1 which is rigidly attached to a floating floor with an equivalent mass m_2 installed on rubber mounts with stiffness k_2. However, caution should be exercised to ensure adequate damping is provided in the floating floor and that there are no common natural frequencies present between the floating floor system and equipment isolation mounted to it (ASHRAE 2013, ASHRAE 2015).

EVALUATING THE NOISE TRANSMISSION TO AN ADJACENT SPACE

There is an ordinal relationship between noise exposure and annoyance which in turn causes social and health problems. Introducing exposure-response relations which are being used in strategic or health impact assesments for estimating long-term annoyance to these insulation performance criteria help designers to develop lower cost and highly efficient insulation systems. Directives in many countries and codes written by independent technical institutes lay down and define specific noise emission limits mostly for environmental noise problems (Nathan 2015). Despite these valuable efforts, there has been little interest on improving exposure-response relationships considering people living in occupied residential and office spaces.

Once the transmitted noise levels are calculated from the total noise produced within the plant room, the noise transmitted can be calculated by substructing the sound reduction of the separating structures. The noise levels that are obtained in the adjacent occupied space should be lower than the legal requirements (or at least from the guideline values given by WHO in 1999). However, in most cases, these limit levels are also given as single numbered equivalent noise levels that give no indication of which frequency components may be the source of non-compliance. Therefore, single-number weighting curves should be used.

The human response to vibration inside the occupied space can be evaluated by making measurements and following ANSI S2.71 and ISO Standard 2631-2. From the designers perspective, vibration isolators transmissibility values can be used eliminate vibration induced noise in the design phase. The transmissibility of the isolators can be selected as 10% for nonsensitive, %5 for sensitive installations and 3% for critical installations. The static deflection of the isolation system can be calculated once the maximum allowable transmissibility has been decided (Simmons 2007).

CONCLUSION

Properly accomplished acoustic insulation and vibration isolation measures within the design phase is the most inexpensive way to prevent objectionable vibration and noise. The general scheme described above is based on following up international codes and common engineering principles. Eventhough decoupling structural components of a plant room to avoid flanking sound and vibration transmission while properly mounting the equipment to rigid housekeeping pads and leaving clearances for service installations for seismic safety considerations creates a dilemma, it has been shown that both sound and vibration transmission values can be taken into account within the design phase.

ACKNOWLEDGMENTS

This project has been funded by Ulus Yapı Tesisat Malzemeleri A.Ş.

NOMENCLATURE

- R = airborne sound transmission loss
- q = surface mass
- ω = radian frequency
- ρ = density
- c = speed of sound
- f = frequency
- s = dynamic stiffness
- S = surface area
- k = stiffness
- T = transmissibility
- I = vibration isolation efficiency

Subscripts

- 0 = gas
- D = double partition
- I = first partition
- II = second partition
- d = disturbing
- n = natural
- S = single partition
- l = gap between double partitions
- C = composite structure
- FF = floating floor
- HP = housekeeping pad

REFERENCES

Bies, D.A., C.H. Hansen. 2009. *Engineering Noise Control – Theory and Practice 4th Edition*. USA: CRC Press.

Ver, I.L., L.B. Beranek. 2006. Noise and Vibration Control Engineering. USA: Wiley.

Schaffer, M.E. 2011. *A Practicle Guide to Noise and Vibration Control for HVAC Systems,* 2nd edition (SI). USA: ASHRAE.

Goesele, K. 1980. Prediction of the Sound Transmission Loss of Double Partitions (without Structureborne Connections). *Acoustica* 45:218–227.

Mak, C.M., Z. Wang. 2015. Recent advances in building acoustics: An overview of prediction methods and their applications. *Building and Environment* 91:118–126.

Kurra, S. 2012. Comparison of the models predicting sound insulation values of multilayered building elements. *Applied Acoustics* 73:575–589.

Nathan, W., Peris, E., Condie, J., Woodcock, J., Brown, P., Moorhouse, A.T., Waddington, D.C., A. Steele. 2015. Development of a social survey for the study of vibration annoyance in residential environments: Good practice guidance. *Applied Acoustics* 87:83–93.

Rao, R.R. 2004. *Mechanical Vibrations,* 4th ed. USA: Pearson Prentice Hall.

ASHRAE. 2013. *ASHRAE Handbook—Fundamentals.* Atlanta: ASHRAE.

ASHRAE. 2015. *ASHRAE Handbook—HVAC Applications.* Atlanta: ASHRAE.

Simmons, R. 2007. Vibration isolation. *ASHRAE Journal.*

Energy Savings of Windows with Shutters in Hot and Humid Climates

Albert Al Touma[1]

Nesreen Ghaddar, PhD[2]
Member ASHRAE

Kamel Ghali, PhD[3]
Member ASHRAE

ABSTRACT

Energy consumption has gained tremendous attention in the purpose of finding alternative reduction methods. One of these methods is the installation of shutters on the external side of the windows to deteriorate the effect of outdoor conditions. The aim of this paper is to assess the effect of shutter placement on windows in terms of energy savings for different shutter positions in hot and humid climates such as in Saudi Arabia. An EnergyPlus simulation has been conducted on a typical residential building abiding by the Saudi Arabian construction standards with a Window-to-Wall Ratio (WWR) of 13.5% to find its thermal energy demand. The simulation has been validated with results from literature. Upon the validation of the results, a case study of the same villa but with different WWRs was conducted to extrapolate the usefulness of shutter placement into larger residential spaces for different shutter positions ranging from fully-open to fully-deployed. Results have shown that 74.4% and 79.1% reductions in total annual energy demands could be achieved for the cases of 20% and 30% WWRs, respectively, when the shutter is fully-depoloyed.

INTRODUCTION

Energy consumption is increasing annually and more awareness is brought into this field, especially as non-renewable resources are being harshly exhausted. The global responsibility to provide means of reducing energy consumption has attracted lots of researchers. This understanding has led to a deceleration in the annual increase in energy consumption from +2.0% over the past few years to 0.9% in 2014 (BP Global 2015). Among all the energy-consuming sectors, building alone account for 40% of the world's energy (Ghiaus et al., 2004).

Building in the Kingdom of Saudi Arabia (KSA) consume 65% of the total electricity production, which is much higher that the world average during 2010 (Said et al., 2003). Aldossari et al. (2013) reported that around 51.1% of the total KSA electricity consumption is attributed to residential buildings, 66% of which is consumed by air-conditioning (AC) systems. This is why providing new methods for reducing the energy consumption in residential buildings in KSA has become necessary.

In buildings, the building envelope is what surrounds the spaces from the outer environment. It consists of walls, roof, windows, materials, insulations, and so on. Extreme care has to be taken when designing the building envelope due to its importance on the lifecycle of the building. Among all these components, windows are known to be the most critical due to their thermo-physical properties that are easily overcome by outdoor conditions; significant amount of heat and solar radiation enter the space during hot seasons causing an increase in the cooling load (Hassouneh et al., 2010). For instance, windows were shown to contribute by around 32% to 39% of the space total

[1] Graduate Student, Department of Mechanical Engineering, American University of Beirut, email: aga32@mail.aub.edu
[2] Professor, Department of Mechanical Engineering, American University of Beirut, email: farah@aub.edu.lb, corresponding author.
[3] Professor and Department Chair, Department of Mechanical Engineering, American University of Beirut, email: ka04@aub.edu.lb

cooling load (Chasar et al., 2002).

The space cooling load due to windows is actually caused by the direct solar radiation transmitted through the windows and the heat caused by solar absorption that is leaked later into the space. Many researchers have considered diminishing the effect of each of these two causes. Some considered changing the building orientation and adapting it so that windows are hit by the least possible amount of solar radiation (Vanhoutteghem et al., 2015). It was found that this method could actually decrease the building energy consumption, but yet it is not enough and should be accompanied by an enhancement in the walls heat transfer coefficient (U-value) to decrease the amount of heat dissipated into the space. Others have considered the placement of solar films on the window outer surface (Li et al., 2015). Reductions in electricity consumptions of 4.1%, 1.3% and 10.5% were found when a solar film is placed on a 12 millimeter (0.47 in.) clear glazing window on the west façade only for an office, mall and hotel rooms, respectively. They have also found that using absorptive material in colors such as bronze or grey actually leads to reductions in the window visible transmittance. On the other hand, methods of reducing heat passage through windows have been studied by using multiple-pane windows. Arici et al. (2015) considered replacing double-pane windows with triple-pane and quadruple-pane windows for the hot seasons of Turkey and found 50% and 67% reductions in energy consumptions, respectively.

Means of reducing the thermal heat passage and the radiation transmissivity into the space simultaneously was also studied. For instance, Tsikaloudaki et al. (2015) studied the energy performance of windows in the warm regions of Europe and found that by decreasing their U-value from 3.2 $W/m^2 \cdot K$ (0.56 $Btu/hr \cdot ft^2 \cdot °F$) to 0.72 $W/m^2 \cdot K$ (0.13 $Btu/hr \cdot ft^2 \cdot °F$) and their solar transmittance from 0.76 to 0.40, the cooling energy index decreased from 320 kWh/m^2 (101.44 $kBtu/ft^2$) to 190 kWh/m^2 (60.23 $kBtu/ft^2$) for an office space and from 290 kWh/m^2 (91.93 $kBtu/ft^2$) to 150 kWh/m^2 (47.55 $kBtu/ft^2$) for a residential space. Furthermore, many researchers have considered the installation of blinds, awnings and screens on the outer side of the windows for the same purpose. These materials are usually opaque and cover some parts of the window, thus reducing the surface hit by solar radiation and serving as another layer against heat passage. Tzempelikos et al. (2005) studied the effect of window shading design and control on the building cooling demand. The shading device used was roller shades and reductions in cooling demand from 4700 MJ (4.45 MBtu) to 2300MJ (2.18 MBtu) were seen when the shading and lighting control strategies were implemented. This reflects the importance of shutter placement on reducing the spaces energy demands.

In this paper, the thermal energy demand of a typical residential villa built according to the Saudi contruction standards and subjected to the climate of the Kingdom is found through an EnergyPlus simulation and validated with similar work found in literature. Then, a case study considering the placement of shutter with different positions ranging between fully-open and fully-closed for different Window-to-Wall Ratios (WWRs) is conducted to quantify the energy savings for these different cases.

METHODOLOGY

In order to study the effect of shutter deployment over the window surface area, a building simulation tool had to be used. EnergyPlus V8-1, which is a whole building energy simulation software, has been used to find the thermal energy demand of a villa built according to the KSA construction standards for a WWR of 13.5%. The results are compared with literature for verification. Upon the validation of the results, the base-case was modified in a case study to add a shutter on the external side of the window and simulate both heat and transmitted radiation energies caused by windows, and hence calculate the energy reductions for cases of shutter 25% closed, 50% closed, 75% closed and fully-closed. Since the energy conservation requirements for residential buildings in KSA recommend that the WWR ranges between 15% and 25% (Saudi Building Code, 2007), the same procedure was repeated for the cases of 20% and 30% WWRs in order to extrapolate the usefulness of shutter placement to larger glazing areas.

Building Charateristics

A typical residential villa in KSA is considered in the simulation. The villa is two-storeys high with a floor area of

262.5 m^2 (2825.53 ft^2) and a height of 7 m (22.97 ft). The WWR was 13.5% and the windows used were 6 millimeters (0.24 in.) thick made of clear glass. The walls, roof and floor construction materials are described in Table 1 and their properties are shown in Table 2.

Table 1: Building Construction Materials (Alaidroos et al, 2014).

Parameter	Value
External Wall Materials	Starting from outside: 20 mm (0.79 in.) plasterboard, 200 mm (7.9 in.) concrete hollow block and 20 mm (0.79 in.) plasterboard.
Roof Materials	Starting from outside: 10 mm (0.39 in.) built-up roofing, 150 mm (5.91 in.) concrete roof slab and 12.7 mm (0.50 in.) plasterboard.
Floor Materials	150 mm (5.91 in.) slab on grade

Table 2: Properties of the Construction Materials (Ahmad, 2004).

Material	Thermal Conduvtivity [W/m·K], [Btu·ft/hr·ft^2·°F]	Specific Heat [J/kgK], [Btu/lb·°F]	Density [kg/m^3], [lb/ft^3]
Plasterboard	0.797 (0.46)	840 (0.20)	1682 (105.01)
Hollow Concrete Block	1.038 (0.60)	840 (0.20)	977 (60.99)
Concrete Slab	1.219 (0.71)	840 (0.20)	2122 (132.47)
Built-up Roofing	0.17 (0.10)	1460 (0.35)	1100 (68.67)

In addition, the window properties are shown in Table 3.

Table 3: Properties of the Window.

Parameter	Value
Solar Transmittance	0.775
Solar Reflectance	0.071
Visible Transmittance	0.881
Visible Reflectance	0.080
Thermal Conductivity	0.9 W/m·K (0.52 Btu·ft/hr·ft^2·°F)

Moreover, the villa had 30.5 W/m^2 (9.67 Btu/hr·ft^2) of total internal loads as a combination of lighting and miscellaneous demands. Lastly, the cooling set-point temperature was 22°C (71.6°F). Note that the villa has been assumed to be a single block with no internal partitions.

Weather Conditions

The weather file that has been used for the simulation is the Saudi Arabia .epw file available on the EnergyPlus website for the city of Riyadh. The city is located at a latitude of 24.6° north and a longitude of 46.7° east with an elevation of around 600 m (1968.5 ft) above the sea level (Zell et al., 2015). The average daily temperature, direct solar radiation and diffuse solar radiation are shown in Figure 1. The average daily temperature ranged between 14°C (57.2°F) and 37°C (98.6°F) during the months of January and July respectively, where the peak temperature reached 44°C (111.2°F) during the month of June. On the other hand, the direct solar radiation intensity ranged between 4590 Wh/m^2 (1455.02 Btu/ft^2) and 7600 Wh/m^2 (2409.19 Btu/ft^2) per day in December and June whereas the diffuse solar radiation intensity ranged between 1200 Wh/m^2 (380.40 Btu/ft^2) and 2900 Wh/m^2 (919.30 Btu/ft^2) per day during December and July, respectively. The peak direct and diffuse solar radiations were 1100 W/m^2 (348.70 Btu/hr·ft^2) and 450 W/m^2 (142.65 Btu/hr·ft^2) per hour at noon in June. In comparison with the published weather data for the central area of KSA, these values were very similar to those collected experimentally (Zell et al., 2015).

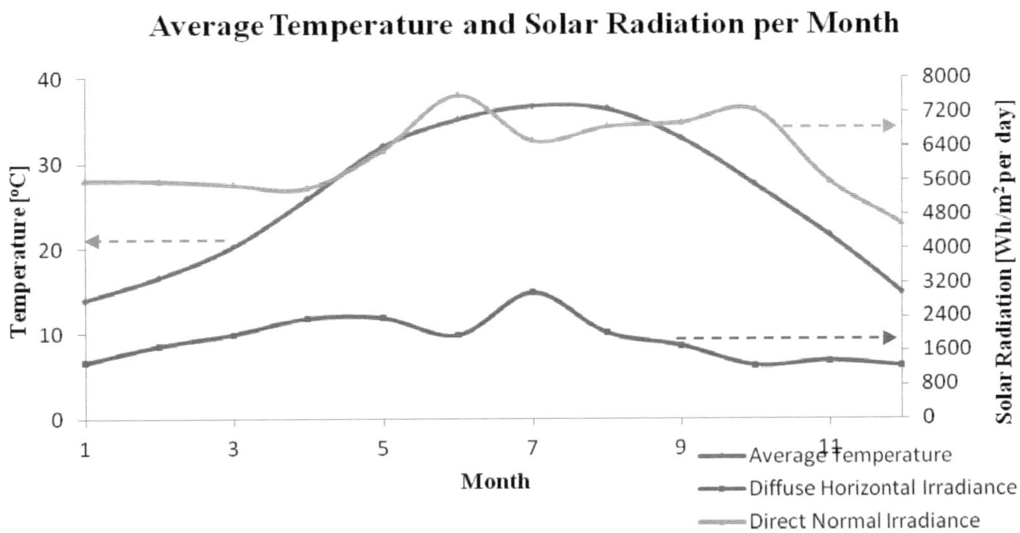

Figure 1: Average daily temperature, direct and diffuse solar radiation in Saudi Arabia.

RESULTS AND DISCUSSION

The building total thermal energy demand found on EnergyPlus turned out to be 249.8 kWh/m² (79.19 kBtu/ft²), which was found to be very close to the 228 kWh/m² (72.28 kBtu/ft²) value found by Alaidrous and Krarti (2014) for exactly the same building. The percentage difference between the two results was as low as 8.7%, where the error could be attributed to the assumption of single block villa with no internal partitions. Hence, this represented a validation of the simulation that was conducted and based on which the shutter was added to check the energy savings caused by its deployement.

Case Study

After validating the base-case simulation conducted on EnergyPlus, the effect of deploying the shutter is now studied. Since EnergyPlus has no exact component that replaces or acts as a shutter, an alternative method was used; the original window surface area is reduced by the percentage of shutter deployment and is altered by the design of a new opaque window that replaces it but has the characteristics (U-value) of the original window, the shutter and the airgap combined. Similar to the shutter, the new designed window has zero solar and visual transmittances. The same procedure was used for different shutter closeness percentages. The shutter properties used in the simulation are shown in Table 4.

Table 4: Properties of the Shutter.

Parameter	Value
Solar Transmittance	0
Solar Reflectance	0.72
Visible Transmittance	0
Visible Reflectance	0.79
Thermal Conductivity	205 W/m·K (118.90 Btu·ft/hr·ft²·°F)

The procedure followed for the calculations of energy savings starts by using windows of the base-case, simulating the building total annual thermal energy demand per meter squared of floor area and finding the amount caused by windows by summing the heat gain energy and transmitted solar radiation energy. After the shutter is added, the energy caused by windows is found again and reductions in the building total annual energy are calculated. Results are shown in Table 5.

Table 5: Energy Savings Due to Shutter Deployment for WWR of 13.5%, 20% and 30%.

WWR [%]	Shutter Closeness Percentage	Building Total Annual Energy Demand [kWh/m²], [kBtu/ft²]	Annual Energy Caused by Windows [kWh/m²], [kBtu/ft²]	Percent of Heat Gain Energy Through Windows [%]	Percent of Radiation Energy Transmitted Through Windows [%]	Percent Decrease in Annual Energy Demand [%]
13.5%	0 %	249.8 (79.19)	143.3 (45.43)	82.9	17.1	-
	25 %	212.2 (67.27)	105.7 (33.51)	83.1	16.9	15.1
	50 %	179.4 (56.87)	72.9 (23.10)	83.3	16.7	28.2
	75 %	146.6 (46.47)	40.1 (12.71)	83.8	16.2	41.3
	100 %	108.1 (34.27)	1.6 (0.51)	100.0	0.0	56.7
20%	0 %	288.5 (91.45)	216.9 (68.76)	81.3	16.9	-
	25 %	235.4 (74.62)	163.8 (51.92)	83.3	16.7	18.4
	50 %	182.1 (57.73)	110.5 (35.03)	83.5	16.5	36.9
	75 %	128.3 (40.67)	56.7 (17.97)	84	16.0	55.5
	100 %	73.9 (23.43)	2.3 (0.73)	100	0.0	74.4
30%	0 %	329.8 (104.55)	264.6 (83.88)	79.5	20.5	-
	25 %	258.9 (82.07)	193.7 (61.40)	75.8	24.2	21.5
	50 %	174.4 (55.28)	109.2 (34.61)	75.1	24.9	47.1
	75 %	109.7 (34.77)	44.5 (13.11)	69.4	30.6	66.7
	100 %	68.8 (21.81)	3.6 (1.14)	100	0.0	79.1

The base-case building total annual energy demand for a WWR of 13.5% was 249.8 kWh/m² (79.19 kBtu/ft²) where the direct load from the windows was 143.3 kWh/m² (45.43 kBtu/ft²). This represented around 57% of the annual energy compared to other loads caused by other components from the building envelope as well as internal loads that contributed by 106.5 kWh/m² (33.76 kBtu/ft²). This energy decreased when the shutter was being deployed until it reached savings of 56.7% for the case of fully-deployed shutter. Larger savings were seen for the cases of 20% and 30% WWRs; the annual energy demands were 288.5 kWh/m² (91.45 kBtu/ft²) and 329.8 kWh/m² (104.55 kBtu/ft²) and the window shares were 216.9 kWh/m² (68.76 kBtu/ft²) and 264.6 kWh/m² (83.88 kBtu/ft²), respectively. These demands decreased significantly when shutters were being deployed until they reached savings of 74.4% and 79.1% for fully-closed shutters.

CONCLUSION

This study highlights the importance of shutter placement in reducing building energy loads. EnergyPlus simulation tool has been used to find the thermal energy demand of a typical residential villa in the Kingdom of Saudi Arabia with and without shutters placed on windows. Significant energy savings of 56.7%, 74.4% and 79.1% were found for a fully-deployed shutter for the cases of 13.5%, 20% and 30% window-to-wall ratios. Despite its extreme usefulness in reducing energy loads, shutters' full deployment often sacrifices the visual and thermal comfort of the occupants and increases the electrical lighting demand of the spaces. For this reason, using shutters to cover some parts of the glazing areas without shutting the entrance of natural lighting would ensure both energy savings and thermal comfort of the occupants.

REFERENCES

Ahmad A. Energy simulation for a typical house built with different types of masonry building materials. *Arabian Journal for Science and Engineering*. 2004 Oct 1;29(2):113-26.

Aldossary, N., Rezgui, Y., and A. Kwan. 2013. Domestic energy consumption patterns in a hot and humid climate: A multiple-case study analysis. *Applied Energy* 114: 353-365.

Alaidroos, A., and M, Krarti. 2014. Optimal design of residential building envelope systems in the Kingdom of Saudi Arabia. *Energy and Buildings* 86: 104-117.

Arici, M., Karabay, H., and M. Kan, 2015. M. Flow and heat transfer in double, triple and quadruple pane window, *Energy and Buildings*, 86: 394-402.

BP Global. 2015. *BP Statistical Review of World Energy*, 64th edition.

Chasar, D., Moyer, N., Rudd, A., Parker, D., and S. Chandra. 2002. Measured and simulated cooling performance comparison; insulated concrete form versus frame construction, *American Council for an Energy Efficient Economy*.

Ghiaus, C., and C. Inard. 2004. *Energy and environmental issues of smart buildings. A handbook for intelligent building*, p. 26-51.

Hassouneh, K., Alshboul, A., and A. Al-Salaymeh. 2010. Influence of windows on the energy balance of apartment buildings in Amman, *Energy Conversion and Management*. 51: 1583-1591.

Li, C., Tan, J., Chow, T. T., and Qiu, Z. 2015. Experimenal and theoretical study on the effect of window films on building energy consumption. *Energy and Buildings*.

Said, S., Habib, M., and M. Iqbal. 2003. Databse for building energy prediction in Saudi Arabia, *Energy Conversion and Management* 44: 191-201.

The Saudi Building Code National Committee, *The Saudi Building Code (SBC)*—Section 601: Energy Conservation. Riyadh, Kingdom of Saudi Arabia, 2007, pp. 196.

Tsikaloudaki, K., Laskos, K., Theodosiou, T., and D. Bikas. 2015. The energy performance of windows in the mediterranean regions. *Energy and Buildings*, 92: 180-187.

Tzempelikos, A., and A. Athienitis. 2005. The effect of shading design and control on building cooling demand. "*International conference "Passive and Low Energy Cooling for Built Environment*", pp. 953-957.

Vanhoutteghem, L., Skarning, G., Hviid, C., and S. Svendsen. 2015. Impact of façade window design on energy, daylighting and thermal comfort in nearly zero-energy houses, *Energy and Buildings*.

Zell, E., Gasim, S., Wilcox, S., Katamoura, S., Stoffel, T., Shibli, H., Engel-Cox, J., and M. Al-Subie. 2015. Assessment of solar radiation resources in Saudi Arabia, *Solar Energy*, 119: 422-438.

Energy Performance Contracting for Lebanon

Charles Blaschke IV
ASHRAE Member

Henry Jeha

ABSTRACT

Lebanon is a country that has a heavy dependence on energy imports in order to meet its domestic demand. Although it is a relatively small country, it rivals countries like Mexico and Brazil in electric power consumption per capita at 3,102 kWh/person. The Lebanese economy has a very high energy cost to GDP ratio, which translates to more money being spent on energy. Prior to 1975, more than 40 percent of Lebanon's produced power was hydroelectric; however, due civil war, the majority of this infrastructure was damaged. Today, Lebanon depends on rehabilitated Heavy Fuel Oil (HFO) and Light Fuel Pol (LFO) power plants to satisfy its demand; a strategy that has, thus far, proved insufficient in satisfying the country's energy demand. A large portion of this demand comes from existing commercial and residential buildings and developments. This accounts for upwards of 60 percent of the electricity consumption in the country. Building owners and businesses throughout the region, and the world, are starting to realize that energy consumption needs to be reduced before tackling supply side problems. The concept of Energy Performance Contracting (EPC) is one of the most effective mechanisms that can be used to tackle this problem. An EPC is an innovative solution that enables the reduction of energy consumption and demand using paid-from-savings, performance based mechanisms. The owner of a building does not pay upfront costs, and only pays a portion of the savings once it is generated. Deployed on a country-wide scale, this model can significantly aid deficient capacity by dealing with demand side consumption. In most cases, an EPC lays the groundwork for making energy efficiency upgrades to buildings and assets at no cost to the owner. The capital expenditure required to implement these upgrades are paid for throughout the life of the contract by the savings generated; a structure that ensures payment that is based solely on performance. A strong working mechanism between Lebanese government bodies, building owners, energy services companies and finance providers can help reduce the impact of buildings on the environment, while reducing energy demand and increasing asset profitability and value.

INTRODUCTION

As countries start to become more developed, and people move into their urban centers, the need for sustainable buildings and infrastructure is becoming increasingly important. Safe, healthy and comfortable spaces are required to live and work. These requirements almost always come with a price, financially and environmentally. Over the course of the last fifty years, buildings have been designed, built and operated without giving much consideration to their environmental impacts. In Lebanon, specifically, the pace of development has been hindered due to political and financial constraints. Energy and; consequently, tariffs in the country are not cheap. With increasing demand for energy due to population and development growth, and the insufficient power generation capacity of the country, Lebanon's demand side consumption must be the center of focus and analysis. The critical demand culprit is existing buildings. These are responsible for the majority of energy consumption in the country, with the utility company having to cycle regional power supply at different times of the day just to satisfy the demand of these buildings, usually located in major cities such as Beirut. This risks leaving a large amount of the population with no power; causing regular blackouts throughout different towns. Thus, it has become vital to upgrade these existing buildings in order to promote efficient operation and, thereby, reduce the reliance on energy supply.

In order to upgrade buildings, and deploy the latest technology on a large scale, public, private and non-government organizations throughout the country need to collaborate and cooperate. Currently, a strong regulatory framework does not exist to support Energy Services Companies (ESCOs) and the EPC model. The absence of such a body hinders the model's growth, does not instill confidence in the market and does not safeguard against fraud and wrongdoing. In many cases, the primary obstacle for building owners when choosing to upgrade their assets is the capital investment that is required to accomplish energy efficiency goals, compounded with the lack of financing options available. There are many ways to finance energy efficiency projects, either through capital budgets, loans or tax incentives; however, most owners are not willing to bear the interest and tax incentives are non-existent in Lebanon. Energy Performance Contracting is a unique and proven method that can be applied on a large scale and implemented in the majority, if not all, of the existing buildings in Lebanon. With proper planning and execution, EPCs can be a key driver in achieving energy reduction milestones with no cost to the government or building owners.

THE PROBLEMS

Lebanon has undergone a stable growth in development and rebuilding over the last two decades after the civil war. This has had a huge impact on energy demand and consumption throughout the country. This growth will continue into the future with the development of more high-rise buildings in Beirut. The current status of development has resulted in existing buildings that are inefficient and have poor energy performance. As a result of this inefficiency, there is a high energy cost to the building owners and the environment. Over the summer, and more specifically on the coast, weather conditions require a large amount of energy to satisfy the ventilation and cooling demands of these buildings. In order to keep up with the growing economy and pace of development, there should be sufficient energy that drives this growth.

From a design and operations aspect, corners are being cut in order to save on capital budgets. There is no thought given to the impact that this mentality has on annual operations costs. There is a big gap in building technology, and while it may be cheaper to save money during the initial parts of design and construction, it inevitably results in a higher spend throughout the lifetime of the asset. In addition, the lack of knowledge and information about energy, efficiency and how to save is further contributing the problem.

Energy Cost has a Huge Impact on Business

The high energy use per capita has a massive impact on the profitability of building owners and citizens. The current utility rates are high relative to most countries in the region due to the fact that there is not enough electricity being supplied, which forces the population to resort to private diesel generators. These diesel generators are inefficient, bad for the environment, and result in a high average tarrif cost. When comparing population size and GDP, the resulting cost from these tariffs make up a large proportion, if not the largest proportion, of a building owner's annual expenditure. This, coupled with the inefficient operation of the buildings, results in a big impact on the overall profitability of the asset. In addition, standards of comfort are not met, which results in an uncomfortable and unhealthy environment throughout the buildings. In addition, these same utility rates are the main obstacles facing the growth of the industrial and agricultural sectors. These industries are unable to match and compete with overseas importers of manufactured materials and consumer goods.

Owners are Reluctant to Invest in Energy Upgrades

Even when equipped with this knowledge, owners are still reluctant to invest in energy efficiency measures and upgrades for their buildings. There are many valid reasons as to why owners would not want to invest in energy upgrades for their buildings. Given the current economic situation in Lebanon, and the world, a scenario that hinders the allocation of budgets and capital for investment in energy upgrades exists. This, together with the lack of education, comfort and confidence in energy efficiency projects results in a failure to act by the owners. Every day that passes results in insurmountable lost savings.

Access to Finance for Upgrades

Assuming that owners were well versed in the matter of energy efficiency upgrades, and that the intent was there, there is still no viable solution for owners to finance these projects. The United States provides incentives for energy efficiency upgrades. This is something that is not practiced at all. Furthermore, banks do not see the financial or environmental benefit of energy efficiency projects. Because of that, these financial institutions are not ready to make funds readily available for this specific purpose. This results in the owners having to contemplate self-finance, which almost always results in them forgetting about the project altogether.

THE SOLUTION

An Energy Performance Contract is an agreement between an owner and an ESCO that allows the ESCO to finance and implement energy efficiency measures at no cost to the owner. This model enables the building to start saving energy almost immediately. The main advantage of an EPC is that the ESCO bares all the risk of the upfront capital cost and only gets paid based on performance.

What is Energy Performance Contracting?

An EPC is an innovative model that provides a flexible solution for building owners. There are many ways that an EPC can be structured; however, the most common structures are Guaranteed and Shared Savings contracts. These two types of contracts have the same fundamental goal; however, they are drastically different when it comes to ESCO payment. In both cases, the contracts are structured so that the capital expenditure of the project is paid for from the energy cost savings achieved over the lifetime of the contract. Buildings that utilize this model undergo targeted engineering design and modifications that are solely for the purpose of energy reduction. The two types of contracts will be discussed later. EPCs consist of the elements shown in Figure 1.

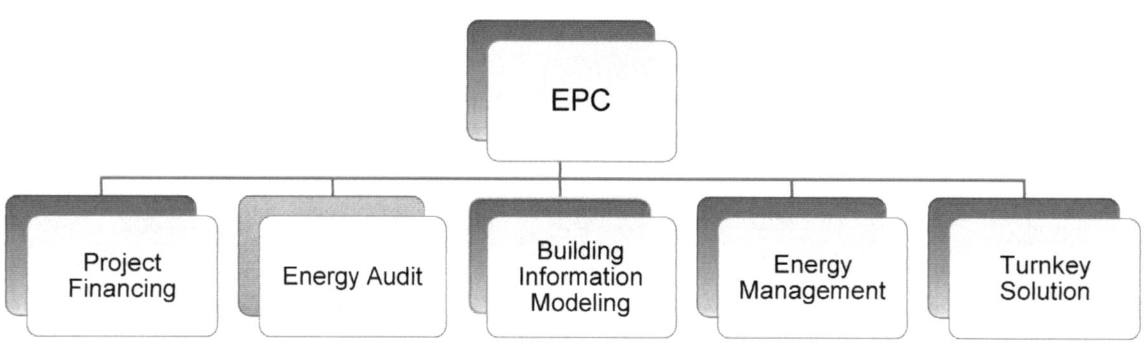

Figure 1 Elements of an Energy Performance Contract.

Since a building owner is already having his or her needs met by their existing facilities management company or engineering team, an ESCO will only serve to complement the services provided. The ESCO adds consulting, contracting, legislation, technology, finance and energy management experience, while looking at the facility as a whole.

Project experience coupled with contractor relationships and leveraging software and technology, such as Building Information Modeling (BIM), results in a complete solution for the building owner.

How Does an Energy Performance Contract Work?

In an EPC, the ESCO engineers a turnkey solutions based on the characteristics of the project, equipment in place and feasibility of implementing certain measures that are identified during the audit. First, the client's requirements need to be understood. If the building owner wishes to invest in his or her asset, then a Guaranteed Savings EPC is signed which hold the ESCO responsible for guaranteeing an annual savings figure. On the other hand, if the owner does not wish to invest his or her own money in the asset, the ESCO provides financing under a Shared Savings EPC in return for a portion of the annual savings over the duration of the project. Once this is identified, an Energy Audit is conducted in order to identify existing systems, understand current consumption traits and operational parameters and to gain a general idea into the possibility of implementing pre-defined energy efficiency measures.

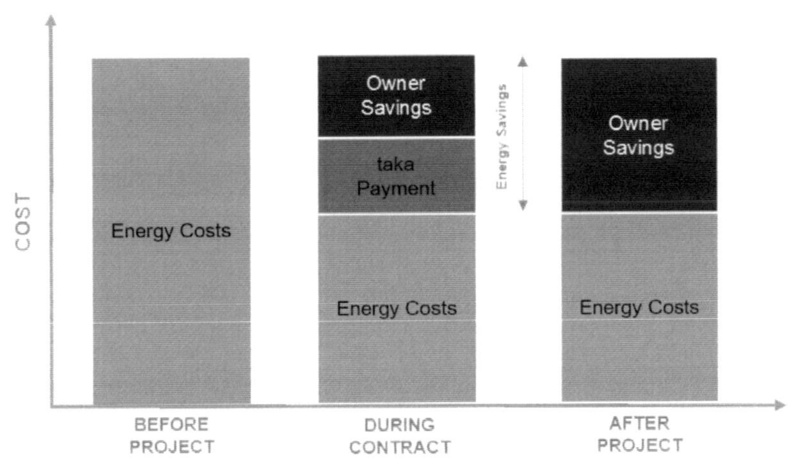

Figure 2 Typical Project Cash Flow Chart.

Next, the ESCO implements the selected and approved measures by engineering, designing, installing and commissioning the upgrades over the length of the contract to meet the contractual performance goals. Finally, once all the selected measures are implemented, the ESCO proceeds to manage the entire facility, from an operations standpoint, for the duration of the contract. This is done to ensure performance goals or guaranteed figures are met. This is done through the process of Measurement and Verification (M&V) to verify how much energy is being saved on a monthly, quarterly and annual basis. Most owners tend to prefer the Shared Savings EPC due to the fact that they do not have to worry about financing the project themselves and can start realizing savings almost immediately after the operations period begins.

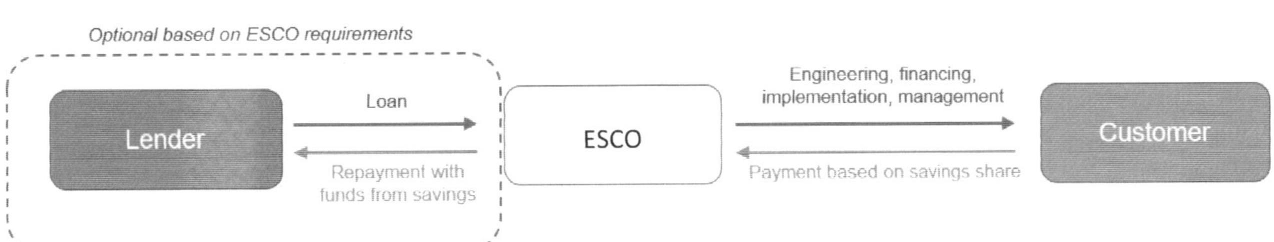

Figure 3 Shared Savings EPC Structure.

If, however, an owner chooses to self-finance, the ESCO guarantees the project energy and cost savings, which will be sufficient to pay for the project over the term of the contract. That is, the owner is guaranteed to receive payback on his or her investment and the ESCO is required to perform in order to avoid paying the difference of any shortfall in savings. At the end of the contract, the building is handed back over to the owner, who then benefits from all the savings along with an upgraded and optimized building. Most of the buildings in Lebanon can expect contract lengths of less than five years to satisfy the capital expenditure of efficiency projects.

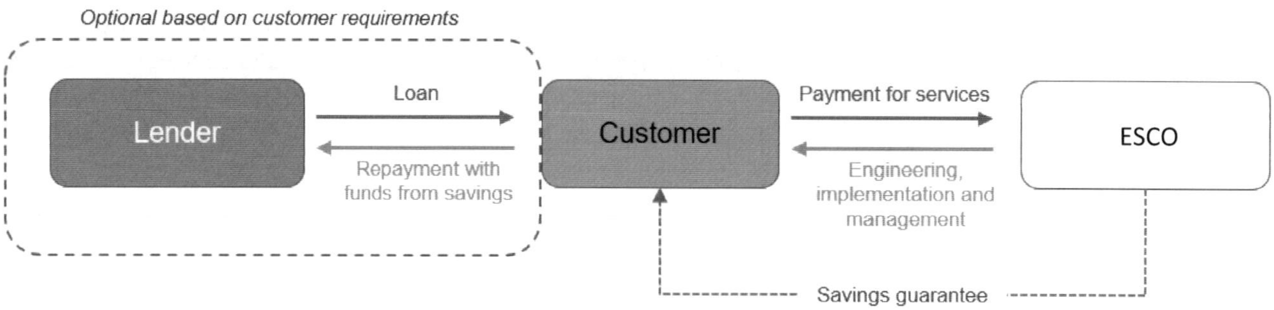

Figure 4 Guaranteed Savings EPC Structure.

Advantages of Energy Performance Contracts

Energy Performance Contacts aid building owners and facility managers in meeting energy efficiency, renewable energy, water conservation and environmental impact targets by streamlining private-sector funding and expert technical experience. The mechanism provides several benefits, described below and shown in Figure 5.

1. Aligned owner and ESCO interests to provide high quality equipment, services and support.
2. A healthier, safer working and living environment.
3. Economic development that ensures building efficiency improvements and equipment upgrades without upfront capital costs.
4. Provides energy security without risking capital expenses.
5. Guarantees energy and related operational and maintenance economic savings.
6. Environmental stewardship in energy efficiency, renewable energy, water conservation and reduced greenhouse gas emissions.

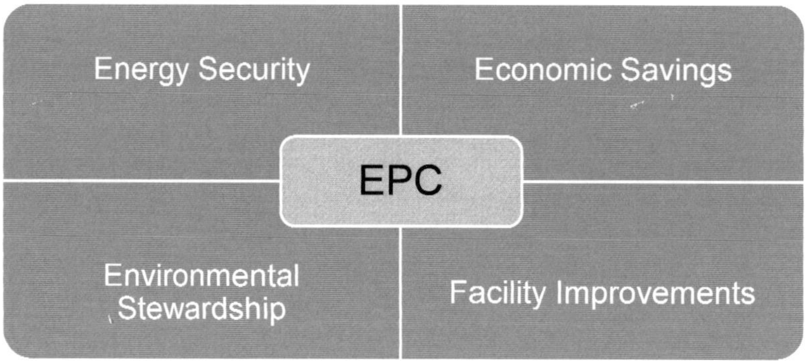

Figure 5 Benefits of an Energy Performance Contract.

Impact

Applying the concept of Energy Audits and EPCs for residential and commercial buildings will result in a huge impact on the overall consumption of the country as a whole. For example, a sample size of approximately two-hundred buildings in Beirut, with an average annual consumption of 1 million kWh each spends close to 150,000 dollars on electricity alone. Energy efficiency upgrades using EPCs will result in 9 million dollars of savings across this sample, and allow a further 60 million kWh to satisfy existing demand without the need to build additional power plants. This is equivalent to 42 million kilograms of Carbon equivalent reduction.

UTILIZING ENERGY PERFORMANCE CONTRACTS

United States

Energy Performance Contracts have been widely used for many years throughout the United States, with great success. The Department of Energy created the Federal Energy Management Program (FEMP) with the aim to assist the federal government departments and operators in securing and completing energy efficiency projects through EPCs. As of May 2011, Energy Performance Contracts worth 3.9 billion dollars have been awarded through the program, spanning over five-hundred projects. Collectively, these projects have saved approximately 33 trillion British Thermal Units (BTU) of energy on an annual basis; a total of 13 billion dollars in energy costs. Using EPCs, government agencies upgraded their buildings and reduced their energy consumption without any capital investment or tax incentives. Many state and local governments across the United States have also benefited from using the EPC model as a mechanism to upgrade thousands of public buildings. Today, private and public utility companies are offering residential and commercial incentives for energy efficiency upgrades and retrofits.

United Arab Emirates

The U.A.E. has recently become more involved in sustainability and energy efficiency. More specifically, Dubai has set a 30 percent reduction goal by 2030. In order to achieve this, a "super-ESCO" called Etihad Energy has setup a retrofit program, with the backing of the government, to make Dubai a leading example of energy efficiency in the region and the world. Etihad enables the EPC market in Dubai by developing projects for over thirty-thousand buildings. A group of ESCOs in Dubai have been involved in multiple projects throughout the city, with excellent results being reported from a savings forecast and guarantee standpoint.

Lebanon

Lebanon can follow suit by using the United States and United Arab Emirates as examples. With new projects and continuous development on the horizon, there will be a greater need for energy to be supplied in order to satisfy the demands of the population. This demand will result in a strain on the government budget and strategic goals, while also adding to the already existing problem of supply side deficiency. Lebanon, as a country, does not have the ability to building power plants on a large scale across the country in time to meet these rising demands; therefore, the demand side issues must be tackled through the use of EPCs first. Buildings are consuming more than half of Lebanon's energy, and more than 75 percent of these buildings are eligible for an upgrade. These upgrades will result in a minimum average of 42 percent energy reduction throughout each building, which will result in an approximate energy demand and consumption decrease of 30 percent throughout the country.

Lebanon's total electricity net consumption in 2012 was 13 billion kilowatt-hours (kWh). A 30 percent decrease in consumption would amount to 3.9 billion kilowatt-hours in savings; a total of approximately 500 million dollars. Currently, there is no EPC market in Lebanon. This is due to the lack of demand for energy related services in buildings and the lack of knowledge by the owners with regards to energy consumption and cost patterns. Energy prices have traditionally been extremely high, and do not look like they will be reduced any time soon. This, coupled with the supply deficiency, means that owners need to find solutions outside of the norm to ensure continuous energy supply to their buildings. It is no surprise that the large amount of energy and money spend on existing buildings is robbing the country of its precious resources and its residents' hard-earned income; both of which could be better allocated and managed.

Lebanon's Sustainability and Energy Sector

The correct steps are being taken when it comes to energy efficiency; however, not in the direction that has the greatest impact. A minority is beginning to realize that the status quo needs to be changed in order to prolong development and wellbeing throughout the country. The Lebanese Green Business Council (LGBC) was enacted to tackle the issue of green building and sustainability in Lebanon. While, this has raised awareness with regards to existing buildings and their issues, it has not gained the traction seen in other countries such as the United Arab Emirates. The same can be said about the Lebanese Center for Energy Conservation (LCEC). Regardless of how much awareness is being raised, owners are still not interested in upgrading their buildings and paying for it out of their own pockets.

Currently, Lebanon's implementation with regards to energy efficiency is focused solely on lighting retrofits and solar generation on a small scale. Street lights are being replaced with solar powered LED ones in order to maintain and provide lighting during blackouts and throughout the night. Solar hot water heating for the residential sector has also exploded; however, may residents are reluctant to pay the relatively high price.

Results

Country wide implementation of Energy Performance Contracts in Lebanon would be very beneficial to the country in many sectors. Estimates provided by Frost & Sullivan (2011) show that the energy performance contracting market in the Middle East earned revenues of around 80 million dollars, and is expected to grow annually at a rate of 4.1 percent. The rise of this industry will have a great impact on the nation, from an economic point of view. Furthermore, ESCOs that promote EPCs in Lebanon will serve to create jobs for the many skilled and unskilled workers and contractors around the country.

DISCUSSION

Local ESCOs and implementation contractors will help to bring new knowledge to the market about energy consumption traits and cost associated with energy. In addition, the acceptance of the EPC model will serve to create a new market that was non-existent in Lebanon. Energy upgrades through performance contracts will spur a growth in job creation and help satisfy the energy demands of the nation, while reducing the strain on the already scarce supply. This will allow the government to focus on future capacity planning without inflated consumption demands.

A true "green economy" will begin to form that consists of a skilled labor force, expertise and market knowledge. With cooperation and support from the government bodies in Lebanon, as well as the existing committees and councils formed, the framework and regulation surrounding EPCs can be created and promoted throughout the entire country. As EPCs becomes mainstream in the market, the private sector will begin to become more involved, and the expertise, technology, implementation, knowledge and results will grow exponentially. As owners become familiar and comfortable with the process, ESCOs can provide ongoing support and energy management to continuously drive down energy consumption.

Finally, as new products and technologies begin to emerge and penetrate the market, another implementation phase can begin in which ESCOs upgrade the existing buildings to further optimize the facility, while reducing energy consumption and cost.

CONCLUSION

Energy Performance Contracting is an innovative model that implements energy efficiency measures and upgrades for existing buildings, which will provide services to businesses or owners at a lower cost. The history and proven results of EPCs demonstrate how it is a viable solution for existing building owners' energy consumption problems. EPCs allow for 100 percent project financing at no cost whatsoever to the owner. This enables projects to be completed with no capital cost, while receiving immediate benefits in the form of reduced energy consumption and shared cost savings with the ESCO. Cash flow and asset value increase through the reduction of operational costs and additional equipment improvements, respectively. In addition, the existing buildings are operated according to industry prevailing code, thus promoting a better and healthier environment for the tenants.

There is a huge potential and need to improve existing buildings before thinking about building bigger and better buildings. The money is locked inside these existing buildings and the correct regulatory framework, awareness and procedures in place will push the private sector to accelerate and exponentially grow, while producing rapid results. Buildings such as residences, offices, hotels, factories and hospitals are all capable of benefitting from the Energy Performance Contract model. There are signs of progress in the local market, with focus beginning to shift towards green, sustainable and efficient practices; however, most of it is not being applied due to the financial constraints. With assistance from the government and private building owners, ESCOs wishing to operate in Lebanon can make an immediate and positive impact on the energy usage and profitability of the available resources, and reduce overall energy consumption in the country.

If the government bodies in Lebanon can set up the regulatory framework, then Energy Performance Contracting can be a scalable strategy that will immensely aid Lebanon in reducing its overall energy consumption and demand from existing buildings. This will help delay, or completely eradicate, the need for larger supply capacities and hundreds of millions of dollars needed to lay the infrastructure for such capacity. Energy Performance Contracts lays the groundwork for a triple win situation involving building owners, the government and the environment.

ACKNOWLEDGMENTS

The material and claims made in this paper are based on actual energy efficiency projects performed by taka solutions, an ESCO that operates primarily in Dubai, United Arab Emirates. Any opinions, findings and conclusions or recommendations expressed in this paper are those of the authors.

REFERENCES

Badelt G. and Yehia M. The way to restructure the Lebanese electric power sector: a challenge for transitional management. Energy Policy 2000, 28:39-47

Fardoun, F., Ibrahim, O., Younes, R., and Louahlia-Gualous, H. Beirut. 2012. Electricity of Lebanon: Problems and Recommendations.

U.S Department of Energy, (2011, February 04). Energy Savings Performance Contracts. Retrieved March 18, 2011, from U.S Department of Energy: http://www1.eere.energy.gov

World Bank. LEBANON: Social Impact Analysis For the Electricity and Water Sectors. MENA Knowledge and Learning. Quick notes series, Number 14, November 2009b

Development of a Model for the Prediction of Indoor Climate to Enhance Design Tasks in Southern Climates

Ayman Bishara, PhD **Helge Kramberger-Kaplan, PhD**

ABSTRACT

In planning practice for high performance buildings comparatively simple and reliable tools for the quantification of indoor climate conditions (temperature and humidity) are rare, whereas established tools (e.g. TRNSYS, ENERGY PLUS) are very comprehensive and require detailed user knowledge in operation. In the present paper, the development, implementation, verification, and exemplary application of a new thermal-hygric model for solving building-climatic challenging planning tasks is presented. The use of thermal insulation in buildings under hot climatic conditions, for energy saving in summer, has been verified in southern climates and the implementation into a user-friendly software, named CLIMT (Climate-Indoor-Moisture-Temperature) is shown. For the model calibration, a case study in Germany was chosen, in which interior and exterior temperature and relative humidity values were measured. Even though the building shows the peculiarity of an irregular user behavior, very good accordance between measured and calculated values could be achieved. In a further step, the application of the developed program CLIMT is demonstrated by means of a specific example: the hygrothermal optimization of a church building design, located in a hot humid climatic zone (Latakia, Syria). As part of this design, several challenging boundary conditions had to be considered. Besides specific building physical and building climatic requirements (high humidity and high temperatures), also structural requirements had to be taken into account. Furthermore, the new church building had to assimilate with the historically grown environment and the specific building typology traditions of the region. The optimized building design, facilitating a damage-free operation and low building energy consumption, is presented in detail and essential optimization steps are discussed. The represented model allows for a simplification of building climatic design, even in challenging southern climate conditions, through the implementation of a user-friendly software and is therefore particularly suitable for use in architectural planning offices.

INTRODUCTION

In southern countries, the energy consumption of buildings is steadily rising due to an increasing cooling demand. In addition, the requirements for a comfortable room climate are increasing. For the design, planning and reconstruction, user-friendly and easy to use design tools can contribute to a consideration of building climatic criteria, already in the architectural planning phase. To quantify indoor climate and energy consumption of buildings, and optimize the hygrothermal behavior of components, a number of software tools are already available. However, they are mostly very comprehensive and extensive knowledge about the software operation is necessary.

To quantify the influence of the individual construction and climate measures during the design and planning phase, models are needed that allow the calculation of the indoor climate, depending on outside climatic stress, building cubature, details of construction, material parameters and usage regime. An additional application for these models is to review the damage freedom of constructions. Thus, the application of materials can be optimized, their limits identified, and new materials with improved thermal and hygric properties can be developed. The interdisciplinary coordination between architecture, civil engineering, building technology and materials science is a prerequisite for this optimization process. This allows a holistic approach in the design, planning and execution, and requires a solid professional knowledge of the planning process.

The main aim of this work is to develop, implement and verify an adequate precision hygrothermal space model,

Dr. Ayman Bishara is a researcher in the department of building physics, Dr. Robert-Murjahn-Institute, Ober-Ramstadt, Germany. **Dr. Helge Kramberger** is head of Dr. Robert-Murjahn-Institute, Ober-Ramstadt, Germany.

which is also applicable to Mediterranean and tropical conditions. The model should be applicable for the preliminary and further design phases and accordingly should include a clear input and output option of all necessary parameters.

MODEL DESCRIPTION

The developed simplified model and program CLIMT (Climate-Indoor-Moisture-Temperature) can be used for the calculation of the indoor climate components consisting of air temperature and air relative humidity (compare Glaubitz and Schneider 2006). The program CLIMT is user friendly and suitable for the normal professional use. Figure 1 shows a basic situation in order to quantify the indoor climate in a building. At first, we need the hourly values of the outdoor climate: temperature, relative humidity, direct solar radiation, diffuse solar radiation and long wave radiation. The latter has been calculated by using the first climate parameters. Secondly, the geometry of the building, the construction of the building parts and the thermal and hydric material parameters must be given. Finally, all parameters of use – heating and air conditioning, inner heat and moisture sources, ventilation and the shadow equipment -are necessary (Häupl, Bishara and Hansel, 2010, Bishara 2011). Before the energy and moisture balance was formulated, a set of exterior climate components (here the climate of the German city Essen) has been measured: wind direction in °, wind velocity in m/s(ft/s) , precipitation in mm/h (in/h), air pressure in kPa (psi), temperature in °C (°F), relative humidity in %, direct radiation in W/m² (Btu/h·ft²), diffuse radiation in W/m² (Btu/h·ft²), luminosity in Lux (fc), building long wave radiation and sky long wave radiation in W/m² (Btu/h·ft²).

Additionally, a model for the calculation of long wave radiation heat loss by means of the measured climate components mentioned above has been developed (Bishara 2011). The basic equation is the STEFAN-BOLTZMANN law (3). The ε-value of the building surface is about 0.96, the ε-value of the sky (1) is dependent on the temperature T, the relative humidity ϕ and the sky covering scale n (2). n can be calculated by the relation VT of the diffuse to the whole solar radiation in a defined time interval from 9am to 3pm (day), (2). For the time steps from 3pm to 9am (night) a linear approximation V_N (2) has been used (Konzelmann 1994). Finally, the whole long wave radiation is given by (3).

$i = 0, 1....364, l = -3, -2....3, k = -8,-9......9,\quad j_T(i, l) = 12+24.i+l,\ j_N(i, k) = 24.i+k$

i: Index of day, l index of hours from 9am to 3pm, k index of hours from 3pm to 9am, j index of all years hours

$$^{\varepsilon}T(i,l)=(\phi_T(i,l))^{0.3}\cdot\left[\left(\frac{T_T(i,l)}{400}\right)^{0.5}\cdot\left[1-(n_{T(i,l)})^4\right]+(n_T(i,l))^4\right]^{0.75}.0.95 \qquad (1)$$

$$n_T(i,l)=V_T(i,l)^{1.15}.1.075-0.084,\quad n_N(i,k)=V_N(i,k)^{1.15}.1.075-0.084 \qquad (2)$$

$$G_{langT}(i,l)=\sigma.T_T(i,l)^4.(\varepsilon_T(i,l)-\varepsilon_B),\quad G_{langN}(i,k)=\sigma.T_N(i,k)^4.({}^{\varepsilon}N(i,k)-{}^{\varepsilon}B) \qquad (3)$$

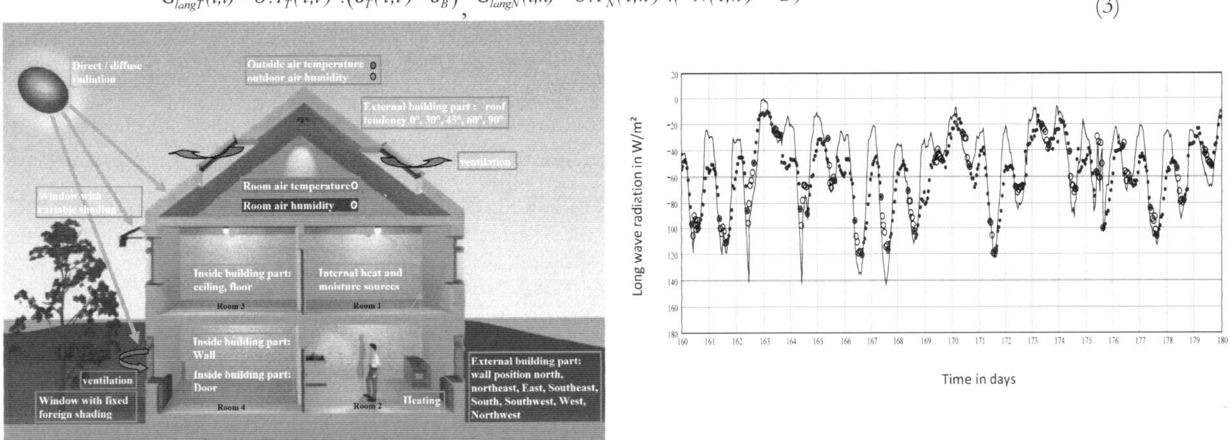

Figure 1 (a) Test house construction with outdoor climate loading and given use regime. (b) Comparison of the TRY Essen long wave radiation (thin line) and the calculated long wave radiation (equations (1) to (3), small points 3pm to 9am, big hollow points 9am to 3pm)

Modelling of energy balance

The model used has been described in (Bishara 2011, Häupl 2008) (compare also the models in Häupl and Bishara 2010, Brown 1990). The heat capacity of the room air is very low, which means that C_L is approximately zero. The heat capacity of the inner storage mass is C_i and that of the outside mass is C_e. The heat transport terms are $T_{W'}$ (transmission for walls and roofs without transition coefficients), T_F (transmission for the windows including the transition coefficients), S_W (indirect radiation transport through the walls and roofs), S_F (direct radiation through the windows), $\ddot{U}_i$, $\ddot{U}_e$ (transition on the surfaces), and L (heat transfer value by ventilation). Hence, we obtain the following simple system of energy balance equations for the calculation of the outer surface temperature (4), for the inner surface temperature (5) and the indoor air temperature (6).

$$S_W = T'_W.(\theta_{oe}-\theta_{oi}) + \ddot{U}_e.(\theta_{oe}-\theta_e) + C_e.\frac{d\theta_{oe}}{dt} \tag{4}$$

$$0 = (L+T_F).(\theta_e-\theta_i) + \ddot{U}_i.(\theta_{oi}-\theta_i) + \frac{J}{2} \tag{5}$$

$$C_i.\frac{d\theta_{oi}}{dt} = S_F + T'_W.(\theta_{oe}-\theta_{oi}) + \ddot{U}_i.(\theta_{oi}-\theta_i) + \frac{J}{2} \tag{6}$$

The inner surface temperature for the time steps (one hour) j =1, 2....8760 is given by the solution of the differential equation system (4) to (6):

$$\theta_{oi,j+1} = \theta_{oi,j} + (\theta_{oiLIM,j} - \theta_{oi,j}).(1-exp(\beta_j.t)) \tag{7}$$

If the temperature is known at the time, then step j, the temperature at the time step j+1 can be calculated by a simple exponential time function. The limit temperature $\theta_{oiLIM,j}$ contains only the transport terms, the radiation gains and the inner heat sources; the time coefficient βj contains the transport terms and the heat capacities of the building.

$$\beta_j = \frac{-E_j}{2} - \sqrt{\left(\frac{E_j}{2}\right)^2 + B_j}, \quad E_j = \frac{\ddot{U}_i}{C_i.(L_j+T_F)}.\left(\frac{1}{\frac{1}{\ddot{U}_i}+\frac{1}{L_j+T_F}}\right) - \frac{T'_W+\ddot{U}_e}{C_e} - \frac{T'_W+\ddot{U}_i}{C_i}, \quad B_j = \frac{T'_W+\ddot{U}_e}{C_i.C_e}.\left(\frac{1}{\frac{1}{\ddot{U}_i}+\frac{1}{L_j+T_F}} + \frac{1}{\frac{1}{T'_W}+\frac{1}{\ddot{U}_e}}\right) \tag{8,9,10}$$

Finally, the indoor air temperature for each time step is given by

$$\theta_{i,j+1} = \theta_{i,j} + \left[\begin{array}{l}\theta_{e,j} - \theta_{i,j} + \\ \dfrac{\dfrac{\left(\frac{S_{Fj}}{\ddot{U}_e}+\frac{S_{Fj}}{T'_W}+\frac{S_{Wj}}{\ddot{U}_e}\right)}{(L_j+T_F)}}{\frac{1}{\ddot{U}_e}+\frac{1}{T'_W}+\frac{1}{\ddot{U}_i}+\frac{1}{L_j+T_F}} + \dfrac{\dfrac{\left(\frac{1}{\ddot{U}_e}+\frac{1}{T'_W}\right)}{(L_j+T_F)}.\frac{J_j}{2}}{\frac{1}{\ddot{U}_e}+\frac{1}{T'_W}+\frac{1}{\ddot{U}_i}+\frac{1}{L_j+T_F}}\left(1+\frac{\ddot{U}_i}{L_j+T_F+\ddot{U}_i}\right) + \dfrac{\frac{J_j}{2}}{L_j+\ddot{U}_i+T_F}\end{array}\right].\left(1-exp\left(-\beta_j.t\right)\right) \tag{11}$$

This solution (11) contains all heat transport coefficients T, Ü, L, the solar radiation values S, the indoor heat sources J and in β_j the heat storage coefficients C in a logical clear manner. Moreover, heating power can be integrated.

Modelling of moisture balance

The moisture balance can be formulated in the same way as the heat energy balance. The driving potential is the vapor pressure. The moisture storage term depends on the hygroscopicity and the μ-value of the materials of all inner surfaces. The model for determining the room temperature can be transferred analogously to calculate the water vapor pressure or relative humidity of the room air in the day- or annual cycle in natural outdoor climate and any use of space, depending on the building parameters. The moisture model for calculating the indoor climate is comparatively easy, because no moisture is transported by transmission through the comprehensive construction (unless rain penetrates the component) and there is no "humidity radiation "(Bishara 2011). The important variables (see Figure 3) are: vapor pressures p_{De}, p_{Se} and relative humidity ϕ_e for the outside and p_{Di}, p_{Si} and ϕ_i for the inside, moisture storage stream of the room air, the walls, the ceilings and the furnishings dm_{SPL}/dt or dm_{Spi}/dt, of the airborne moisture flow dm_L/dt, the moisture transfer current $dm_{Üi}/dt$ and the moisture source current dm_{QU}/dt. (Moisture flow stored from the room air (12) and from the walls, ceilings and furnitures (13), moisture flow transmitted through ventilation (14), moisture flow emitted and released from the surface (15) and released from the inner moisture sources (16)).

$$\frac{d}{dt}m_{SPL} = C_{LF}\cdot\frac{d}{dt}p_i \; , \quad \frac{d}{dt}m_{SPi} = C_{Fi}\cdot\frac{d}{dt}p_{oi} \qquad (12, 13)$$

$$\frac{d}{dt}m_L = \frac{V_L.n_L}{R_D}\cdot\left(\frac{p_e}{T_e}-\frac{p_i}{T_i}\right) , \quad \frac{d}{dt}m_{Üi} = Ü_{iF}\cdot(p_{oi}-p_i) , \quad \frac{d}{dt}m_{QU} = m_{ptV}.V_L \qquad (14, 15, 16)$$

The energy flow model in section 2 (equations (4) to (6)) is obtained analogously for the formulation of moisture flows (outer surface moisture balance (17), indoor air humidity balance (18), interior surface moisture balance (19)).

$$\frac{d}{dt}m_T = 0 \qquad (17)$$

$$C_{LF}\cdot\frac{dp_i}{dt} = \frac{V_L.n_L}{R_D}\cdot\left(\frac{p_e}{T_e}-\frac{p_i}{T_i}\right)+Ü_{iF}\cdot(p_{oi}-p_i)+m_{ptV}.V_L \qquad (18)$$

$$C_{iF}\cdot\frac{dp_{oi}}{dt} = Ü_{iF}\cdot(p_{oi}-p_i) \qquad (19)$$

Instead of the temperatures, the water vapor pressures act as the driving potentials. The moisture balance equation on the outer surface is not relevant for indoor air humidity and can therefore be omitted (17). However, the moisture storage capacity of the indoor air C_{LF} in kg/Pa (lb/psi) in the moisture balance equation for room air (18) must be considered. It follows from the gas equation (20)

$$C_{LF} = \frac{V_L}{R_D.T_i} \qquad (20)$$

The transmission terms T are omitted. Analogous to the specific heat ventilation flows L are the specific (now related to the vapor pressure) ventilation moisture flows L_{Fi} and L_{Fe} in (18)

$$L_{Fi} = \frac{V_L.n_L}{R_D.T_i} \; , \quad L_{Fe} = \frac{V_L.n_L}{R_D.T_e} \qquad (21a, 21b)$$

The moisture transfer value from room air /inner surfaces in kg/sPa (lb/s.psi) is given by:

$$Ü_{iF} = \sum_{j=1}^{n} 7{,}9.10^{-9}\,\frac{kg}{Pa}\,\frac{K}{Ws}.h_c.A_{Wij} \qquad (22)$$

The humidity flow of the sources dmp/dtdV= m_{ptV} in kg/m³ (lb/ft³) h is completely (not just half like in (5)) released into the air. In the moisture balance equation for the enclosure surface (16) the terms T and S are deleted. The same applies to the term J/2, because moisture is not transferred by radiation. Moisture storage in the enclosure surface or furnishings C_{iF} in kg/ms² = Pa (psi) is for a step load with t_{sp} as penetration time of the hygric signal.

$$C_{iF} = \frac{1}{2}\cdot\sqrt{\frac{t_{sp}}{\pi}}\cdot\sqrt{\frac{\delta_L}{\mu}.\rho_W.\frac{w_h}{p_s}}.A_{oi} \qquad (23)$$

The balance equations (19) are inserted into equation (18), which leads to a simple differential equation for calculating the time-course of water vapor pressure poi on the inner surface and later for the room air p_i. The solution for the (j +1)-st (selected here a total of 8760 hours time steps for a year) period is (compare Hansel 2011):

$$p_{oi,j+1} = p_{oi,j} + \left(p_{oi,LIM,j} - p_{oi,j}\right).\left(1 - exp(-\beta_{Fj}.t)\right) \quad (24)$$

Herein means $p_{oi,LIM,j}$, water vapor pressure at the surface after an infinitely long time, poi,j is the pressure at the beginning of the j-th time step, $p_{oi,\,j+1}$ is the pressure at the end of the j+1 st time step and β_{Fj} is the time constant for humidity adjustment within the j-th time step. It approximately follows that

$$p_{oi,LIM,j} = p_{e,j} + \frac{m_{ptV}.R_D.T_i}{n_{Lj}}, \quad \beta_{Fj} = \frac{1}{C_{iF}}.\left[\frac{1}{\frac{R_D.T_i}{V_L.n_{Lj}} + \frac{1}{\ddot{U}_{iF}}}\right] \quad (25, 26)$$

To determine the partial pressure of water vapor of the outside air at time j, the saturation pressure must be calculated depending on the outside air temperature (see table 1)

$$p_{se,j} = 610,5.\left(e^{\frac{17,26.\theta_{e,j}}{273,3+\theta_{e,j}}}.\Phi\left(\theta_{e,j}\right) + e^{\frac{21,87.\theta_{e,j}}{265,5+\theta_{e,j}}}.\Phi\left(-\theta_{e,j} - 10^{-6}\right)\right) \quad (27)$$

A step function Φ, which is used here and in the following relationship, is 1 for positive arguments and 0 for negatives. With the measured values of relative humidity outside ϕ_{ej} it follows that the water vapor pressure is given by

$$p_{e,j} = \phi_{e,j}.p_{se,j} \quad (28)$$

The inertia of a moistening or drying process β_{Fj} is determined by the storage capacity of the indoor air $V_L n_{Lj}/R_D T_i$, the storage capacity of the enclosure surface area and of the furnishings C_{iF} and the moisture transfer resistance $1/\ddot{U}_{iF}$ of the room air to the inside surface. Finally, we can determine the vapor pressure of the room air:

$$p_{i,j+1} = \frac{p_{o,j+1}.\ddot{U}_{iF} + p_{o,j+1}.\frac{V_L.n_{Lj}}{R_D.T_i}.3600}{\frac{V_L.n_{Lj}}{R_D.T_i}.3600 + \ddot{U}_{iF}}, \quad p_{si,j+1} = 610,5.\left(e^{\frac{17,26.\theta_{i,j+1}}{273,3+\theta_{i,j+1}}}.\Phi\left(\theta_{i,j+1}\right) + e^{\frac{21,87.\theta_{i,j+1}}{265,5+\theta_{i,j+1}}}.\Phi\left(-\theta_{i,j+1} - 10^{-6}\right)\right) \quad (29, 30)$$

From the water vapor pressure (29) and the saturation pressure (30) in the room it follows that the second major component of indoor climate, the inside relative humidity at the time j+1 is:

$$\phi_{i,j+1} = \frac{p_{i,j+1}}{p_{si,j+1}} \quad (31)$$

From the values for room air temperature and air humidity at time j, the values at time step j+1 can be calculated by a simple loop. No iteration is required. The heating power J(j) can be given manually or independent of the outside or inside climate. The same applies to ventilation rate n(j) and shading z(j). The intensity of internal heat and moisture sources are based on the use of space.

The validation of the calculation modules has been realized under laboratory conditions.

VALIDATION OF THE MODEL BY A CASE STUDY

The test house is a so-called bath house in South Brandenburg, operated by the University of Applied Science in Cottbus (figure 2 (a)). This bath house is used as a holiday flat for families and it is equipped with an electric heating system. The outdoor and indoor climate and the usage regime has been measured. The indoor climate values are compared with the calculated values. For the calculation we have used the Windows program CLIMT (Hansel 2011), which is based on our MathCAD model, with a user-friendly surface. Figure 2 (b) shows an example of the input masks. The following figures give the calculated and measured results (indoor air temperature and indoor air relative humidity) for the bath house in South Brandenburg.

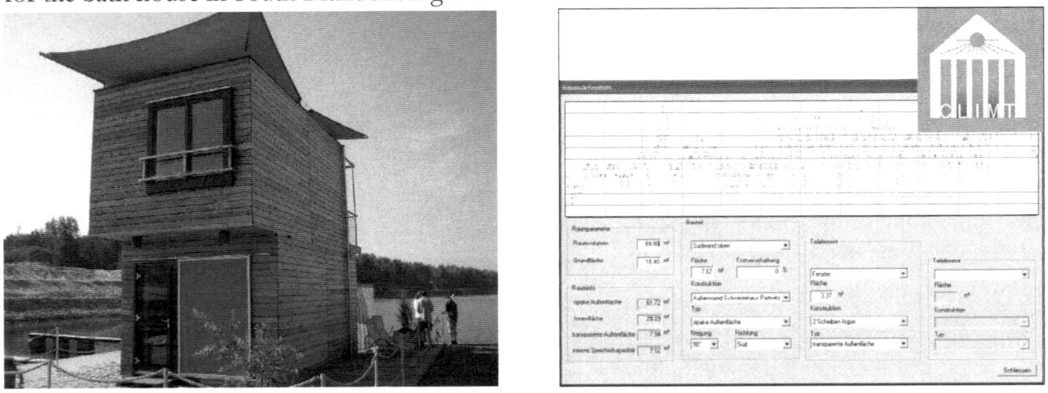

Figure 2 (a) bath house In South Brandenburg. (b) CLIMT Input data for room parameters

In January and February, a slight increase in heating guarantees 16°C (60.8°F). In March and April we have only sporadic use, later (until October) we have regular occupancy by guests. In November the heating operation was switched off (the house was under repair). In December, a temperature of only 6°C (42.8°F) is necessary for frost protection. The measured and calculated values correspond amazingly well (figure 3(a)). A similarly satisfactory correspondence was noticed for the calculation results of temperature by the TRNSYS (Glaubitz and Schneider 2006) program. Also the correspondence between the measured and calculated (CLIMT) values is satisfactory. Of course, if the temperature decreases, then the relative humidity increases (figure 3(b)).

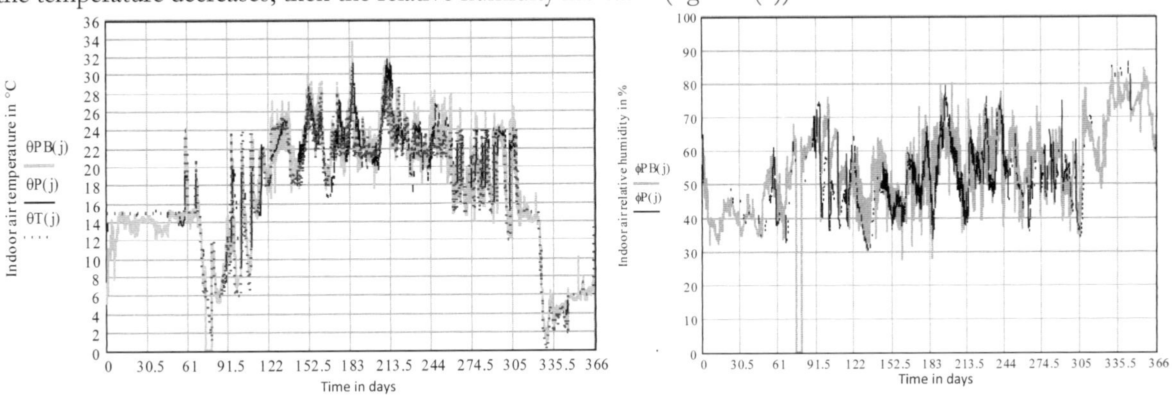

Figure 3 (a) Room air temperature of the bath house in 2008: Calculated by TRNSYS (pointed line), calculated by CLIMT (black line) and measured values (grey line). (b) Room air relative humidity of the bath house: calculated by CLIMT (black line) and measured values (grey line)

NEW BUILD DESIGN OF A CHURCH IN ALRWDA USING CLIMT SOFTWARE

A new church building was planned to be designed in Alrwda, Syria. This project deals with an architecturally sensitive building, which is historically and liturgically very consistent. Several boundary conditions, such as structural constraints, extreme weather conditions and highly irregular user behavior had to be taken into account, in addition to the specific architectural design criteria. The church has an area of 500 m^2 (5381.96 ft^2) and a total volume of 6000 m^3 (211888 ft^3) (figure 4). The aim is to design the church in such a way that the indoor room temperature is kept below 28 °C (82.4°F) in summer and above 13 °C (55.4°F) in winter, without any technical equipment (air conditioning, heating, etc.) installed (Bishara 2011). Already in the tendering stage, a building construction based on reinforced concrete was specified. This was due to limited financial resources, as well as limited available craft skills. Therefore, the selected architectural and structural variants are based on the premise of reinforced concrete. The church is to appear like a traditional building in spite of modern reinforced concrete structure. For this reason, the external walls of the church are covered with natural stone and the interior walls are covered with lime plaster. The concrete dome and barrel vaults of the roofs are covered with tiles and the inside ones are also plastered. Thus, the thermal mass is immediately adjacent to the interior, and an improvement of the interior climate is given. Additionally, the exterior stone cladding increases the thermal mass of the whole building.

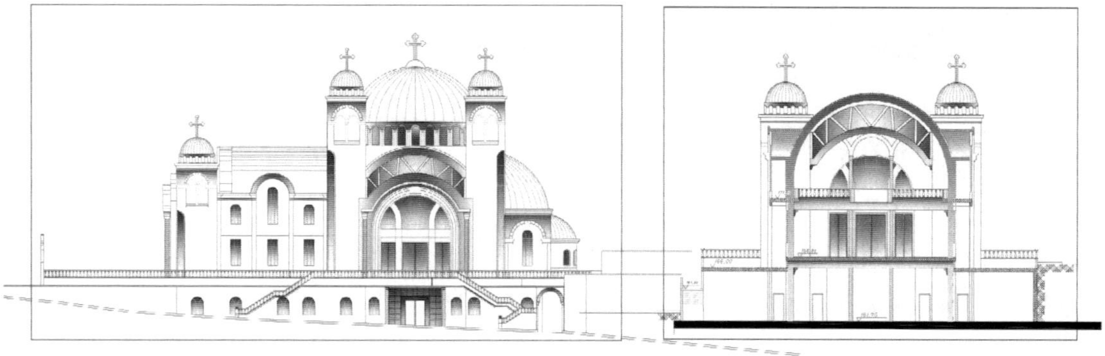

Figure 4 (a) south façade of the Church in Alrwda. (b) Vertical section on the southern facade

Design variants

For heat insulation, wood fiber boards were chosen as insulation for a basic church design variant. The wood fiber boards are made of untreated, domestic softwood and contribute to an effectively insulated and vapour permeable construction design of walls and roofs. The characteristics of this system are: high compressive strength and permanent hydrophobic properties. Additionally, the high density 250 kg/m³ (15.61 lb/ft³) of the insulation boards contributes to particularly good summer heat protection. In our laboratory, we measured thermal conductivity to be 0.051 W/mK (0.354 But in/h.ft².°F) and the water vapour diffusion resistance factor to be 5 (μ-value).

The basic design variant consists of the following features: roof insulation, brightly-painted inner walls, colored double glazing (a stained glass pane and clear glasspane), and intensive night ventilation in summer. Additionally, a second variant was investigated, possessing the same properties as the basic variant, but without roof insulation. In a third variant, all interior walls are painted dark. For better comparison, the room air temperature was also calculated for a worst case. This variant consists of single glazing, no roof insulation, dark interior walls, and basic ventilation.

Results and discussion

Figure 5 (a) shows the annual course of room air temperature and relative humidity level in the basic variant. The room air temperature is between 12.5°C (54.5°F) and 26.6°C (79.88°F). Room air humidity varies from 40 to 95%, the average value is 67%. Figure 5 (b) shows the course of a year for room air temperature for all variants described above

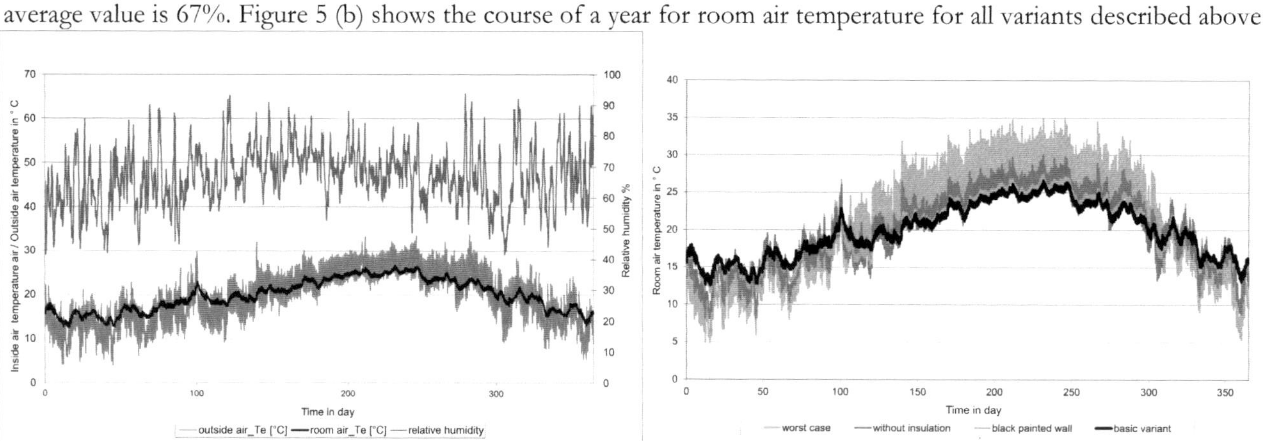

Figure 5 (a) Annual course of inside air temperature and inside relative humidity level for the basic variant. (b) Course of a year for room air temperature for all variants

In Variant 2 (without insulation), the maximum temperature is 30.0 °C (86.0°F) and the minimum temperature 8 °C (46.4°F), compared to 26.6 °C (79.88°F) and 12.5 °C (54.5°F) in a basic variant. For 38 hours in 7 days in a year the temperature is above 30 °C (86.0°F). In the basic variant, the inner church walls are painted white, as solar absorption of the inner surfaces will be positively affected. About 30-40% of the radiation which penetrates through the windows is reflected by the white walls. In variant 3, dark interior walls were prepared to demonstrate the influence of this contrary effect. In this case, the maximum temperature is 27.5 °C (81.5°F) and varies 528 hours to 29 days a year between 26 °C (78.8°F) and 30 °C (86.0°F). The worst variant shows, as expected, a very high temperature of 35.1°C (95.18°F) in summer and a very low room air temperature of 4.9 °C (40.82°F) in winter.

Table 1 summary of results for all variants

	Internal air temperature			Hours			
	Maximum °C (°F)	Minimum °C (°F)	Mean value °C (°F)	> 30°C (> 86°F)	26-30°C (78-86°F)	20-26°C (68-78°F)	>10°C (> 50°F)
Basic variant	26,6 (79.88)	12,5 (54.5)	19,8 (67.64)	0	95	4408	0
Without insulation	30,0 (68.0)	8,1 (46.58)	19,8 (67.64)	38	1307	3149	340
Dark interior wall	27,5 (81.5)	11,5 (52.7)	20,0 (68.0)	0	528	3687	0
Worst case	35,1 (95.18)	4,9 (40.82)	21,3 (70.34)	1134	1991	1740	985

SUMMARY

On basis of simplified energy and moisture balance of a building, a model and program CLIMT for the practical calculation of indoor air temperature and relative humidity has been developed. The results have been compared with TRNSYS and with measurements, achieved from a case study in Germany. All results correspond very well.

Additionally, the climatic optimization of a new church design was performed by the use of CLIMT. Various simulation studies lead to a hygrothermally optimized church construction design. The final design consists of natural ventilation (no heating, no air conditioning and no ventilation system) and provides maximum indoor temperatures of 27 °C (80.6°F) during summer and minimal indoor temperatures of 13 °C (55.4°F) during winter. For approximately two-thirds of the year, the indoor air temperature reaches the comfortable range of 20 °C to 26 °C (68.0 to 78.8°F). The hygrothermal building model and program CLIMT, presented in the present paper, allows for the verification of the influences of the outside climate and the construction techniques as well as the building use on the quality of the interior climate. Thus, it establishes a variety of building climatology rules.

NOMENCLATURE

T` = Heat transfer value

U` = Specific heat transfer value of constr

δ_L = Heat transition value

S = Solar radiation value

R = Gas constant

m = Effective storing mass

n = Air change rate

REFERENCES

Bishara, A. 2011 *Bauklimatische Simulationsverfahren zur Lösung von Entwurfs-, Planungs- und Sanierungsaufgaben in Syrien, Dissertation,* Dresden: University of Technology

Brown, G. 1990. *The BRIS Simulation Program for Thermal Design of Buildings and Their Services.* Energy and Buildings

Glaubitz, J. and Schneider, J.2006. *Handbuch zu TRNSYS 16*, Hamburg: HWAK

Häupl, P., Bishara, A. and Hansel, F. 2010. CLIMT - model and program for the prediction of the indoor air temperature and indoor air relative humidity. *CESBP (first Central European Symposium on Building Physics), Cracow 2010*

Häupl, P. 2008. *Bauphysik – Klima, Wärme, Feuchte, Schall – Aktiv in Mathcad*: 550 S. Berlin: Ernst & Sohn.

Hansel, F. 2011. *Dokumentation CLIMT*, BTU Cottbus, Senftenberg, Germany http://www.hslausitz.de/fakultaet4/lehrgebiete/bauphysik/climt.html

Konzelmann, T. et al. 1994: *Parameterisation of global and long wave incoming radiation for the Greenland ice sheet.* Global Planetary Change Vol. 9, p. 143-164

Reassessing the Numerous Proposed and Existing U-values for Lebanon

Philip H. Saleh

ABSTRACT

Since 2005 a number of publications have proposed different U-values to be used in Lebanon in order to reduce the buildings' energy demand for cooling and heating. This research considers those different values from the two editions of the Thermal Standard for Buildings in Lebanon (2005 & 2010) and Lebanon Center for Energy Conservation LCEC guidelines (2014), in addition to recommended U-values from similar worldwide climates. In the second part of the paper, dynamic thermal simulation software (EDSL TAS) is used to test the proposed U-values in conjunction with typical local construction materials and using Bayrouth weather files (Meteonorm 7). The tridimensional model used for the simulations is based on a typical existing building with characteristics kept constant throughout the comparative study. Furthermore the same internal heat gains and patterns for occupancy and appliances, as well as window opening areas and schedules are also kept the same in all the simulations. For each case, the four main cardinal orientations are tested. The research compares the overall yearly energy demand for mechanically heated and cooled buildings, in order to assess which U-values will give the lowest energy demand under varying levels of internal heat gains.

INTRODUCTION

The concern to reduce buildings energy reliance on fossil fuel for internal comfort is, in today's social media language a "trending topic". The triggering incident for this concern can be traced back to the 1973 first OPEC crisis that resulted in an attempt to reduce dependency on fossil fuel. Later, the environmental impact of global warming and its close relationship with fossil fuels' CO_2 emission as the main culprit, consolidated the concern with fossil fuel consumption, especially considering that the built environment global end use emissions account for an estimated 40 to 50% of overall energy consumption.

Construction codes and bio-climatic design source books (Nicol et al, 2012; Szokolay, 2004; Yannas 1994; Littlefield, 2007) promote qualitatively and quantitatively the use of low U-values, or highly insulated construction, for temperate European or north American climates, whereas they include marginal notes with limited quantitative data stating that the use of thermal mass will greatly affect the thermal performance of any construction by reducing its internal temperature fluctuation. This reductive effect is further enhanced by its combination with night time, cooling ventilation. Kiel Moe's (2014) 'Insulating modernism', attacks all the buildings' insulation practice by going into a detailed historical overview on how insulation emerged in the refrigeration industry and was imposed into the architects as an ideal solution, without leaving any room for re-questioning its efficiency. He carries on with numerical demonstration of the drawback of insulation. He concludes that the alternative is the use of thermal mass, yet he only refers to limited built examples without any quantitative demonstration of their efficiency.

Moving to Beirut, Lebanon, whose coastal climate is defined by hot and humid long summer without precipitations, and warm short winters (TSB, 2010 & Kottek et al; 2006), its building stock is almost entirely made out

Philip H. Saleh PhD candidate, Architecture & the Built Environment; University of Westminster, London, UK.

of stone, concrete or related and combined construction material, all fitting within the definition of heavy weight construction. Based on the above described benefits of thermal mass, one would expect to find a very good example of low energy performing buildings. Yet it appears that for Beirut's internal summer comfort is highly reliant on mechanical cooling.

Furthermore, a number of local environmental bodies are active in the construction regulations field, yet working in parallel, and accordingly have recommended different U-values and energy benchmarks. This paper starts by listing all the available U-values (W/m^2K), Table 1, from the different local sources, including the foreign ones used as a reference for appropriate local values along with those from similar climatic zones in a range of countries. The paper continues by listing different energy benchmarks ($kWh/m^2.y$), Table 2, as well as proposed limits of internal gains from occupants (W/m^2) and the maximum energy per year from lighting ($kWh/m^2.y$)Table 3 (LCEC, 2014). The paper then, carries on the first round of parametric energy simulations using all the values listed above onto a typical apartment building rotated to check the four cardinal orientations. In a second round, the internal gains are exaggerated to check the resilience of the systems.

List of different bodies

The following list shows the different bodies whose published values are considered herein: Both the Thermal Standard for Buildings in Lebanon (TSB) (Republic of Lebanon 2005 & Order of Engineers 2010); and the National Energy Efficiency and Renewable Energy Action (NEEREA) (LCEC, 2014). In addition to those, Agence Libanaise pour la Maitrise de l'Energie et l'Environment (ALMEE) published The Thermal Insulation Market in Lebanon (Comair et al, 2011) where they interpreted the thickness of the insulation based on the TSB 2010. Also within the introduction of the TSB (2010) it is stated that data were collected from a variety of sources including: the TSB 2005; ASHRAE 90.1.2004; the RT2005 French Thermal Regulation, and the Tunisian Thermal Regulation ZT1, 2008. For both French and Tunisian case, only values relative to the Mediterranean zone (south of France and North-East of Tunisia) were considered (FFEM et al, 2010; French Republic, 2006 a&b) whereas for ASHRAE 90.1, zone 2's values are considered (ASHRAE 90.1;2007). Finally based on the climate classifications of Koppen-Geiger (Kottek et al, 2006) California falling within similar Beirut climate has its U-values shown herein.

Envelope U-Values

Table 1 combines all the U-Values for each of the external walls, the roof and the windows, from the above mentioned list. The values are considerably different from one source to the other, and hence the roof U-values range from 0.1 to 0.75 W/m^2K, the external walls from 0.18 to 1.62 W/m^2K. In both cases the lowest values are the LCEC guidelines (2014), which did not specify any value for the windows. Otherwise those windows U-Values range from 1.81 to 6.2 W/m^2K which encompass triple glazing with low-e; to single panel windows, as well as the intermediate double glazing. When the source gave the values in imperial units for U-factor, a conversion factor of 5.678 is used to change into SI units to U-Values (ASHRAE 2013).

Table 2 shows the limited proposed cooling and heating energy benchmarks from sources where available. The "business as usual" as defined by the LCEC (2014) is for any typical building without any energy consideration, it is set for residential at 118 kWh/m^2 per year out of which only 3 are for heating. On the other hand the benchmark value for a building to start to be considered as energy efficient is 80 kWh/m^2 per year.

Table 3 shows the only available proposed internal gains from occupants, as well as the proposed yearly energy from lighting (LCEC; 2014).

Heavyweight Construction

The term thermal mass refers to any material that has the capacity to absorb, store and release heat (Littlefield,

2007; Szokolay, 2004). Concrete, masonry walls and stone finishes are high density material and at the same time with high thermal capacity, hence the term heavy weight refers to that type of construction, which is the main construction type in Lebanon. Szokolay (2004) defines heavy weight or thermal mass as a building with a density of 400kg/m³ and above. A calculation from a typical residential concrete local building in Beirut will give values above 500kg/m³

Table 1 All U-Values from different local and foreign sources. The Bold Numbers show similar values within different references.

Title	Year	Roof	Walls (W/m².K)	Window	Notes	Reference Number
ASHRAE 90.1.2007 (Zone 2 A,B)	2007	0.27	0.70	4.26	Window U-Value for up to 40% Wall Area	2;3
Thermal Standard for Buildings	2005	0.57	2.1	6.2		19
RT2005 H3[1]	2006	0.34	0.45	2.6		9;10
Tunisia ZT1[2]	2008	0.75	1.2	6.2	Low Window to Floor area Ratio (WFR)	8
			1.1	6.2	Medium to High WFR area ratio	
			0.8	3.2	Very High WFR area ratio	
Thermal Standard for Buildings	2010	0.71	1.6	5.8	Till 25% WFR area ratio	18
			1.6	4	For 26 -35% WFR	
			1.26	3.3	For 36-45% WFR	
Thermal Insulation Market in Lebanon [3]	2011	(3.2cm)	(1.2cm)	N.A.	Data given in thickness	6
		0.58-074	1.59-1.62		Calculated U-Value	
2013 Residential Compliance Manual California Energy Commission [4]	2014	0.14-0.17	0.71	1.81	Maximum 20% Window to Wall area Ratio Maximum 5% Window to Wall area Ratio for West Orientation	4
International Energy Conservation Code [5]	2015	0.15	0.71-0.44	1.98	Values of Roof & Walls converted from R-values, converted to SI. U-Values of Windows converted to SI	1
LCEC Guidelines on Preparing Technical Proposal for Non-Certified High Energy Performance Building	2014	0.1-0.15	0.18-0.31	N.A.		13

[1] French Thermal standards for H3 Zone: Mediterranean area of south France
[2] Tunisian Norms for Private Buildings ZT1 zone which is the Mediterranean area of North East Tunisia
[3] Calculated U-Value based on XPS insulation thickness. Density 26-75Kg/m3 and R=0.026-0.037 W/m.K
[4] All U-values converted to metric value by multiplying by a conversion factor 5.678
[5] R-Values Converted to U-Value by U=1/R ; Values of Window given in U-Value

Table 2 Yearly energy values in kW.h/m^2/year standards and benchmarks for residential and offices from different sources.

Title	Year	Residential Standard	Residential Benchmark	Office Standard	Office Benchmark	Reference Number
RT2005 H3*	**2006**		80**			9;10
Lebanon Thermal Standards	**2010**		80		85	18
LCEC Guidelines on Preparing Technical Proposal for Non-Certified High Energy Performance Building	**2014**	118	80	107	75	13

* French Thermal standards for the H3 Zone : Mediterranean area of south France
** Based on fossil fuel heating (as opposed to electrical heating which has higher value)

Table 3 Internal gains from occupants and maximum yearly energy from lighting (LCEC 2014)

	Internal Gains from occupants (W/m^2)	Maximum Yearly Energy from Lighting (kWh/m^2)
Residential	5	13
Offices	14	17

Thermal Mass Behavior

Thermal performance of heavy weight construction is described as taking longer to warm up when exposed to solar gains and slower to respond to temperature variation (Littelfield, 2007; Szokolay, 2004). In the same way it takes longer to cool down and loose the extra heat. This stored heat will dissipate during the night provided there is the opportunity to do so due to diurnal temperature swing and adequate ventilation. This situation is ideal to maintain a comfortable indoor temperature when the temperature difference between day and night is considerable. Furthermore Szokolay (2004) considers it a very practical free-running solution to keep internal conditions within the thermal comfort zone. Littelfield (2007) and Szokolay (2004) both add that the practical effect of thermal mass lies in its surface area rather than in its thickness hence larger exposed area have more impact than effective thicknesses, and as such 5 to 12 cm are enough to provide a 24 hour cycle of heat absorption and release. Those thicknesses are available in any reinforced concrete construction, in any of the elements, slabs, walls and even some floor finishing; with not less than 5cm of uninterrupted thickness of solid material in/on any wall. Within the 2013 Residence Compliance Manual (2014) issued by the California Energy commission, it is stated that a 20 to 30cm concrete wall will have a combined dampening thermal time delay between 6 and 10 hours. Koch-Nielsen (2002) states that the process of cooling down thermal mass, or the heat dissipation, cannot be achieved without night time ventilation for hot and dry climates. This ventilation could be either natural or mechanically induced. He does not specify however hot and humid climates, but specifies that for warm and humid zones with little temperature fluctuation, thermal mass should not be applied; instead well ventilated light weight construction is the better choice of construction materials.

Building sample

For this study, the building that is taken as a prototype model for dynamic thermal simulation is located in the Ain er-Remmeneh area, on the outskirts of Beirut, and based on the users/owners testimony, it is one of the first residential and commercial developments to be built there in the late 1960's. Constructed of 5 floors, with two apartments on each, and a commercial ground floor, with a south-west main orientation for the living areas and blank exposed walls to its eastern and western façades. The building is made out of concrete slabs, plastered hollow concrete block walls, and all the windows have wooden frames. Each apartment is 110sqm and is made out of two bedrooms, one living and dining area, kitchen, two WCs and one entrance functioning as a small family living. The apartment is occupied by a family of four.

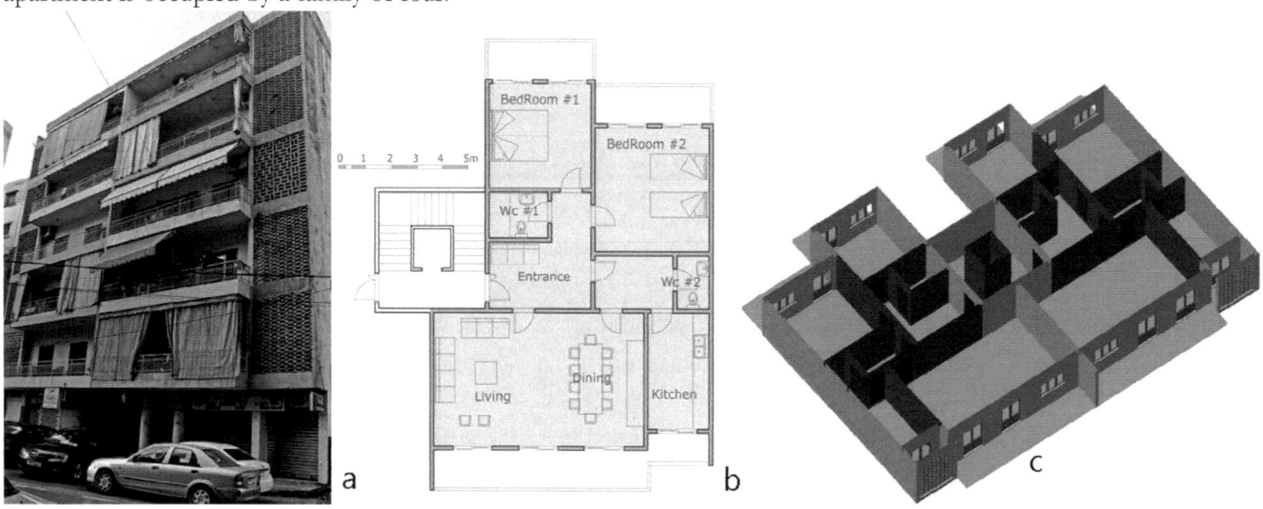

Figure-1 (a)The Building used as a model for the simulation; (b) Typical plan of one apartment; (c) 3D axonometric of the entire floor showing both apartments.

The apartment has individual split A/C units one in each bedroom and one in the living-dining area, which is seldom used, whereas the bedrooms' A/C are used during the night, when the main EDL power is available, hence not during the regular black outs when the neighborhood generator is the sole power supply. In the living area a fan and the open windows are used to provide some comfort throw ventilation from the summer heat and humidity.

This particular apartment has been monitored for ambient temperature and surface temperature from late August 2015 till early October 2015. The building is used as a typical generic model for the dynamic thermal model and simulation, over which the different U-values are tested.

Thermal Simulation

The EDSL TAS 9.3.2 thermal simulation software is used with the Bayrouth weather file 2000-2009 (Meteonorm 7). Four typical floors are modeled, each with the two adjacent apartments and cooling and heating yearly values per area are shown herein for the third and fourth floors, with the fourth considered as the roof or top floor and third as intermediate floor. Furthermore a total of four fully developed models with each having a different orientation (where the orientation is that of the main living areas) are simulated and analyzed.

The annual loads for cooling and heating are calculated for intermittent mode only when users are there, and occupancy is based on the observed apartment users' living patterns and remained unchanged in all the different simulations. Cooling temperature is based on 24°C set-point and 50% RH threshold; as for heating it is based on 20°C. The internal gains from lighting, users and equipments are the same throughout. For the LCEC values, 3 sets are done each with different window U-values from 6.41; 4.2 and 1.98 W/m^2K

Results

Run #1 Figure 2 shows all the values from the different orientation and runs for cooling and heating annual loads. The cooling load is always considerably higher than the heating load, averaging three to four time larger. Whereas the roof has higher heating and cooling values than the intermittent floor by less than 10%. As for orientation, the lowest values are always for the living areas with a north orientation followed by the south, and both East and West show almost the same values with very little difference. The highest values for a north orientation are 37 and 39 kWh/m^2 for intermediate and roof floor respectively, using the TSB 2005 U-values. Whereas the lowest values are from the LCEC with 4.2 W/m^{2}K window U-values reaching 27 and 29 kWh/m^2. The south orientation will again have its lowest values with the same LCEC U-values at 35 and 37 kWh/m^2, and its highest values with the TSB 2005 at 49 and 52 kWh/m^2.

When it comes to east and west orientations, the highest values are 55 and 58 kWh/m^2 this time from the TSB 2005 intermediate floor and the base case for the roof. Whereas the lowest values 37 & 39 kWh/m^2 are still with the LCEC with 4.2 W/m^{2}K windows U-values.

All achieved values are less than the 80 kWh/m^2 benchmark given by the local sources. Nevertheless, in order for the comparison to be feasible, one should note that major parameters that are intrinsic part of energy simulation are missing from the sources. Those are the specific weather files, the internal gains and the schedules of users, equipments and lighting. Also the number of users, the air exchange rate and volumes and finally the notion whether the cooling and heating are intermittent or continuous. All those missing factors render the comparison unsubstantial.

Run # 2 A second set of runs is performed to check the impact of increasing internal gains while pointing out the lack of benchmark values within the numerous local sources. For this scenario, the TSB (2010) and the LCEC (2014) models are considered as the base case and the overall internal gains are raised to 2.5 times, then, 5 times then 10 times greater, as shown in the table 4. Without a reference to assess whether the 10 times values might be over exaggerated or not, a house with considerable solar exposure will receive around noon time only an average of 600 to 800 W/m^2 summing up to a daily total well above the 1150W all internal gains combined. Yet what should be noted here, is the critical point where low U-values might not perform well, or at least as good as expected when internal gains are considerably high. In this case the percentage difference between the TSB and LCEC are decreasing with the increase of the internal gains. Starting at 40% for the base case at 53 and 38 kWh/m^2 for the TSB and LCEC respectively, it reaches negative 1% when internal gains are ten times higher than the initial base case.

What should be noted here is the efficiency of high insulation versus low insulation when there are no limits for internal gains, those could be from more users, different sort of physical activities, equipments and lighting as well as from solar gains.

CONCLUSION

With so many references for U-values and tentative energy benchmarks, a lot of missing valuable data are still hindering the local sustainable energy research scene. Theoretically, and based on thermal simulation software, the lower the U-Value the lower the energy. It seems that internal gains do have a considerable effect on reverting the impact of highly insulated buildings. Also major books and reference on sustainable design refer to the positive effect of adding thermal mass to any structure in order to reduce the cooling energy loads. Taking this into account, along with the overwhelming exposed heavy weight concrete structure in Lebanon, some contradiction exists: the simplest is that thermal softwares are not accurately simulating the performance of thermal mass. And if so extensive research should be focused on practical and comparative experiments to check the accuracy of the above. More so, one scholar (Moe; 2014) is strongly advocating the use of thermal mass, instead of high insulated buildings.

ACKNOWLEDGMENTS

The author wishes to thanks Dr. Rosa Schiano-Phan & Dr. Collin Gleeson for their support and helpful comments on an earlier draft of the work .

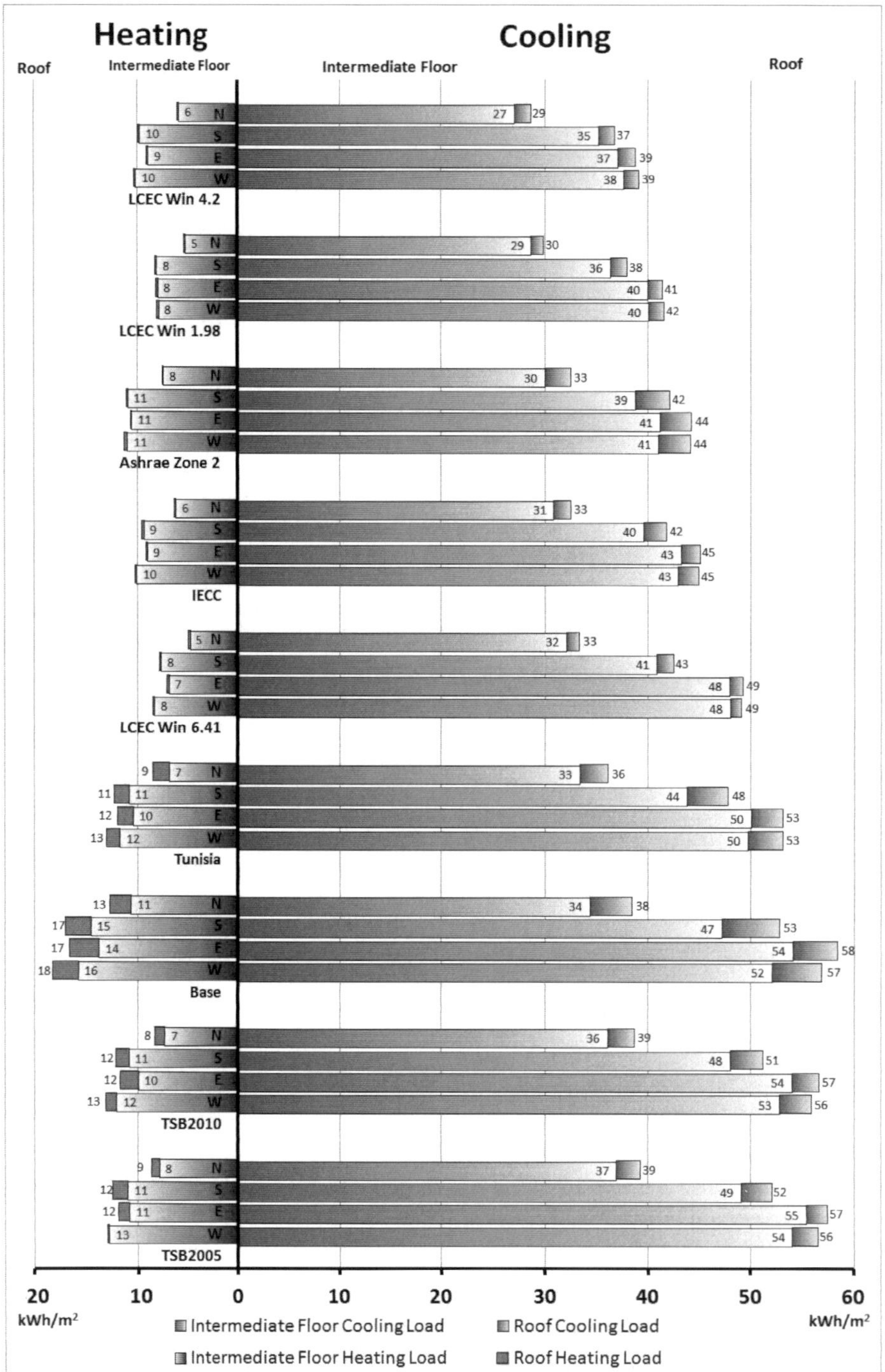

Figure 2 Overall summary of the cooling and heating load based on the different U-values from local, regional and international sources for a 110sqm residential apartment in Beirut.

Table 4 showing in the first two columns the increasing internal gains in total yearly and daily per area, followed by the corresponding energy values for both the LCEC & TSB along with the percentage difference between both.

	Internal Gains		Energy Cooling Values		
	Total kWh (per Year)	Daily Total (W/m^2)	LCEC Base (kWh/m^2)	TSB Base (kWh/m^2)	% difference (LCEC/TSB)
Base	4622	115	38	53	40%
x2.5 Internal Gains	11556	288	71	85	19%
x5 Internal Gains	23352	582	147	154	5%
x10 Internal Gains	46225	1151	323	321	-1%

NOMENCLATURE

TSB = Thermal Standard for Buildings in Lebanon
LCEC = Lebanon Center for Energy Conservation
NEEREA = National Energy Efficiency and Renewable Energy Action
RH = Relative Humidity
ICEC = International Energy Conservation Code
EDL = Électricité du Liban
OPEC = Organization of Petroleum Exporting Countries

REFERENCES

[1]***2012 International Energy Conservation Code*** 2011, International Code Council, Inc, U.S.A.

[2]ASHRAE Standard 90.1 2007, ***Energy Standard for Buildings Except Low-Rise Residential Buildings***, I-P Edition

[3]ASHRAE 2013, ***Fundamentals***, IP editions

[4]California Energy Commission 2014, ***Residential Compliance Manuel*** *for the 2013 Building Energy Efficiency Standards*, Title 24; Part 6.

[5]CIBSE 2006, ***Environmental Design CIBSE Guide A***, , London.

[6]Comair F.G., Mourtada A., Chebad S., Matar T. & Sebrini C. 2011, ***Thermal Insulation Market in Lebanon***, ALMEE.

[7]EDSL TAS, **Environmental Design Solutions Limited,** Version 9.3.2 www.edsl.net

[8]FFEM & ANME 2010, ***La Mise en Place de la Réglementation Thermique et Énergétique en Tunisie***.

[9]French Republic 2006 (a), ***Caractéristique Thermiques des Bâtiments Nouveaux et des Parties Nouvelles de Bâtiments***.

[10]French Republic 2006 (b), ***Réglementation Thermique 2005 des Bâtiments Confortables et Performants***.

[11]Koch-Nielsen, H. 2002, ***Stay Cool, a design guide for the built environment in hot climates***, James & James Ltd.

[12]Kottek, M., Grieser, J., Beck, C., Rudolf, B. & Rubel, F. 2006, **"World Map of Koppen-Geiger Climate Classification updated"**, *Meteorologishe Zeitschrift*, vol. 15, no. 3, pp. 259-263.

[13]LCEC 2014, ***National Energy Efficiency and Renewable Energy Action (NEEREA)***, RCREEE; Banque du Liban

[14]Littlefield, D. 2007, ***Metric Handbook, Planning and Design Data***, 3rd edition, Elsevier Architectural Press.

[15]Meteotest 2014, ***Meteonorm***, 7.

[16]Moe, K. 2014, **Insulation Modernism**, Birkhauser

[17]Nicol F; Humphreys M; Roaf S; 2012, ***Adaptive Thermal Comfort, Principles and Practice***, Earthscan; Routledge

[18]Order of Engineers and Architects 2010, ***Thermal Standard for Buildings in Lebanon***, Order of Engineers and Architects of Beirut; LIBNOR; ECOTECH engineering; ADEME; ALMEE; LGBC; ASHRAE, Beirut.

[19]Republic of Lebanon 2005, ***Thermal Standard for Buildings in Lebanon***, UNDP/GEF and MPWT/DGU, Beirut.

[20]Szokolay, S.V. 2004, ***Introduction to Architectural Science: The Basic of Sustainable Design,*** Elsevier Architectural Press.

[21]Yannas S; 1994, ***Solar Energy and Housing Design***, Volume 1, the department for Enterprise, Architecture Association

ENHANCING FUTURE ENERGY PERFORMANCE OF MIXED MODE COOLING IN BEIRUT

Jessica Daaboul[1] **Kamel Ghali**[2]**, PhD** **Nesreen Ghaddar**[3]**, PhD**

Member ASHRAE *Member ASHRAE*

ABSTRACT

In this paper an assessment of the use of mixed mode ventilation is performed using dynamic simulation software with a typical office building as a test case. The study is performed under future weather data to make sure the design will last through the building lifecycle. Future weather data of Lebanon is generated using Climate Models to perform the future simulations. The model assessed is simulated using current construction practices, 19% energy savings are expected in the future. To increase the future energy saving potential of the system, thickening of walls and including 5 cm (0.16ft) insulation was studied which decreased the energy consumption further and made the mixed mode process save 24% in the future. Then double glazing was used instead of single glazing for the building which lowered the energy consumption and the future savings became 23% against the base case. These two measures proved efficient in the present and the future. The best way to enhance the use of mixed mode ventilation in the present and the future is then to use a combination of those two measures.

INTRODUCTION

Increased attention is being paid on the fact that climate is warming at an alarming rate, in fact the Intergovernmental Panel on Climate Change expects a global temperature increase of 1K to 7K (1°C to 7°C) in the coming 100 years (Wang et al. 2014). The International Energy Agency has exposed data on primary energy consumption, growing at a rate of 2% per year (Lombard et al. 2008). HVAC is one of the pivotal elements of energy consumption in buildings and reaches 50% of the total buildings energy consumption (Lombard et al. 2008). With increasing temperatures, this energy consumption might increase due to the link of HVAC energy consumption to the outdoor temperatures. This triggers the necessity to reduce HVAC energy consumption in buildings. Offices are one of the most energy intensive constructions (Lombard et al. 2008) and thus need to be considered for sustainable design.

Mixed mode cooling is a passive cooling technique considered as an example of sustainable designs. This technique consists of using natural convection whenever possible for space cooling in the aim of reducing energy consumption. Considerable research has been carried out with regards to this ventilation principle to prove its reliability (Alves et al. 2015, Schulze et al. 2013, Tuohy et al. 2007, Rijal et al. 2008). However these researches have been mostly concerned with application of the system in the present. Will this system remain as efficient in the Middle East and countries of the Mediterranean region in the future? Daaboul et al. 2016 studied mixed mode ventilation in Beirut Lebanon under the present and the future weather data, generated a calibrated model for the office, and showed that the energy saving potential of the system will decrease in the future. The aim of this study is to enhance

[1] Graduate Student, Department of Mechanical Engineering, corresponding author email:jgd05@mail.aub.edu.
[2] Professor and Department Chair, Department of Mechanical Engineering, Bechtel Building, Room 324 email:ka04@aub.edu.lb.
[3] Professor, Department of Mechanical Engineering, Room 322 email: farah@aub.edu.lb.

the effectiveness of mixed mode ventilation in the future for the purpose maintaining the sustainable characteristic of the mixed mode system through the future in the area of Lebanon. In an office building comfort should be necessarily provided during working hours. As the mixed mode system's performance is highly dependent on the outdoor conditions, favorable outdoor conditions during working hours would reduce HVAC energy consumption. Similarly, favorable outdoor conditions during non-working hours might be transferred to the inside and stored till working hours in the presence of a good heat storing building envelope. Also, heat gain from the outside might be significantly reduced all day long using good insulation materials, this can help making use of the convection part of the summer breeze (using wind and wall openings) while opting out the walls heat gain from the outside (using walls insulation). Improvements on typical office building envelope materials are suggested to enhance the use of the mixed mode approach and leave it energy efficient through the future. To enhance the building envelope materials, one pivotal envelope material with major effect on energy consumption is the external walls material. Yassine et al. 2012 studied the external walls material effect on mechanical ventilation in Lebanon and suggested the use of double walls with insulation to enhance mechanical ventilation use, and proved the advantage of this wall composition over the single wall composition commonly used in Lebanon. Annan et al. 2015 also suggested the use of double wall with insulation in Lebanon in her study on natural ventilation. In this study the mixed mode cooling approach will be simulated using the single wall, and insulated double wall patterns as external walls to deduce the configuration with the most energy saving potential in Lebanon under future weather data. The use of double glazing is also assessed to further enhance the performance of the mixed mode system. The target will be reached through simulation of a typical office building in Beirut Lebanon under the mixed mode approach in the present and the future.

METHODOLOGY

In this study, mixed mode cooling is studied as a sustainable design that limits HVAC energy consumption while maintaining indoor thermal comfort. Dynamic simulation software was used to conduct this simulation.

The principle behind calculating natural ventilation rates in the software lies on studying the flow characteristics of air through openings using fluid dynamics equations. To this end, needed are the sources of resistance to airflow through openings. In addition to the wind velocities and windrose associated with the country.

Moreover, the system is simulated under future 2050 weather data to assess its future performance. The future weather data used for the simulation are predicted using HADCM3 model. The future performance of the system is improved by making enhancements on the envelope materials such as walls and windows. The above mentioned aims will be attained following the below steps:

1.1.1. Windows and AC control algorithm:

Mixed mode cooling is an alternation between natural ventilation and mechanical cooling, it helps maintaining an acceptable indoor temperature while consuming less energy by simple switching between windows operation and heat pump operation. This combination is beneficial in two ways: it keeps the occupants in contact with their outdoor environment which limits their expectations of a stable indoor temperature and helps them adapt to its conditions, it extends the applicability of natural ventilation throughout the year by controlling windows and AC.

Wang et al. 2015 and Ezzeldine et al. 2013 suggested guidelines to be followed for setting windows and HVAC system operation algorithm in mixed mode buildings. The window control algorithm applied in this study was previously suggested by Daaboul et al. 2016 in the study on mixed mode cooling in Lebanon as per the guidelines summarized in figure 1. Note that a proportional bandwidth of 2°C is used to linearly open or close windows each time ±1°C divergence of indoor temperature from the natural ventilation set-points is detected. In mixed mode system and during AC operation in occupied hours, 20% dissatisfaction (Olesen et al. 2004) is allowed. Especially that the difference between the natural ventilation operation temperature and the AC system operation temperature should not be high to avoid temperature jumps while alternating between systems.

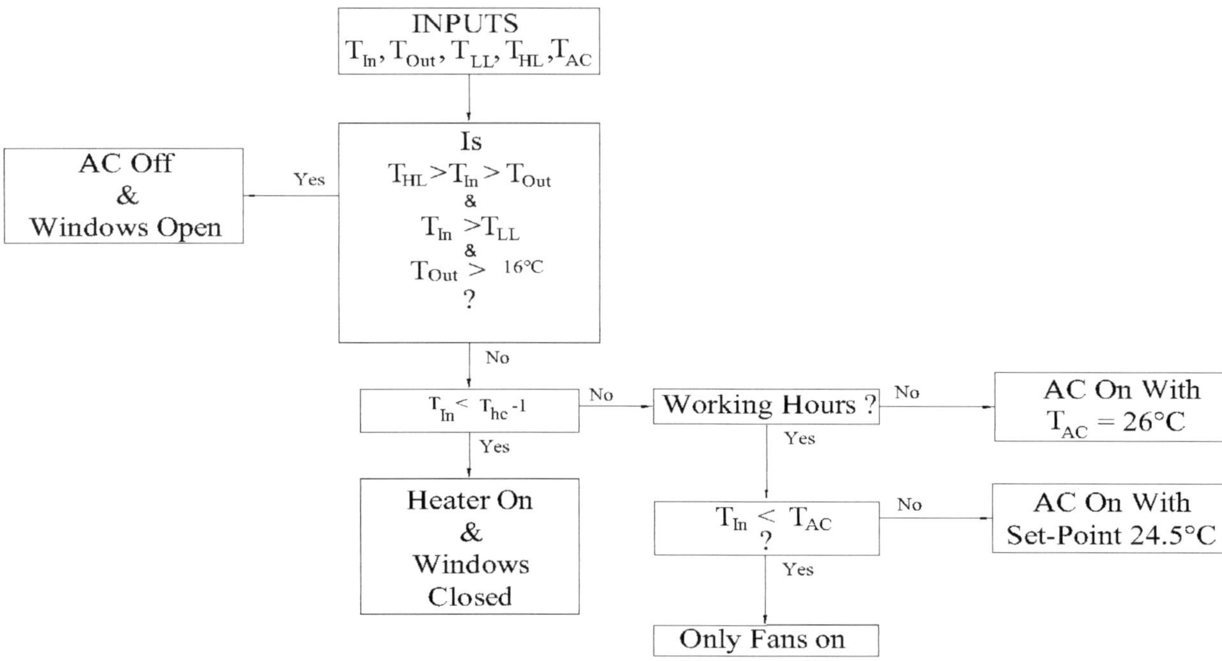

Figure 1: Window operation algorithm.

1.2. Generating weather data:

In our study, the current weather files used for the simulation represent the Typical Meteorological Year (TMY) weather data for the country. A TMY weather file is a weather file describing the average climate of a country derived from previous measured weather data using the Finkelstein-Schafer method over a period of 10 to 30 years. The future 2050 weather data of the country is estimated using the IPCC Climate Model. This model generates monthly means of future climate parameters based on climatological observations and on outputs from transient coupled atmosphere-ocean GCM simulations. To down-scale these monthly means into hourly weather data and form a weather file "Climate change world weather generator" is used. This is a morphing tool that generates future hourly weather files based on the morphing method developed by Belcher et al. 2005. The "Climate change world weather generator" was used in previous studies that performed future weather data simulations. (Holmes et al. 2011, Yi et al. 2014)

CASE STUDY

1.3. Prototypical Office Building description (Base Case)

The typical office model used in this paper is chosen from Daaboul et al. 2016 calibrated office model in Beirut Lebanon and served as a base case to test the applicability of mixed mode cooling. The actual office follows the typical building features used in the country; it is located on the fifth floor of a 25 years old building in Beirut, Lebanon. The features of the office are as follows: (i) Open space with an area of 832 m2 (8956 ft^2) and a height of 3.7m (12.1ft). (ii) SE to NW building orientation. (iii) 60% double glass on the south-west façade and 11% single glass on the north-east façade. (iv) Windows are closed all the time even though operable.

The office operation follows a schedule of 8h per day all over the year during the weekdays, lights and plug loads follow the schedule of the workers, computer servers remain in operation day and night and HVAC system follow a low level operation during out-of-hours. The office's construction materials and glass properties are summarized in table 1. The AC system used is a direct expansion (DX) unit used all year long with AC set-point of 23°C in the summer during the working hours, 26°C during out of hours operation, and 22°C for the winter during the working hours. The supply flow rate is 5200 L/s (68630 Gal/min) of which 300 L/s (3959 Gal/min) are exhaust.

ENERGY SIMULATION

RESULTS AND DISCUSSION:

The mixed mode system together with the window opening algorithm and AC operation strategies were simulated using dynamic simulation software with the weather data file of Beirut Lebanon for the present and the future. Then Energy Conservation Measures (ECM) were studied to enhance the mixed mode cooling performance in the present and the future. By reducing walls heat gain and favoring natural convection through altering some building envelope features and construction practices. To that end simulation results were assessed for thermal comfort and system energy consumption.

MM: Mixed mode system energy consumption:

The stated building features were input into the software under present weather data to assess the feasibility of the system in the present. The building envelope materials are included as per Table 1, the lighting power density used was 13 W/m2 (1.2 W/ft^2), the electrical equipment heat dissipation rate used was 6.5 W/m2 (0.6 W/ft2) as per the previous study's calibration result, the AC and windows operation algorithm is defined as per Figure 1. The schedule of people and equipment is defined as per the actual case (refer to section 1.3).

The simulated total annual HVAC electricity consumption of the base case appeared to be 47.45 MWh in the present and becomes 32.7 MWh (1.1 x 10^8 BTU) when the mixed mode system was introduced. This represents 31% energy saving against the base case. Figure 2 summarizes the monthly energy consumption comparison between mixed mode and the base case. As shown in Figure 2, due to the use of natural ventilation intermittently, the energy consumption decreased significantly, however during the peak summer no significant energy savings were witnessed and that is due to the fact that high outdoor temperatures limit the use of natural ventilation in the summer. Note that Ezzeldin et al.'s 2013 study in the Middle East and North Africa region resulted in higher energy saving potential of the mixed mode system since it was based on an optimized design where the glazing percentage and envelope construction were optimized for a low annual energy demand (Ezzeldine et al. 2013). A better building envelope would reduce HVAC energy consumption. With mixed mode cooling, thermal storage and thermal insulation both act in favor of energy savings. To elaborate, thermal storage helps shifting the favorable night time conditions to working hours, and thermal insulation helps making use of wind velocities while eliminating the unlikely effect of heat gain through wall. The envelope parameters should be optimized in this study using the recommended optimal features for the area to enhance the energy saving potential of the mixed mode system.

Table 1: Base Case Construction Materials (based on Daaboul et al. 2016 study)

Construction Element	Construction Materials (in to out)
External Walls	Plaster-Concrete Block-Plaster
Windows	Single Glass With 0.8cm thickness
Windows SE façade	Double Glass With 1.2cm Internal Cavity
Roof	Wood Wool-Mineral Fiber-Aerated Concrete-Asphalt Mastic Roofing
Ground Floor	Carpet-Reinforced Concrete-Insulation-Reinforced Concrete

Mixed mode was then simulated for the future (Figure 3) and the results show that the energy savings in the future against the base case (also projected to the future) became 19% which is still significant even though lower than the decrease in the present, the present savings are higher than the future savings due to the considerable energy consumption in the future during the summer. The decrease in energy consumption in the future is in line with Daaboul et al. 2016 paper results. Daaboul et al. 2016 suggested enhancing envelope materials of the base case in the study for better future results.

Figure 2 shows the electricity consumption of the base case Vs mixed mode. As the proposed system operates on mixed mode cooling, the electricity savings shown in figure 2 are principally linked to two foundations: the savings caused by people's adaptation to the environment due to the exposure to the outdoor conditions, and the electricity savings due to the heat loss to the environment by natural ventilation. It is necessary thus to assess the heat loss to the environment by air exchanges and natural ventilation for a better assessment of the contribution of the adaptation (ASHRAE 55 adaptive comfort model) to energy savings.

Taking the cooling system's COP into account (COP=2.6), and keeping in mind that the winter heat loss is more a burden than an advantage on energy saving (winter month heat loss should not be accounted in the electricity savings calculation) a total electricity saving amount of 11.3MWh per year is obtained due to natural ventilation. Thus out of the 14.7MWh per year electricity savings incurred due to the use of the mixed mode system, 3.45MWh savings are due to people's adaptation to the outdoor environment.

Not to mention that no natural ventilation or mixed mode cooling would be achieved without people's adaptation, as constant temperature and relative humidity conditions are impossible to get based on mixed mode.

COMPARISON BETWEEN ECM:

ECM 1:

ECM 1 was performed through enhancing the envelope material in favor of mixed mode conditioning by transforming the single external wall configuration of the mixed mode case into a double wall configuration with insulation. The U value of the wall was 1.79 W/m^2K prior to adding insulation and becomes 0.36 W/m^2K (0.06 $BTU/(h \cdot ft^2 \cdot °F)$) using the proposed configuration. This wall composition was suggested by Annan et al. 2014 and Yassine et al. 2012 for Lebanon and is in line with the recommendations of the Order of engineers for Lebanon nonresidential buildings in Beirut (U-value less than 1.26 W/m^2K; 0.22 $BTU/(h \cdot ft^2 \cdot °F)$) (OEA 2010). The resultant total energy consumption became 29.3 MWh (9.99 x 10^7 BTU) which represents 39% energy saving against the base case and 9% decrease against the mixed mode case.

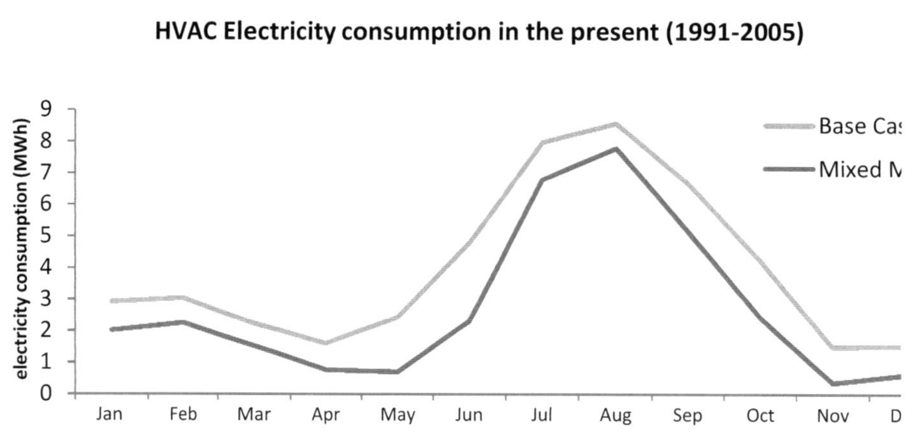

Figure 2: Energy Consumption of ECM vs. the Base Case in the Present

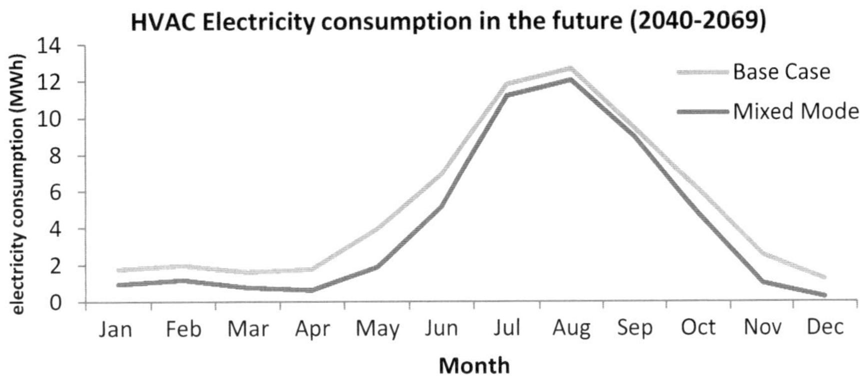

Figure 3: Energy Consumption of ECM vs. The Base Case in the future

The energy savings are achieved all over the year since this wall configuration increased the walls heat capacity by doubling the masonry walls and the walls resistance by inserting the insulation which lowered the need for cooling in the summer and heating in the winter.

ECM1 was then simulated for the future (figure 3) and the results show that the energy savings in the future against the base case (also projected to the future) became 24%. These savings are more significant than the energy saving potential of MM (the mixed mode without enhancement) in the future, the summer energy consumption decreased against MM due to lower heat gain through walls and the winter energy consumption decreased against MM due to warmer air entering the space during the naturally ventilated days.

ECM 2:

ECM 2 was performed by enhancing the glazing material of the mixed mode case in favor of MM by transforming the single external glass configuration into a double glazed window configuration. The resultant total energy consumption became 31 MWh (1.06×10^8 BTU), which represents 35% energy saving against the base case.

The energy savings are achieved more importantly during the summer due to the lower external heat gained in that season. This measure also decreased the energy consumption both while AC is on and likewise extended the comfort hours without AC, this makes it feasible for future use.

ECM2 was then simulated for the future and the results show that the energy savings in the future against the base case (also projected to the future) became 22% which is more significant than the decrease in the present, the summer energy consumption decreased due to lower heat gain through walls and the winter energy consumption decreased due to warmer air entering the space during the naturally ventilated days.

ECM 3:

ECM 3 was performed by enhancing the walls and glazing material of the mixed mode case in favor of MM (bare mixed mode conditioning) by Combining ECM2 and ECM3. The resultant total energy consumption became 28 MWh (9.5×10^7 BTU), which represents 40% energy saving against the base case. Figure 4 shows the monthly energy consumption of the suggested ECM3 vs the monthly energy consumption of the base case. As shown in figure 4 the energy saving potential of the system increases all over the year and even in the peak summer months.

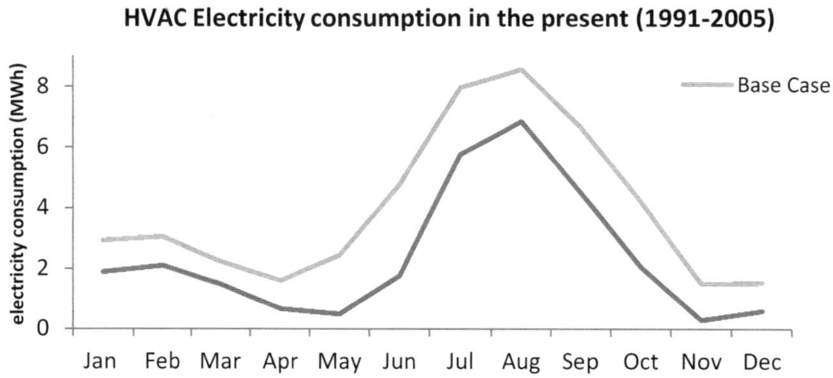

Figure 4: Energy Consumption of ECM 3 vs. The Base Case in the present

28% savings against the base case were achieved under future weather data when ECM 3 was implemented. Simulating ECM3 under future weather data resulted in important enhancement over the base case in the future with future energy savings close in percentage to the present energy savings of the mixed mode system. This proves that the ECM 3 played an important role in maintaining the sustainable characteristic of the mixed mode system and maintaining its energy saving potential through the future.

CONCLUSION:

By operating the same office under mixed mode system and enhancing some common building envelope features, huge savings could be achieved. These features affect positively the energy efficiency sustainability of the building. The energy consumption of the office decreases by 40% in the present and by 28% in the future when ECM 3 (enhancing both walls and windows) was used. This makes the energy saving potential of the system in the future (28%) close in value to the energy saving potential of MM in the present (33%). The two other ECM were also beneficial for the present and the future due to their increased range of applicability during the year. The following opportunities were identified for future studies around the subject:

- Perform CFD on the naturally ventilated space during peak natural ventilation day to provide detailed information about the indoor thermal environment and improve model prediction.
- Assess the life cycle cost of the mixed mode system.

NOMENCLATURE:

T_{OUT} Outdoor temperature.
T_{UL} Adaptive cooling upper limit set-point defined in the ASHRAE adaptive comfort Model.
T_{LL} Adaptive heating lower limit set-point defined in the ASHRAE adaptive comfort model.
U-value Overall heat transfer coefficient [W/m^{2}K]; [BTU/ (h °F ft2)].

REFERENCES:

Alves, C. A., Duarte, D. H., & Gonçalves, F. L. 2015. Residential buildings' thermal performance and comfort for the elderly under climate changes context in the city of São Paulo, Brazil. Energy and Buildings.

Annan, G., N. Ghaddar, K. Ghali. 2014. Natural ventilation in Beirut residential buildings for extended comfort hours. International Journal of Sustainable Enegry (2014) DOI:10.1080/14786451.2014.972403.

Belcher, S., J. Hacker, D. Powell. 2005. Constructing design weather data for future climates. Building Serv. Eng. Res Technol. 26,1 (2005) 00. 49-61.

Daaboul, J., K. Ghali., N. Ghaddar., 2016. Assessing mixed mode cooling in lebanon office buildings. (Awaiting publication)

Ezzeldin, S., & Rees, S. J. 2013. The potential for office buildings with mixed-mode ventilation and low energy cooling systems in arid climates. Energy and Buildings, 65, 368-381.

Holmes, S. H., & Reinhart, C. F. 2011. Climate change risks from a building owner's perspective: assessing future climate and energy price scenarios. In Building Simulation.

Lombard, L., P., J. Ortiz, C. Pout. 2008. A review on buildings energy consumption information. *Energy and Buildings* 40 (2008) 394-398.

Olesen, B., G. Brager. 2004. A better way to predict comfort: The new ASHRAE standard 55-2004. ASHRAE Journal, August 2004.

Rijal, H. B., Tuohy, P., Nicol, F., Humphreys, M. A., Samuel, A., & Clarke, J. 2008. Development of an adaptive window-opening algorithm to predict the thermal comfort, energy use and overheating in buildings. Journal of Building Performance Simulation, 1(1), 17-30.

Schulze, T., & Eicker, U. 2013. Controlled natural ventilation for energy efficient buildings. Energy and Buildings, 56, 221-232.

Tuohy, PG., H. Rijal, M. Humphreys, J. Nicol, A. Samuel, J. Clarke. 2007. Comfort driven adaptive window opening behavior and the influence of building design. Proceedings of Building Simulation 2007 10th IBPSA Conference.

Wang, H., Q. Chen. 2014. Impact of climate change heating and cooling energy use in buildings in the united states. Energy and Buildings 82 (2014) 428-436.

Wang, L., S. Greenberg. 2015. Window operation and impacts on building energy consumption. Energy and Buildings 92 (2015) 313-321.

Yassine, B., Ghali, K., Ghaddar, N., Srour, I., & Chehab, G. 2012. A numerical modeling approach to evaluate energy-efficient mechanical ventilation strategies. Energy and Buildings, 55, 618-630.

Yi, C. Y., & Peng, C. 2014. Microclimate change outdoor and indoor coupled simulation for passive building adaptation design. Procedia Computer Science, 32, 691-698.

An Optimum Design of Energy Consumption of a Residential Building in Kuwait

Ilham Saleh, PE **Mohammed Al-Ansary, PhD** **Walid Chakroun, PhD**
Fellow ASHRAE

ABSTRACT

Kuwait is facing an increase in energy consumption and the country is struggling to increase the electricity generating capacity in order to meet the high demand. The main objective of this work is to optimize building envelopes, windows and fenestration systems and energy efficiency ratio (EER) in order to decrease energy consumption. The optimization technique uses Kuwait energy code as a basis, the multi-objective function is to minimize energy consumption and the cost of building materials. Two base cases were optimized further by varying the parameters of Kuwait code. Case 1 involved maximum occupancy and peak load. Case 2 involved minimum occupancy and peak load. Building orientation starting with eight different north angles was applied to both base cases. The most effective parameter was the EER.. Finally, the data was optimized with the payback calculations as a necessary optimality condition, and resulted in five optimized values for each base case. The peak load was determined with the optimized values for each base case and resulted in saving 20.8% of the cost for Case 1 and 15.4 % for Case 2.

INTRODUCTION

Energy consumption per capita in Kuwait is among the highest in the world. This is caused by a high standard of living, harsh summer climatic conditions, which necessitate cooling, highly subsidized energy costs, and rapid economic growth. Kuwait's climate is classified as dry, hot, with long summers, and short winters. As the relative humidity increases, the temperature sometimes exceeds 50 °C in the shade. Climate data is one of the most important variables used to calculate building energy consumption and is well defined by the American Society of Heating, Refrigerating, and Air Conditioning Engineers (ASHRAE) standards, which categorizes countries into zones based on climate and thermal criteria related to cool and hot days. According to climate classification of ASHRAE standard, Kuwait is classified as zone 1b (hot and dry). Kuwait is an oil-rich country located in the Middle East, Western Asia. This small country lies at the far northwestern corner of the Arabian Gulf. Kuwait's production of oil and natural gas reached three million barrels of oil per day in 2015 and is currently working to boost production capacity to about four million barrels of oil per day by 2020 [KOC, Kuwait Oil Company, 2015].

Kuwait's increasing population and the response to it is a main contributor to the problems of energy consumption. Kuwait's population has grown from approximately 1,906,000 people in 2000 to 3,000,000 in 2010. The expected population for 2013 was 3,820,000 [Kuwait Central Statistical Bureau, 2011]. This continuous increase in population, with an average growth rate of 3.3%, contributes highly to the energy consumption in the country. Other factors contribute to this problem of increasing energy consumption as well. One is the significant amount of waste and over-consumption of energy, which adds to the insufficiency of the energy supply. The consumption of electrical

Ilham Saleh is a project engineer in MPW, Kuwait. **Prof. Mohammed Al-Ansary** is a professor in the mechanical department of Kuwait University, Kuwait.

energy per capita was 13,530 KWh during 2013 [MEW Statistical Yearbook, 2014]. In summer, air conditioning systems consume nearly 70% of the peak load demand and 45% of the annual energy consumption [Kuwait Code MEW R-6-2014] as they are the biggest silent energy users in residential buildings. The seasonal demand variation closely follows the climatic cycle, with a summer peak load of 12,060 MW for July alone in 2013 [MEW Statistical Yearbook, 2014]. The total yearly electrical consumption in Kuwait by the end of 2013 was 60,982 Million KWh. The residential (private) sector consumed 41% of the total yearly electrical consumption, representing a high percentage of the total yearly electrical consumption. Energy consumption and carbon dioxide emissions worldwide are increasing at alarming rates. Both low-rise buildings and high-rise buildings are the major users of energy consuming 40-50% of primary energy. Building usage results in approximately 40% of CO_2 emissions.

The building envelope is the parts of a building that shapes the primary thermal barrier between interior and exterior. These parts greatly influence the energy consumption of a building. In this study, the five parameters of U-Wall, U-Roof, U-Window, SHGC, and EER will be focused onto determine the annual energy consumption of a residential building in Kuwait. Thickness of the walls and roof layers will be studied to see their effect on energy consumption. Windows and fenestration systems of a building play an important role in energy saving, therefore, these will be studied. The other significant parameter used to determine the energy conservation on a building is the EER, which has significant bearing on sustainable energy development. In this research, optimization is described as to how to quantify and implement energy performance solutions in residential buildings in Kuwait. This in turn will help to determine where and to what degree of the total annual energy (KWh) is wasted. This study will discuss the extensive range of possibilities for improving the energy efficiency in the residential sector. The main objective of this research is to establish efficient use of energy to achieve and improve energy efficiency and to regulate energy usage in a residential building in Kuwait. In this respect, optimization energy consumption in a residential building is of primary importance, which implies improving energy efficiency and conservation of resources. Specifically, power and cost saving can be achieved simultaneously. Before starting an optimization process in a residential building, data on the building's design parameters should be identified, such as building envelope, windows, fenestration, and heating, ventilating and air conditioning.

Energy conservation presents many opportunities for research and the optimization of the energy consumption of a building has emerged as a major challenge. A number of studies have recently carried out and optimized the most effective parameters of a building on energy saving along with recommendations. A study presented three strategies to improve HVAC system performance for a commercial building located in a hot and dry climate. A significant reduction in energy consumption from 25.3% to 44.2% was achieved [Vakiloroayaa et al., 2014]. Ascionea et al., (2013) studied varying the HVAC systems in Mediterranean climates to achieve energy saving in the building. To analyze the same issue, another important research by Al-Qabandi (2013) presented a parametric study for energy consumption in residential buildings in Kuwait, varying the overall heat transfer coefficients for the walls, roof and windows plus the solar heat gain coefficient of the fenestration with respect to MEW R-6 (Kuwait Code) and ASHRAE 90.2. The results show the effect of the building envelope, glazing and fenestration system on reducing energy consumption by 12%, while the effect of the lighting loads on energy savings was 4%. Turning off the HVAC system in a residential building when residents are away and when rooms are unoccupied could be saved more than 20% of energy consumption [Kanagaraj & Sharmila, 2015].

Experimental designs

The building used in the case study is a residential house located in Kuwait that has a floor surface of 4,305.6 ft2 (400 m2). The building contains an approximate total of 14,262.2 ft2 (1325m2) on three floors and a basement. It is a semi-detached single-family dwelling consisting of a basement, ground floor, first floor, and second floor. The ground floor and the basement each are 4,305.6 ft2 (400 m2). The first and second floors each have 2,825.5ft2 (262.5 m2). The equipment load is 0.62 w/ft2for the basement and the three floors. In addition, the overhead lighting for all the floors is 0.93 W/ft2. The building was completed in compliance to Kuwait Code requirements. The air conditioning system consists of split DX fan coil units and CAV air system.

Test Producer

Two base cases are; Case 1 involved maximum plug load and occupancy and Case 2 involved minimum plug load and occupancy. A comparison will be made between the increase and decrease of each of these five parameters on the two scenarios presented using the Kuwait Code requirements. Based on this, the cost rate (flat price) is assumed to be $0.14 /KWh for the consumer. Moreover, the emissions analysis is set a constant co2factor of 1.41 lb/KWh throughout this study. The main purpose of this work, 1) What is the energy consumption of the same coded residential building in relation to the number of occupants and the plug load?; and 2) What is the optimum energy saving potential of the residential building of each of the five parameters when increased or decreased for the two base cases? In addition, how does the influence of the building orientation, starting with the north angle at zero degrees and rotating progressively 45° (e.g. 0°, 45°, 90°, 135°, 180°, 225°, 270° and 315°), impact the parameters in each base case scenario?. A multi-objective function with inequality constraints, constrains are: the range of the assessed parameters values used is ± 20% of Kuwait code in increment of 2.5%. The data was optimized with the payback calculations as a necessary optimality condition.

Theoretical considerations

Modeling energy consumption was done using an Hourly Analysis Program 4.7 energy simulation software, HAP, and optimizing the results was done using the genetic algorithm module of MATLAB. The assessment of reducing energy consumption will be studied by varying the percentage value of Kuwait Code assessed parameters in increments of 2.5%. The optimization parameters include those of building envelopes, windows, fenestration, and HVAC parameters (i.e. overall heat transfer coefficients, windows SHGC, building orientation, EER, and unit cost).

RESULTS

Table 1: Shows the optimized parameter values of base Case 1 and the effect on the annual energy consumption plus the payback periods.

Parameter	Optimized Value	Parameter Percentage to Kuwait Code	*AEC (KWh)	Reduction Percentages on AEC	North Angle	Payback Periods (Year)
U-Wall	0.08	–20%	300,134	0.5344%	0°& 180°	1.8
U-Roof	0.056	–20%	300,486	0.4203%	0°& 180°	1.1
U-Window	0.557	–5%	301,385	0.1238%	0°& 180°	4.8
SHGC	0.237	–17.5%	297,263	0.9840%	0°& 180°	4.8
EER	8.225	+17.5%	288,134	4.615%	0°& 180°	4.75

Table 2: Shows the optimized parameter values of base Case 2 and the effect on the annual energy consumption plus the payback periods.

Parameter	Optimized Value	Parameter Percentage to Kuwait Code	*AEC (KWh)	Reduction Percentages on AEC	North Angle	Payback Periods (Year)
U-Wall	0.08	–20%	259,390	0.6045%	0°& 180°	1.9
U-Roof	0.056	–20%	259,723	0.4755%	0°& 180°	1.1
U-Window	0.572	–2.5%	260,824	0.0514%	0°& 180°	2.7
SHGC	0.244	–15%	256,923	0.9653%	0°& 180°	4.5
EER	8.05	+15%	249,296	4.560%	0°& 180°	4

*AEC: Annual Energy Consumption (KWh).

Table 3: Monthly energy consumption by HVAC compomnents cost by end user in ($) for the two base cases and the optimum results.

Month	Base Case 1	Optimum Case 1	Base Case 2	Optimum Case 2
January	116	96	104	101
February	164	133	138	145
March	456	366	429	389
April	841	672	814	706
May	1,533	1,217	1,487	1,273
June	1,851	1,463	1,796	1,529
July	2,184	1,730	2,134	1,806
August	2,302	1,825	2,246	1,906
September	1,853	1,472	1,809	1,541
October	1,288	1,024	1,250	1,078
November	564	454	537	481
December	191	157	165	171
Total	13,342	10,611	12,909	11,123

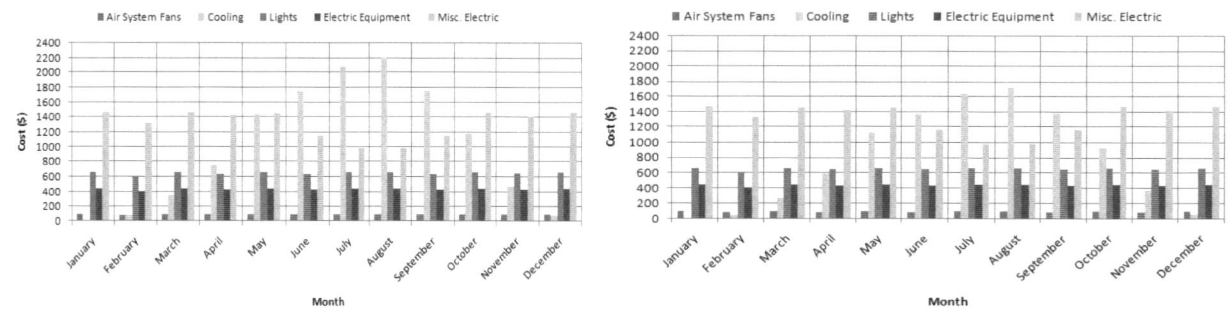

a) Case 1

b)The optimized parameters of Case 1

Figure 1: a) Monthly energy consumption per component by end user in ($) for base Case 1 using Kuwait Code parameters values. b) Monthly energy consumption per component by end user in ($) for the optimized parameters values of base Case 1.

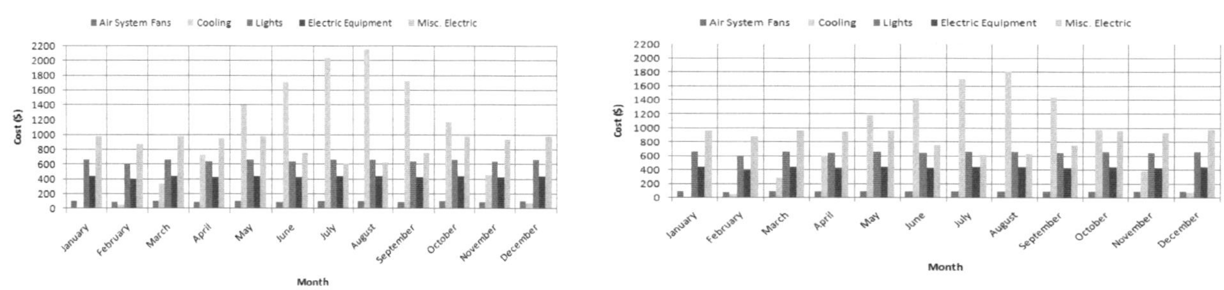

a) Case 2

b) The optimized parameters of Case 2

Figure 2: a) Monthly energy consumption per component by end user in ($) for base Case 2 using Kuwait Code parameters values. b) Monthly energy consumption per component by end user in ($) for the optimized parameters values of base Case 2.

DISCUSSION

Payback period (PBP) analysis was applied as an optimality condition. The consumer benefit is less than five years. Optimization process for the large data was conducted. The PBP calculations for wall roof parameters depend on the insulation layer thickness. The cost increases with an increase in the insulation layer thickness. From Tables 1 and 2, the results show that the optimized values are equal to 0.08 Btu/ft2 h°F (0.454 W/m2 K) and 0.056 Btu/ft2 h°F (0.315 W/m2 K) for the eight north angles for the eight north angles of the building orientation for the range of U-Wall values and U-Roof, respectively. The optimized values were obtained for U-Wall value and U-Roof at –20%, with a north angle equal to zero degrees or 180°. The payback period was approximately less than two years for both assessed parameters. The optimized U-Wall value reduced the annual energy consumption by 0.5% compared to base Case 1. For base Case 2, the optimized U-Wall value has a good reduction percentage of 0.6% in the annual energy consumption. The optimized U-Roof values reduced the energy consumption to 0.42% and 0.47% for base Case 1 and base Case 2, respectively.

The PBP analysis for the glazing depends on the type of glass material, type of frame and thickness of the glass used. Analyzing the window parameter case, the cost of the window increased tremendously when decreasing the U-Window value. The results show that the optimized value was equal to 0.557 Btu/ft2 h°F (3.163W/m2 K) for base Case 1. The optimized value was obtained at –5.0% of the U-Window value. This is because of the high cost of the glass material. The results in Table 2 showed that the optimized value of the U-Window parameter value was 0.572 Btu/ft2 h°F (3.24 W/m2 K) for base Case 2. The optimized value was obtained at –2.5% of the U-Window value. The results of both base cases were at north angle equal to zero degrees or 180° and with payback period was less than five years. It should be noted that the optimized U-Window value of base Case 2 is greater than the optimized U-Window value of base Case 1; this was due to the significant effect of the occupancy and the plug loads. The annual energy consumption of base Case 1 is higher than base Case 2. Subject to this, the optimization results are different. To visualize the effect of using the optimized parameters values in a residential building in Kuwait, in base Case 1, a reduction of 0.12% on the annual energy consumption was achieved when using the optimized U-Window value. On the other hand, only a 0.05% reducing in the annual energy consumption for the optimized U-Window value of base Case 2. The PBP calculations of the SHGC parameter focused on the reflected absorbed cell shade and the shutters. The cost increased with a decrease in the SHGC value. Tables 1 and 2 show the optimized value of the SHGC was 0.237 for base Case 1 and it was 0.244 for base Case 2. The result was obtained from the –17.5% SHGC value for base Case 1 and the –15% SHGC value for base Case 2, both results at north angle zero degrees or 180°. The optimized values of the SHGC parameter have a significant effect in energy conservation. A 0.96% to 0.98% reducing in the annual energy consumption for both base cases was achieved when using the optimized values of the SHGC parameter. SHGC parameter results had higher degree of effectiveness in the annual energy consumption than other previously assessed parameters. The optimized SHGC values had good results in the residential building energy savings.

The PBP calculations for the EER depend on the motor type (the power of the motor) and capacity of the compressor. In addition, the price including maintenance and the spare parts of the cooling loads equipment. The cost increased as EER value increased. Tables 1 and 2 show the optimized value was 8.225 for base Case 1 and it was 8.05 for base Case 2. The result was obtained from the +17.5% EER value of Case 1 and +15% EER value of Case 2 both at north angle zero degrees or 180°. The payback period is less than five years. In both base cases, the most effective parameter was the EER. A reduction percentage of 4.6% on the annual energy consumption was achieved using the optimized value of EER parameter compared to Kuwait code value in both base cases. For base Case 1 the total reduction percentage on the annual energy consumption was approximate of 6.1% using the optimized values of the assessed parameters. Furthermore, the total reduction percentage on the annual energy consumption was 5.3% using the optimized values of the assessed parameters in base Case 2. Using the data of Table 1, the CO2 emissions were reduced by 6.6% approximately for base Case 1. In base Case 2, the CO2 emission has a reduction percentage of 4.5% was applied by the optimized values of the assessed parameters (see Table 2). The reduction of the optimal solution in the CO2 emission of base Case 1 was higher than base Case 2. This is due the higher value of the EER and the lower values of the SHGC and U-Window.

Figures 1 and 2 represent the differences in the energy consumption between the base studys and the optimum results in Tables 1 and 2. A reduction of 20.8% in the peak load was obtained using the optimized values of base Case 1 compared to base Case 1, as shown in Figures 1 (a and b). In addition, a reduction of 18% in the peak load was obtained using the optimized values of base Case 2 rather than base Case 2, as illustrated in Figures 2 (a and b). With the optimized values of base Case 1, implying an orientation towards the north and south, the energy consumption was cost of 1,730$ and 1,825$ of July and August respectively as shown in Fig 1a. However, a peak load of 50.9 (KW) was registered at rush hour 2.00 to 3.00 pm. For the optimized value of base Case 2, it was found that the energy consumption was cost 1,806$ and 1,906$ of July and August as shown in Fig 2. On the other hand the peak load of 52.2 (KW) was noticed at peak time 2.00 to 3.00 pm. This indicates the temperatures falling below 20°C, triggering heating to maintain comfort at the residential building.

Table 3 shows the energy consumption by HVAC components cost in a month ($) for four cases; the two base cases and the optimum results of the two base cases. As a result of peak load reductions was obtained using the optimized values of the both base cases. The energy consumption was cost of 1,730$ in July using the optimized values of base Case 1. It was found to cost 1,806$ using the optimized values of base Case 2 at the same month. In August, the optimized values of base Case 1 was cost of 1,825$ in the energy consumption, and the optimized values of base Case 2 was cost 1,906$, this increase which account for the significant effect of the two factors, occupancy and plug load. The peak load of the optimized assessed parameters values has an energy savings of approximately of 20.8% compared to base Case 1. The peak load of the optimized assessed parameters values has an energy savings of approximately of 20.8% compared to base Case 1. A reduction of approximately 15.5% and 15.1% in the peak load due to July and August, respectively when using the optimized assessed parameters values for base Case 2.

This shows a major reduction was obtained using the optimized assessed parameters values on the north and south sides of the two base Cases. In addition, the good choice of the north angle is protecting some of the indoor spaces from direct solar radiation [Chou, 2001]. Usually, cooling loads cost was lower during the winter season and higher during the three quarters of the year for both base cases as seen in figures 1 and 2. This was due to the lower and higher conductive gains as an effect of the mean outdoor temperatures during the year [Meteotest, 2008].

CONCLUSION

In base Case 1, the increase in occupancy and plug loads fluctuations increases the total energy consumption and emissions by at-most 13% over Case 2. This influences the increase in environmental pollution and harmful effects on human health. Energy consumption is anticipated to increase by increasing the occupancy and the plug load. Lower energy consumption and CO_2 emissions can be achieved by proper choice of glass which delivers a positive impact for the environment in energy savings. Through analyzing the U-Window and SHGC parameters, was found to be more effective is reducing the SHGC of the fenestration system of the residential building, which was critical to performance less energy consumption. A lower SHGC value is advantageous in cooling load consumption.
The U-Window parameter was the most effective value on the incremental initial cost. A lower SHGC value a higher cost material construction. The north angle influence cooling loads of the building. The building orientation position has a big effect on the annual energy consumption with alteration of the U values and SHGC. Where, it has relatively poor effect on the annual energy consumption for building orientation with alteration of the EER value.

The optimized values were found in both base cases at the north angles 0° and 180° for the assessed parameters. The most effective parameter value was found to be the EER for both base cases. It was +17.5% of EER value for Case 1 with approximate saving of 4.6% of the annual energy consumption. Furthermore, it was +15% of EER value for Case 2 with approximate annual energy savings of 4.6%. The optimized value of the U-Roof parameter has the less effect on the annual energy consumption with only little reduction percentage of 0.42% in base Case 1 and of 0.47% in base Case 2. For base Case 1, a reduction of 6.1% in the annual energy consumption was obtained with the optimized values of the assessed parameters. Furthermore, a potential to save up to 5.3% of cooling energy in optimized values of the assessed parameters of base Case 2. The optimized values of the assessed parameters covered the annual costs within less than five years in comparison to Kuwait Code. An energy cost savings of 20.8% in the peak load was obtained using the optimized assessed parameters values for Case 1. In addition, it was an approximately of 15.5% and 15.1% due to July and August respectively, for Case 2.

ACKNOWLEDGMENTS

I would like to express my deepest thanks and gratitude to my advisors Prof. Mohammad D. Al-Ansary and Prof. Walid Chakroun and for this support and encouragement.

NOMENCLATURES

IAQ	Indoor Air Quality
ASHRAE	American Society of Heating, Refrigerating, and Air-Conditioning Engineers
A/C	Air-Conditioning
HVAC	Heating, Ventilating and Air-Conditioning
EER	Energy efficiency ratio
U	The overall heat transfer coefficient
SHGC	Solar heat gain coefficient
OAT	Outside air temperature
PBP	Payback period
KWh	Kilowatt hour
°C	Temperature in degrees Celsius
°F	Temperature in degrees Fahrenheit
CO_2	Carbon dioxide

REFERENCES

Al-Qabandi J. (2012). Parametric study leading to energy conservation in residential houses in Kuwait. Thesis, July, 2012, Kuwait.

Ascionea F., Bianco N., De Mas R. and Vanoli G. (2013). Rehabilitation of the building envelope of hospitals: Achievable energy savings and microclimatic control on varying the HVAC systems in Mediterranean climates, Energy and Buildings, 60, 2013, pp. 125–138.

Chou, S.K. (2001). Seminar Proceedings, Energy Efficiency in Building Design: Building Energy Standard, Building and Construction Authority of Singapore, Singapore (18th April 2001), paper 2.

Kanagaraj, A., and Sharmila, S. (2015). Role of occupancy sensors in energy management. Second International Conference on Science, Technology and Management. 27/9/2015.

KOC, Kuwait Oil Company, 2015.

Kuwait Meteorological Office: Climate Statistics.

Kuwait Central Statistical Bureau, 2011.

Kuwait Code MEW R-6-2014.

Meteotest, (2008). Meteonorm Version 6.0, www.meteotest.ch.

MEW statistical year book, 2014.

Vakiloroayaa V., Samalib B., and Pishghadamc K. (2014). A comparative study on the effect of different strategies for energysaving of air-cooled vapor compression air conditioning systems. Energy and Buildings 74 (2014) 163–172.

Outdoor Air Conditioning of Sports Facility

Essam E.Khalil **Waleed Abdel-Maksoud** **Mohamed E. Ashmawy,**
ASHRAE Fellow

ABSTRACT

In the current study, research on developing air conditioning systems by employing several kinds of technologies to cool the outdoor spaces such as public parks, the holy places and stadiums. The most serious challenge of air conditioning in outdoor spaces is difficult to control the temperature and humidity in the open spaces and also the enormous energy needed by the cooling system. The main task for the analyzed model was to ensure the better method to distribute air to achieve the human comfortable for players and spectators inside the stadium and this study was to investigate the technical and economic aspects of cooling outdoor spaces such as stadium in Qatar. The technical aspect was addressed by developing an empirical mathematical model for the performance of stadium model. The result of the study concludes that the system designed in Qatar stadium for conditioning and distribution air and evaluates results by using Computational Fluid Dynamics CFD code to determine the best way to distribute the air inlets and simulating the flows in air to overcome the high temperature and humidity to reach the human comfort.

INTRODUCTION

In the Middle East, the hot outdoor climate induces a need for cooling to reach the human comfortable. Outdoor air conditioning is rather complicated and its complexity is increasing with the advancement of technology. Nowadays, FIFA World Cup bid is the successful proposal from the government of Qatar to host the 2022 FIFA World Cup and since this moment Qatar proceeded to the Air Conditioning in the stadiums for both the players and spectators and dependence on solar energy to save electric energy that used in cooling system. The major objective of this study is made on simplified simulation model for analyzing the results of Qatar stadium by CFD Software and distribute the air inlets to achieve the best distribution of air and specify influence of human comfortable and study some parameters such as: The effects of ambient temperature, ambient relative humidity, forced air velocity and solar radiation intensity. The simulation results were used to validate the simplified simulation model and find solutions and proposals to achieve the human comfortable. Computational fluid dynamics (CFD) uses numerical methods to solve the fundamental nonlinear differential equations that describe the behavior of fluids and helps to visualize them via graphical simulation. CFD provides exact and detailed information about HVAC design parameters and provides more detailed information about the flow within an occupied zone. CFD is the best solution to provide data and information without making theoretical or experimental methods and saving cost and time.

MATERIAL AND METHOD

Computational fluid dynamics (CFD) has been used to help determine and solve different fluid flow related problems like flow velocity, density, temperature, and chemical concentrations for any area where flow is present in model as well as a wide range of other HVAC applications and get the best distribution, speed and temperature of air to achieve the human comfortable. CFD is a numerical method for calculation for nonlinear differential equations describing fluid flow and is applied in various applications today in order to achieve flawless product designing by combining computational tools and theory of fluid dynamics. CFD Simulation is used for optimization and getting solution in HVAC designs and applications for example, the location and the number of exhausts and diffusers, flow rate and temperature of supplied air for meeting the human comfortable and design criteria. For

Essam E Khalil is a Professor, Waleed Abdel-Maksoud is assistant Professor, Mohamed Ashmawy is a research student, at Mechanical Power Engineering Department, Faculty of Engineering, Cairo University, Giza, Egypt further clarification it can be stated that, CFD simulation helps in design to get solution in special conditions for example, smoke system, ventilation systems, Radiant cooling and study outdoor air conditioning, etc. Lestinen et al [2012] . There are many researches that are focused on using CFD in HVAC especially air conditioning in Qatar stadium, Khalil et al [2016].The study illustrates the influence of different parameters and it is used to determine the diffuser inlet, velocity, temperature, the turbulence model and the importance of radiation in air simulations. The values of these parameters which provide the best solution to achieve the human comfortable. The following figures illustrate the case study (Lekhwiya Sports Stadium).

Description of the Lekhwiya Sports Stadium and HVAC System

Lekhwiya Sports Stadium is based in the city of Doha and exposed to sun rays. The official capacity is 12,000 spectators, and the stadium is located in the Duhail district of the capital Doha. [2015].

Figure 1 Lekhwiya Sport Stadium in Doha [2015]

The strategy of Air conditioning in stadium

To select the locations of the air inlets and air outlets to achieve the human comfortable. Spectators' zone is worked air conditioning systems by three different methods and then it will be inserted in the CFD program to determine the best ways to achieve the human comfortable. So far, CFD is mainly used for the following three domains in the spectators' area, as indicated in Figure 2:

Case (1): Supply the cold air from behind the seats

The first case is to supply cold air from behind the seats with the velocity is 15 m/s [0.825 ft./min] and temperature is 20°C [68° F] with the knowledge that the human skin temperature is about 27°C[80.6 ° F] and the outside air temperature is about 40°C [104° F] and speed 0.5 m/s [0.0275 ft./min] and the humidity of outside air is 70% and the percentage composition of the gases must be taken into account when making calculations and simulation in the CFD program. The following figures illustrate the case (1) of stadium air conditioning.

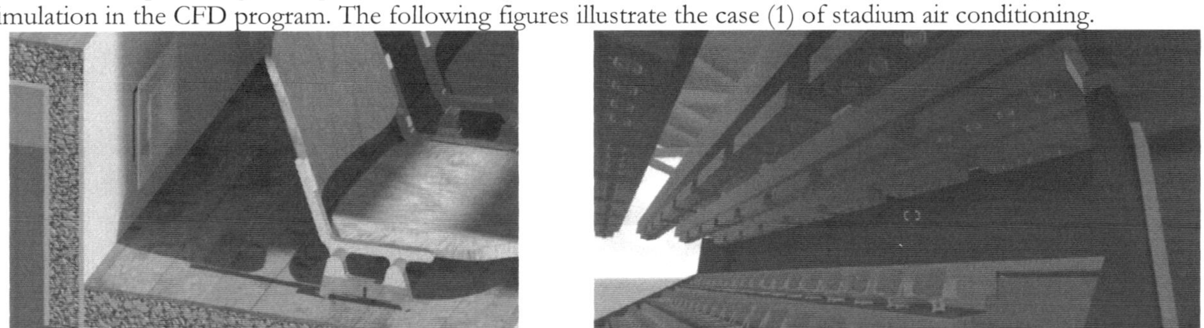

Figure 2 Case (1) Air from Behind the Seats Case (2) Air from above and Return from Behind the Seats

Case (2): Supply the cold air from above and air return from behind the seats

The second case is to supply cold air from above the seats with the velocity is 5 m/s to 25 m/s [0.275 to 1.375 ft./min] and temperature is 18°C [64.4° F] and this difference in the supply air velocity as a result of difference in the form of the stands, where the stands in front row are far from the locations of the jet air nozzles and the stands in last row are very close from the locations of the jet air nozzles so the supply air velocity in the stands in front row is 25 m/s [1.375 ft./min] and the supply air velocity in the stands in last row is 5 m/s[0.275 ft./min]. The human skin temperature is about 27°C [80.6° F] and the outside air temperature is about 40°C [104° F] and speed 0.5 m/s [0.0275 ft./min] and the humidity of outside air is 70%. % and the percentage composition of the gases must be taken into account when making calculations and simulation in the CFD program. The following figures illustrate the case (2) of stadium air conditioning.

Case (3) Supply the cold air from above and air return from behind the seats and radiant cooling in the slab of stands

The third case is to supply cold air from above the seats with the velocity is 5 m/s to 25 m/s [0.275 to 1.375 ft./min]and temperature is 18°C [64.4° F] and in this case will be added another source is the radiant cooling in the slab of stands to absorb the heat and keep the temperature of stands, where the pipes are embedded in the slab concrete and the diameter of pipes are 15mm and the cold water temperature is about 6°C [42.8° F] flows through the pipes. The supply air velocity in the stands in front row is 25 m/s [1.375 ft./min] and the supply air velocity in the stands in last row is 5 m/s[0.275 ft./min]. The human skin temperature is about 27°C [80.6° F] and the outside air temperature is about 46°C [114.8° F] and speed 0.5 m/s [0.0275 ft./min] and the humidity of outside air is 70%. Figure 3 illustrates the case (3) of stadium .

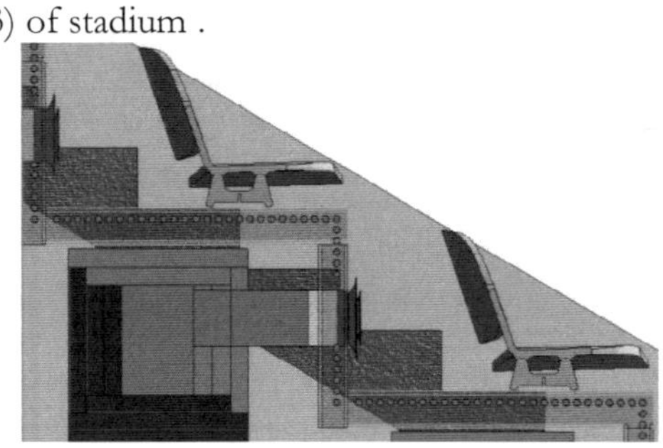

Figure 3 Case (3) Supply the cold air from above and air return from behind the seats and radiant cooling in the slab of stands

RESULTS AND DISCUSSION

The present simulations were conducted with Fluent CFD ANSYS 14 to better understand the air distribution in stadium to achieve the human comfort ; many proposed simulation were used to determine the human comfort. The CFD approach provided simulation results for different design cases in the stadium and in each case, the velocity, temperature, relative humidity are determined and calculated. In the present CFD simulations, the effects of changing the location of air inlets and outlets and using radiant cooling on flow characteristics were obtained and the results of different simulations by **fluent** software, Khalil [2013]were compared and analyzed. In addition, the prevailing boundary conditions in every case were introduced in model. Where it was found that installing supply air from above the seats with radiant cooling could improve the HVAC performance in the Stadium. It also can be seen from the CFD simulation results and installing positions of the air supply can affect the human comfort.

Case (1): Supply the cold air from behind the seats

These results will present by the following figure and note that the distribution inside the stadium and the effects of supply cold air from behind the seats.

The following figures 4, 5 and 6 show the temperature, velocity and Relative humidity contours for this case.

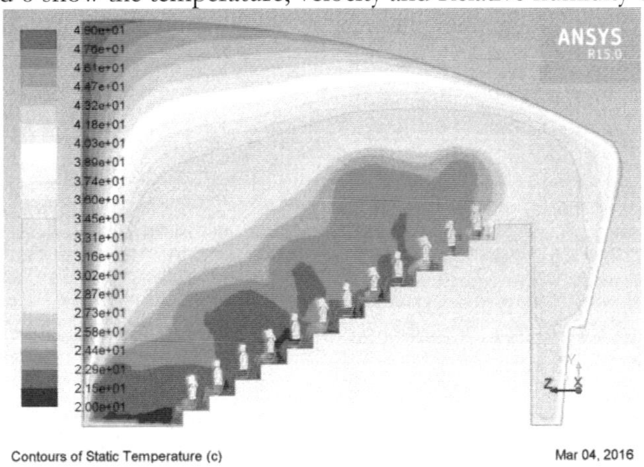

Figure 4 Temperature contours (°C) for supply cold air from behind the seats

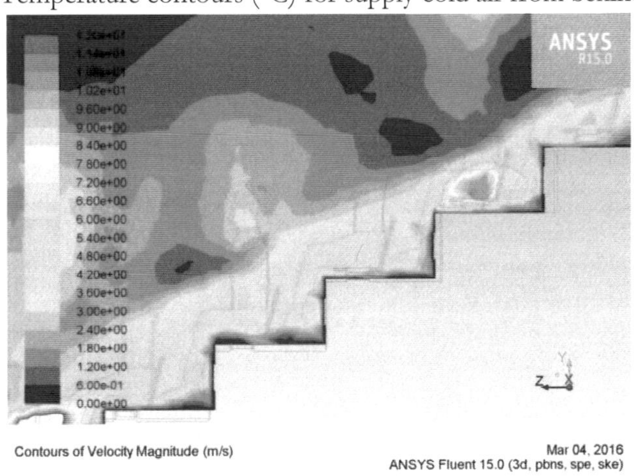

Figure 5 The effect of Velocity on the spectators

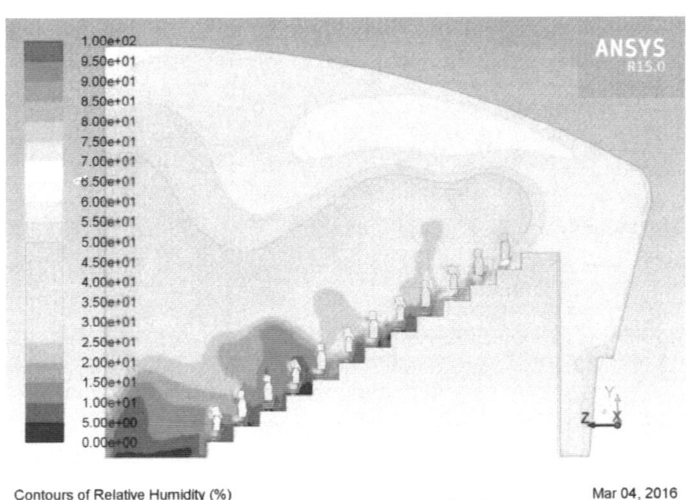

Figure 6 Relative Humidity Contours on the Stands or case1.

The above figure illustrates the results; where the distribution of temperature in the stands is not suitable and note that the temperature is low about 23°C [73.4°F] behind the chairs only. In addition, the chairs and spectators become the obstacle of the cold air movement and shall be used the chairs that are made of materials that do not keep the temperature such as: wood or plastic because the materials that keep the temperature such as: metals may be harmful the spectator's back. Also note that the distribution of humidity, where the humidity is high concentration in areas far from the location of supply air diffusers.

Case (2): Supply the cold air from above and air return from behind the seats

These results will present by the following figures 7,8 and 9 and note that the distribution inside the stadium and the effects of supply the cold air from above and air return from behind the seats. The following figures show the temperature, velocity and relative humidity contours for this case.

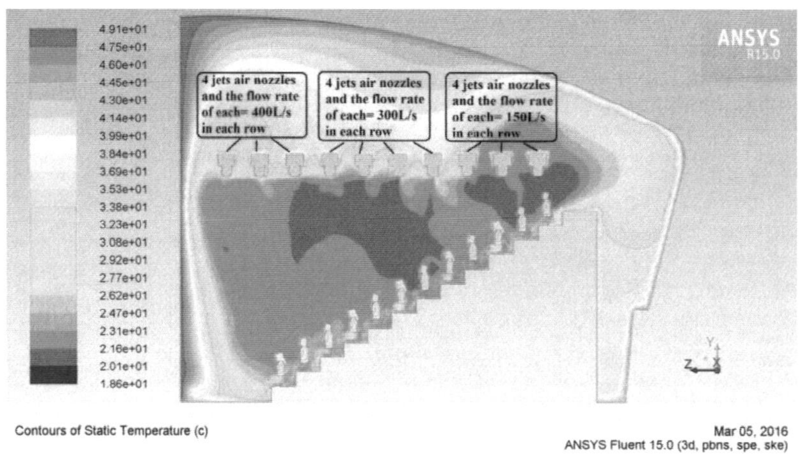

Figure 7 Temperature contours (°C) for supply the cold air from above and air return from behind the seats

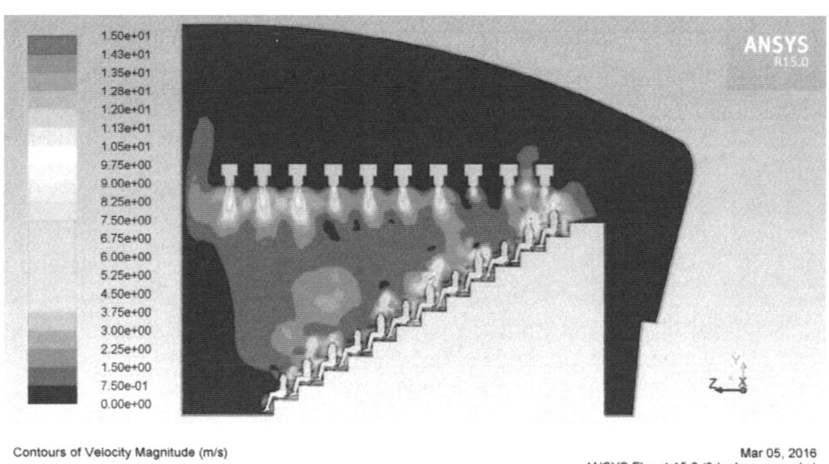

Figure 8 Velocity contours (m/s) for supply the cold air from above and air return from behind the seats

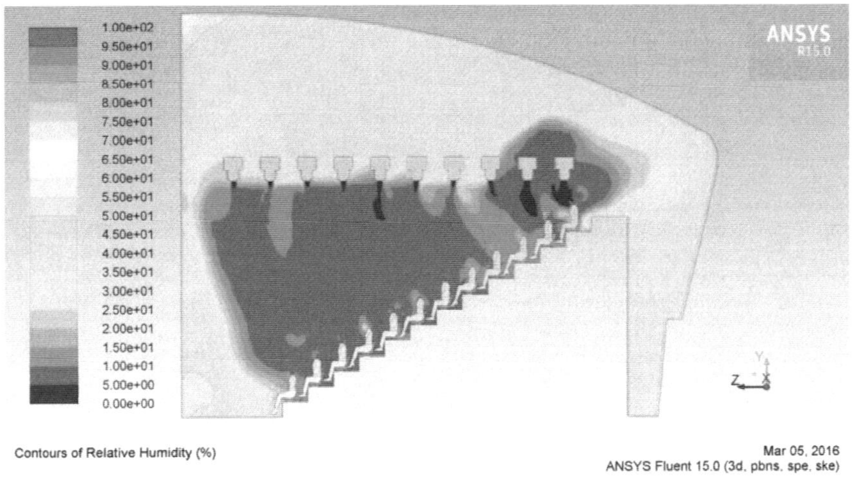

Figure 9 Relative Humidity contours for supply the cold air from above and air return from behind the seats

The above figures illustrate the results, where the distribution of temperature in the stands is more suitable than case (1) and note that the temperature is 24°C [75.2°F] about spectators. In addition, the seats and spectators did not represent direct obstacles for the cold air movement. Also note that the distribution of relative humidity, where the humidity attained low values in areas of seats and stands.

Case (3): Supply the cold air from above and air return from behind the seats and radiant cooling in the slab of stands:

These results will present by the following figure and note that the distribution inside the stadium and the effects of supply the cold air from above and air return from behind the seats and radiant cooling in the slab of stands. The following figures 10, 11, 12 and 13 show the temperature, velocity and relative humidity contours for this case. The above figures illustrate the results, where the distribution of temperature in the stands is more suitable than case (2) and add the radiant cooling in the slab of stands to absorb the heat and keep the temperature of stands, where the pipes are embedded in the slab concrete and the diameter of pipes are 15mm [0.6 inch] and the cold water temperature is about 6°C [42.8°F]flows through the pipes. Note that the temperature is 24°C [75.2°F] about spectators and slab. In addition, the chairs and spectators don't become the obstacle of the cold air movement. Also note that the distribution of humidity, where the humidity is low concentration in areas of stands.

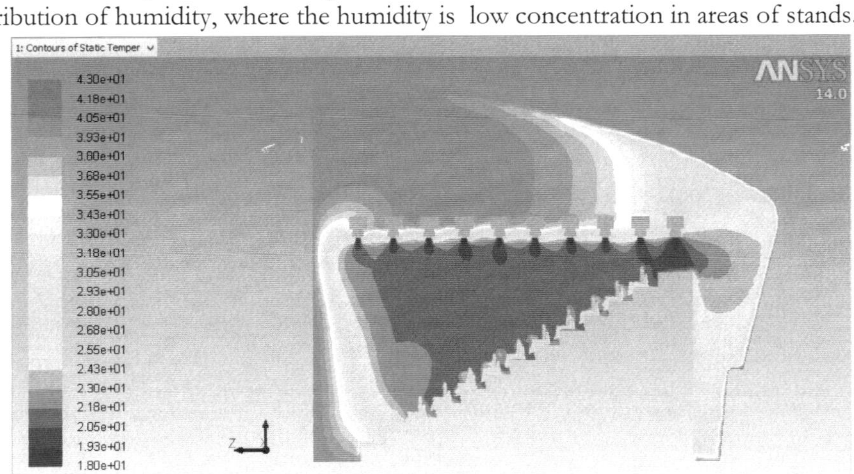

Figure 10 Temperature contours (°C) for supply the cold air from above and air return from behind the seats and radiant cooling in the slab of stands

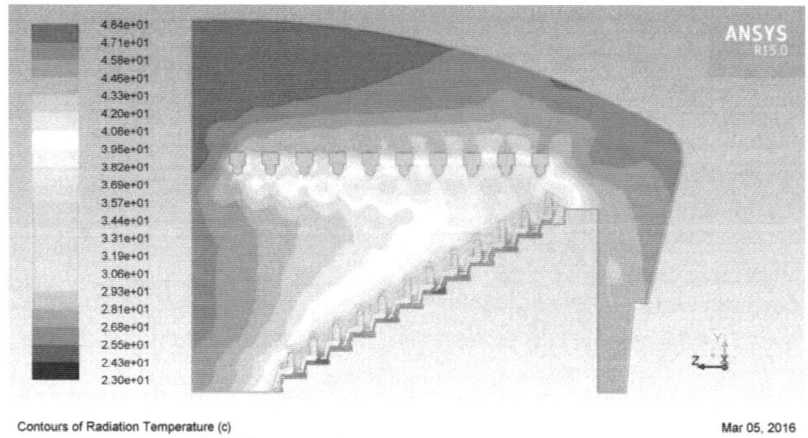

Figure 11 Radiation Temperatures (°C) for Supply the Cold Air From Above and Air Return From Behind the Seats and Radiant Cooling in the Slab of Stands

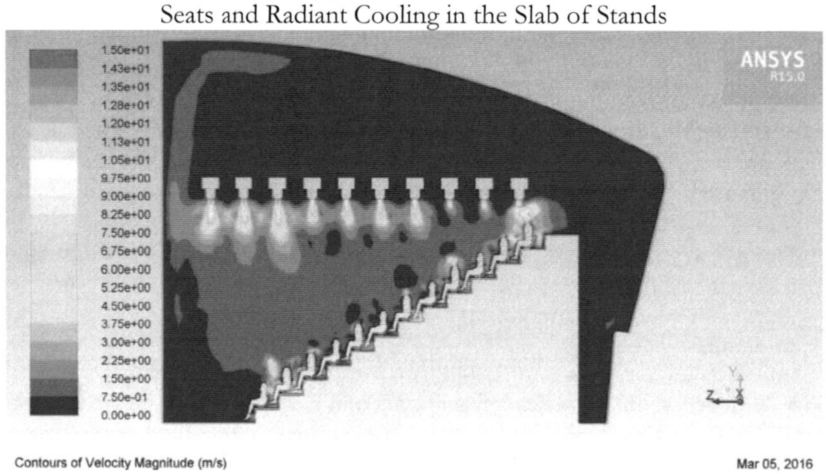

Figure.12 Velocity contours (m/s) for supply the cold air from above and air return from behind the seats and radiant cooling in the slab of stands

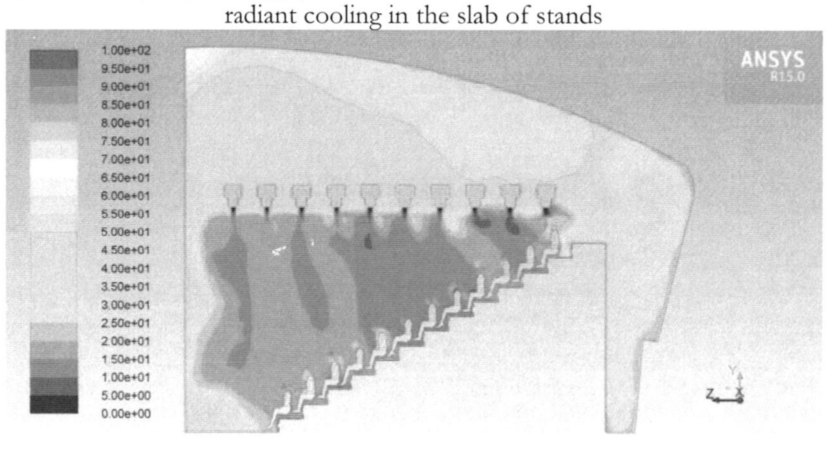

Figure 13 Relative humidity contours for supply the cold air from above and air return from behind the seats and radiant cooling in the slab of stands

SUMMARY of CONCLUSIONS

After presenting the above results, it can be concluded that using Fluent CFD Code aided to improve the air distribution in stadium to achieve the human comfort and many proposed simulation were used to determine the best design solution to achieve the comfort target. The approach provided simulation results for different air-side design cases in the stadium and for each case the velocity, temperature, relative humidity contours were obtained by the model. In the present simulations, with the change of the location of air inlets and outlets and using radiant cooling as alternatives the obtained flow patterns were compared for the different cases. The following points can be concluded:

1. It was found that installing supply air from above the seats with radiant cooling could improve the HVAC performance in the Stadium. It also can be seen from the CFD simulation results and installing positions of the air supply can affect the human comfortable.
2. If the first case (1) is applied, one should use the seats or furniture made from materials that do not store the heat such as: wood or plastic because the materials that store the heat such as metals would be harmful to the spectators' backs.
3. The importance of CFD in HVAC design is realized as it allows engineers to visualize flow velocity, density, thermal impact and chemical concentrations for any region where the flow occurs. This in-turn helps engineers to better comprehend and analyze the problematic areas and suggest the best solutions. CFD is widely used across the construction industry for analysis and design optimization of HVAC air side distribution systems.

NOMENCLATURE

CFD	Computational Fluid Dynamics
t	Time ,s
T	Dry bulb temperature of gas mixture, °C
T"	Temperature fluctuation, °C
u	Instantaneous velocity component in x direction, m/s
v	Instantaneous velocity component in y direction, m/s
w	Instantaneous velocity component in z direction, m/s
U	Axial velocity
V	Computational cell Volume, m^3
Yj	Mass fraction of species j
x, y, z	Cardinal coordinate components

REFERENCE

https://en.wikipedia.org/wiki/Abdullah_bin_Khalifa_Stadium. Accessed 3 December 2015

Khalil, E. E., El-Bialy, E.M., Abdel-Samee W., and Ashmawy, M.E.; "Air Conditioning: Qatar Stadium", Arab Construction World, Vol.34, No.2, pp.35-38, 2016

Khalil, E.E. **Air Distribution in Buildings**, Taylor and Francis, CRC Press, USA, 2013.

Lestinen S., Koskela H., Nyyssölä H., Sundman T. L., Laine T. and Siikonen T., (2012), "CFD-simulation and measurement of indoor environment in a multipurpose arena"; Proceedings of Ventilation 2012, The 10th International Conference on Industrial Ventilation, 17-19 September 2012, INRS, Paris, France.

Optimized Design of Outdoor Cooling Systems for Open Areas in Hot Climate

Adnan Akhdar **Youssef Ghoussoub**

ABSTRACT

The upper concourse is an external area with relatively high occupant density at the east side of Khalifa International Stadium which will host Qatar 2022 FIFA World Cup. Tempering or 'soft conditioning' this area, which is necessary to prevent spectators from experiencing a thermal shock as they enter or exit the cooled bleachers, proved to be a very challenging task, further exacerbated by the severe weather conditions typical for Doha, with ambient temperatures often reaching 47°C (116.6°F) in summer. A detailed Computational Fluid Dynamics (CFD) study has been conducted to investigate and optimize an effective tempering strategy to alleviate hot air wind effects and establish an adequate thermal environment in the open upper-concourse of the stadium with the least energy expenditure. Specifically, the CFD study tests different cooling design options by conducting a detailed investigation of airflow delivery and temperature distribution in the open space while considering external weather parameters such as temperature, humidity and wind speed, as well as detailed internal heat sources such as body heat from occupancy, solar radiation, and heat gains from lighting and equipment. Several design parameters have been considered such as providing shades to protect from solar loads and winds, optimizing the supply air flow rate, speed and temperature of the evaporative coolers, choosing the best air distribution layout and accounting for actual heat loads. The thermal comfort range was defined based on the procedure outlined in Appendix F of ASHRAE 55-2010 which consists of calculating the SET (Standard Effective Temperature) thermal comfort index. A hybrid system consisting of a radiant floor cooling with displacement ventilation at low level, combined with a continuous tent cover with limited shades to maintain views for the spectators, proved to be a solid foundation for guarantying savings in energy costs while establishing adequate environmental conditions in the upper concourse.

INTRODUCTION

Qatar experiences severe weather conditions with temperaures typically reaching 46°C (114.8°F) during the summer season that, if combined with a very humid environment, can create a sensation of temperature that exceeds 50°C (122°F) [Patel, 2013]. Such high temperatures are unsafe and may cause heat shocks to people that have extended exposure to the sun [Bouyer et al., 2007; Nws.noaa.gov., 2005]. In events such as the FIFA World Cup, visiting spectators will be exposed to this extreme and unfamiliar weather [Maughan et al., 2010] for extended periods of time. However, spectators can show a certain tolerance to the adverse climatic conditions prevailing in hot and humid climates [Sawka et al., 2011; Ucuncu et al. 2010] mitigating measures and environmental friendly cooling technologies are integrated in all event-related facilities, including exposed concourses.

However, the realization of these objectives is intricately related to effectively quantifying the extent of hot air infiltration within the open-concourse. Numerous studies evaluating the thermal comfort in semi-outdoor assembly

Adnan Akhdar, M.E., LEED Green Associate is the head of CFD Unit in the Mechanical Department at Dar Al-Handasah. Youssef Ghoussoub, B.E., is a mechanical engineer and a member of the CFD Unit at Dar Al-Handasah.

spaces, such as stadia, have suggested the adoption of designs that naturally promote the ventilation rates in the stadium bowl, in order to reduce the zones of thermal discomfort resulting simultaneously to reduction of the cooling loads in hot and humid climates [Bouyer et al., 2007; Szucs et al., 2009]. The development of any conceptual outdoor cooling strategy requires the adoption of techniques and tools to assess the outdoor space's response to the wind and analyze simultaneously multiple parameters, providing useful information that will assist the decision-making process. In recent years, computational fluid dynamics (CFD) has been playing an increasingly important role in building design. An overview of applications in ventilation [Chen, 2009] has indicated that CFD simulations were a most popular tool used to predict ventilation performance of buildings and that CFD simulations have been widely used to study airflow behavior in the indoor environment, as well as the outdoor environment [Awbi, 1989; Chen & Srebric, 2000; Chow, 1996; Akhdar, 2014]. With the ever-increasing demands for sustainable and energy-efficient buildings, and as expectations of health standards and comfort levels rise, engineers are developing more complex and diversified designs to reduce loads, boost efficiency and utilize renewable resources. For most applications of ventilation and turbulent flows, CFD simulations can simultaneously predict airflow, heat transfer and contaminant transportation in and around buildings with reasonable accuracy; and this detailed information can be used to gain invaluable insight into the underlying physical phenomena. At Dar Al Handasah, Computational Fluid Dynamics (CFD) simulations proved to be an inexpensive yet effective design tool to assist the engineering design process to achieve the desired occupant comfort as well as sustainable development in terms of capital cost and energy consumption for increasing complex and demanding projects.

This paper will illustrate the use of CFD simulations to investigate and compare outdoor cooling strategies to alleviate hot air wind effects and establish an adequate thermal environment in the east upper concourse of Khalifa International stadium with the least energy expenditure.

METHODOLOGY

An extensive number of simulations have been conducted early in the engineering design process to test different design options to provide tempering or 'soft conditioning' for the upper concourse area of Khalifa International Stadium which is an external area with relatively high occupant density. Several options have been proposed to provide adequate environmental conditions for a critical prevailing NNW wind direction while considering external weather parameters such as temperature, humidity and wind speed at typical peak weather conditions and detailed internal heat sources such as body heat from occupancy, solar radiation, and heat gains from lighting and equipments. The selected weather parameters for the analysis are chosen to present severe climatic conditions in Doha after inspecting Doha's long term meteorological data (1983 to 2011) and by considering the position of the stadium's upper concourse. Indeed, prevailing NNW winds are particularly challenging since they sweep through the whole upper concourse from North to South leading to significant hot air infiltration.

Initially, an all-air evaporative cooling strategy using jets at high level was considered. The base simulations indicated that adequate environmental conditions could not be easily achieved for this proposal as the supply was found to be washed away by the hot infiltrating wind and temperatures in the occupant zone were found to be close to ambient. Moreover these base simulations established that wind infiltration was prominent and that partially covering the concourse could help mitigate wind effects and assist the cooling system in achieving the targeted environmental conditions. Indeed, additional simulations indicated that the provision of a continuous tent cover with limited shades to maintain views for the spectators at the north and south sides of the upper concourse, along with a higher flow capacity displacement ventilation system using low level grilles, showed improved results as relatively lower temperatures with a better distribution were achieved. As such, all following simulations adopted this continuous cover at the north and south sides of the upper concourse and seeked to improve supply delivery as the temperature distribution was still uneven.

In a trial to achieve better distribution and in light of the inability of the supply to penetrate deep into the concourse, the base displacement ventilation system was complemented with air supply from rectangular seats along the centerline of the upper concourse to cover a larger area **as shown in Figure 1(a)** below depicting the air supply configuration for Model 1. This combined system of cold air peripheral supply at low level and from rectangular seats **(Model 1)** at the center of the concourse achieved good average conditions along with adequate thermal distribution over the concourse **as shown in Figure 4(a)** below. However, it required a significant amount of cold air supply to cover the whole concourse's area.

Model 2 seeks to minimize the architectural impact by avoiding the use of rectangular seats and still adequately cover the concourse. For this purpose, a combination of high-speed jets at low level (instead of supply grilles) along with high-speed air cannons at high-level -2.5m (8.2 ft) above concourse level- was used at the peripheral wall, **as shown in Figure 1(b)** below depicting the air supply configuration for Model 2. This layout targets to achieve more penetration and mitigate wind infiltration. While this concept succeeded in establishing satisfactory average conditions **as shown in Table 4** below, the thermal distribution was uneven with significant cold spots despite the increased supply flow rate as compared to Model 1 **as shown in Figure 4(b)** below. Moreover, higher air speeds were observed in the concourse especially near the periphery, as well as more mixing and turbulence within the concourse **as shown in Figure 3(b)** below.

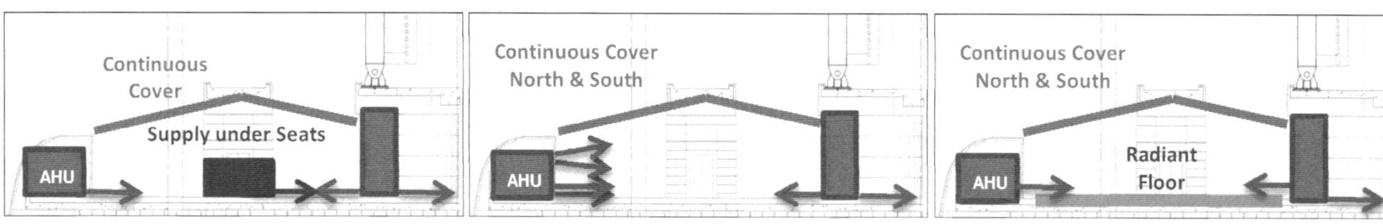

Figure 1 (a) Peripheral displacement ventilation system with cold supply from grilles provided at low level at the periphery of the elevators' block and beneath recntagular seats at the centre of the concourse. A total supply capacity of 244m³/s (517 006.72 cfm) is used, distributed along five supply zones (SE, S, E, NE & N) and under the seats and (b) High speed jets at low level and air cannons at high level, distributed into 5 supply zones along the periphery (S, SE, E, NE & N) with a total supply capacity of 422m³/s (894 167.36 cfm) at 20°C (68°F) and (c) Displacement ventilation system using low level grilles with a total supply of 134.65m³/s (285 307.2 cfm) at 24°C (75.2°F) combined with radiant floor cooling at 24°C (75.2°F).

Considering the limited space available within the skin to house the AHUs, the total supply flow rate needed to be reduced. A hybrid system consisting of a radiant floor cooling with the base displacement ventilation at low level **(Model 3)** was thus proposed and tested in a trial to meet the above constraints **as shown in Figure 1(c)** above, especially that a radiant floor is expected to be very efficient since the concourse is directly exposed to the sun. This hybrid system, combined with the same continuous tent cover, showed great promise and established significantly improved environmental conditions in the upper concourse with drastically reduced supply flow rate as compared to Models 1 and 2 as well as a quasi uniform thermal distribution **as shown in Figure 4(c)** below.

DESIGN CRITERIA & CFD MODEL

The design criteria that the CFD simulations have been based on are **listed in Tables 1 and 2** below:

Table 1. Design Criteria

	Model 1	Model 2	Model 3
Cooling System	Displacement Ventilation + Under-Seats	Periphery Supply using High Speed Jets (Low and High level)	Displacement Ventilation + Radiant Floor @ 24°C (75.2°F)

Wind		NNW @ 5.9m/s (19.36 ft/s)	
Ambient Design Conditions		46°C (114.8°F) DB 26°C (78.8°F) WB	
Supply type	Supply Grilles + Under-seats	High Speed Jets at Low & High Levels	Supply Grilles
Supply Air Temperature	18°C (64.4°F) DB 13.5°C (56.3°F) WB	22°C (71.6°F) DB 16.8°C (62.24°F) WB	24°C (75.2°F) DB 18.5°C (65.3°F) WB
Discharge Velocity	7m/s (22.97 ft/s)	8.6m/s (28.2 ft/s)	3m/s (9.84 ft/s)
Jet Diameter	N/A	Φ250 mm (820.2 ft)	N/A
# Jets	N/A	1008	N/A
Seats type	Rectangular	N/A	N/A
Supply Distribution	5 Supply Zones		Uniform
People Density		0.4 person/m² (0.037 person/ft²)	
Sensible Load		75 W/person (256 BTU/hr.person)	
Latent Load		73 W/person (249 BTU/hr.person)	
Total Flow	244,000 L/s (517 006.72 cfm)	422,000 L/s (894 176.36 cfm)	134,650 L/s (285 307.2 cfm)
Concourse Area		2977 m² (32044.16 ft²)	

Table 2. Supply Distribution

Supply Flow (m³/s)	Peripheral Displacement Ventilation S	SE	E	NE	N	Columns &	Under-Seats	Jets	Total
Model 1	15 [31 783.2]	56 [118 657.3]	13.2 [27 969.2]	50 [105 944]	30 [63 566.4]	25 [52 972]	55.25 [117 068.12]	-	244 [517 006.72]
Model 2	72.5 [153 618.8]	130 [275 454.4]	40 [84 755.2]	120 [254 265.6]	60 [127 132.8]	-	-	-	422.5 [894 173.36]
Model 3			110 [233 076.8]			24 [50 853.12]	-	-	134 [283 930]

A full-scale three-dimensional model of the whole stadium was built as an obstruction inside a wind tunnel of dimensions 4000 m (13120 ft) length x 1750 m (5740 ft) width x 600 m (1970 ft) height **as shown in Figure 2(a)** below in order to accurately establish hot air infiltration into the upper concourse, noting that particular attention was given to capture accurately the geometry of the upper concourse.

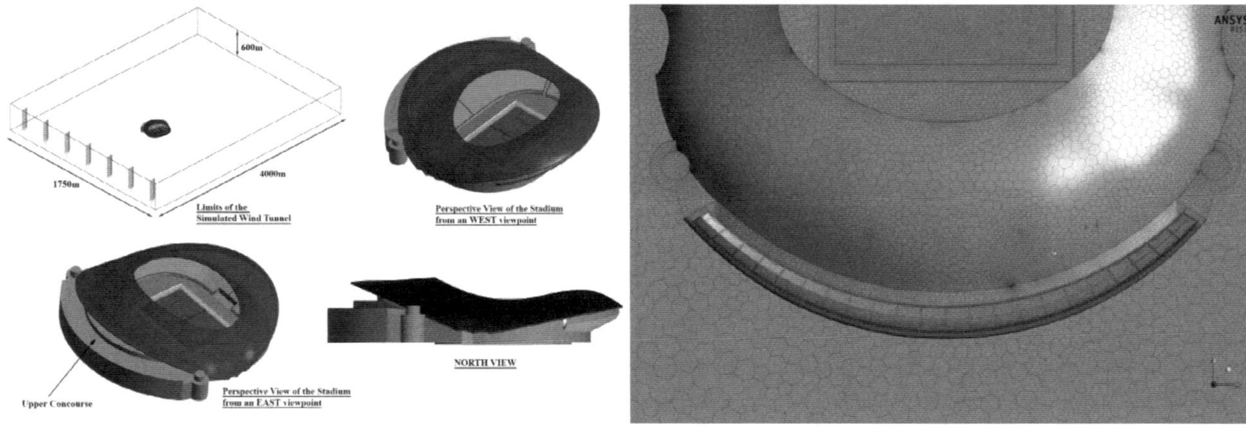

Figure 2 (a) Full-scale 3-D model of Khalifa International Stadium and (b) Simulated mesh for the CFD Model of the Upper Concourse in Khalifa Stadium.

The computational grid consists of approximately 6.4 million polyhedral cells whereas special attention was given to the precise modeling and high grid resolution of the supply jets, as well as the occupant region up to 2m (6.56 ft) from upper concourse floor level. A partial view of the conformal mesh **is shown in Figure 2(b)** below. The minimum cell size used is 0.1 x 0.1 x 0.1 m (0.328 x 0.328 x 0.328 ft) in the occupants' zone and near the diffusers and the maximum size is 12 x 12 x 12 m (39.37 x 39.37 x 39.37 ft) in the uppermost region of the virtual wind tunnel.

In order to define the thermal comfort range, the procedure outlined in Appendix F of ASHRAE 55-2010 is used. This consists of calculating the Standard Effective Temperature (SET) value which combines the temperature, air speed, and humidity values into a single thermal comfort index, where the thermal comfort range is defined **as detailed in Table 3** below. For tempering the uppuer concourse, the thermal comfort range in all simulations is defined as maintaining SET values lower than 30 in the space.

Table 3. SET Ranges for different comfort levels and sensation

SET	Sensation
> 30	hot, unacceptable
25.6 - 30	Slightly warm, acceptable
22.2 – 25.6	Thermal neutrality, comfortable
< 22.2	Cold, uncomfortable

RESULTS

In one hand, installing a continuous tent cover at the north and south sides of the upper concourse helped alleviate severe wind infiltrations. Indeed, **Figures 3(a, b, c)** showing velocity contours and vectors at 1.5 m above concourse finish floor level for Models 1, 2 and 3, respectively, illustrate the fact that relatively low velocities are depicted over a large area of the concourse in all models. In fact, the average velocity in all models is significantly lower than the ambient air speed (5.9m/s) (19.36 ft/s) as **shown in Table 4** below.

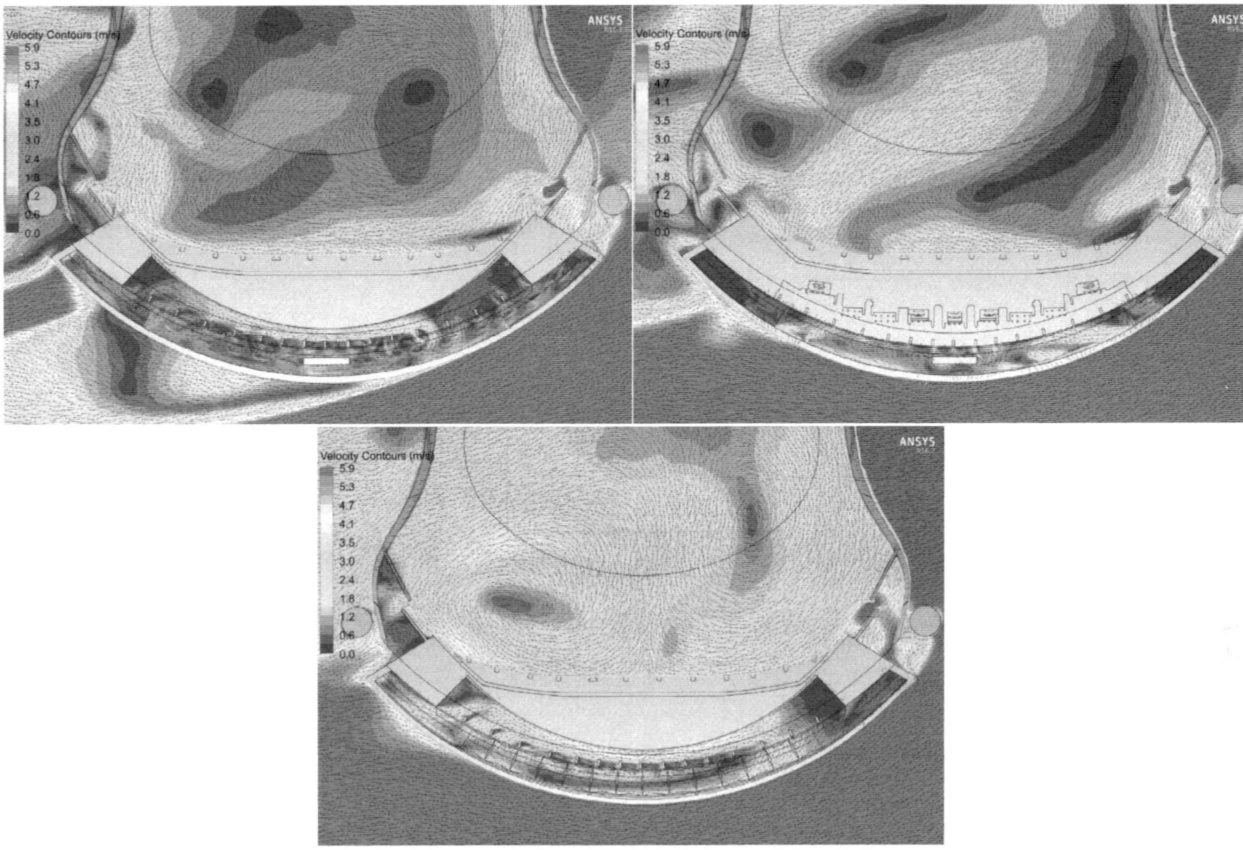

Figure 3 (a) Contours of Velocity magnitude and overlaid velocity vectors at 1.5m Above Concourse FFL for Model 1 and (b) Contours of Velocity magnitude and overlaid velocity vectors at 1.5m Above Concourse FFL for Model 2 and (c) Contours of Velocity magnitude and overlaid velocity vectors at 1.5m Above Concourse FFL for Model 3.

In the other hand, Table 4 below summarizes the average temperature, air speed and SET values at a plane 1.5 m (4.92 ft) above the upper concourse finish floor level. As indicated, Model 3 achieves adequate thermal comfort conditions with significantly reduced flow rate. Moreover, as previously observed, air speeds are significantly lower than ambient (5.9m/s) (19.36 ft/s) and are favorable for people walking about at high DB temperature.

Table 4. Temperature, Air speed, RH and SET at 1.5m above Upper Concourse level areas

Model	NNW Wind @ 5.9m/s (19.36 ft/s)			
	DB (°C) [°F]	RH (%)	Air Speed (m/s) [ft/s]	SET
1	~26.3 [~79.3]	~47	~0.68 [2.23]	~25
2	~26.6 [~79.9]	~45.9	~0.79 [2.6]	~25.5

3	~31.1 [~88]	~39	~1.07 [3.5]	~27.5

Finally, **Figures 4(a, b, c)** below display contours of SET at 1.5m (4.92 ft) above concourse floor level for Models 1, 2 and 3, respectively. These figures clearly denonstarte that a quasi-uniform thermal field is established in Model 3 with no cold or hot spots, despite the reducted flow rate. As such, comfortable conditions are maintained throughout the concourse. Comparatively, while the average conditions are adequate, the thermal variance in Models 1 and 2 is much more pronounced with simultaneous comfortable and un-comfortable regions coinciding with the reported cold and hot spots.

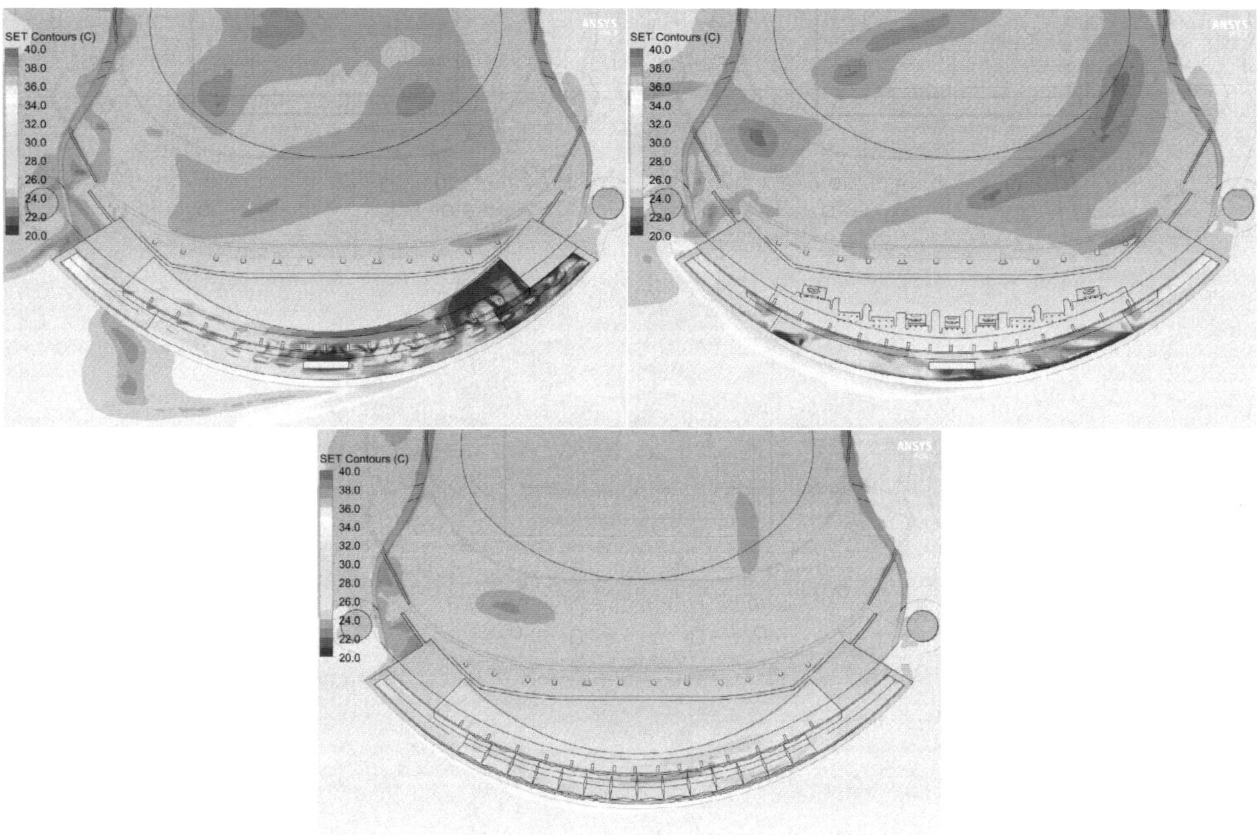

Figure 4 (a) SET Contours at 1.5m above Concourse FFL for Model 1 and (b) SET Contours at 1.5m above Concourse FFL for Model 2 and (c) SET Contours at 1.5m above Concourse FFL for Model 3.

CONCLUSION

Computational Fluid Dynamics (CFD) simulations can provide detailed flow and thermal distributions with excellent accuracy at minimal cost, permitting engineering decisions to be made with considerable precision to optimize HVAC designs for unconventional applications such as outdoor cooling systems. In fact, different outdoor cooling startegies for the open upper concourse in Khalifa International Stadium were investigated using CFD simulations while considering external weather parameters and detailed internal heat sources. A hybrid system consisting of a radiant floor cooling with displacement ventilation at low level, combined with a continuous tent cover with limited shades to maintain views for the spectators, proved to be a solid foundation for guarantying savings in energy costs while establishing adequate environmental conditions in the concourse.

ACKNOWLEDGMENTS

We would like to dedicate a special thank you to the CFD team at Dar Al-Handasah who continues to impress and is successfully achieving design innovations and systems optimization while building up a solid foundation of commitment and technical expertise at all levels. We would also like to thank all design teams for their collaboration, assistance, and valuable contribution.

NOMENCLATURE

CFD = Computational Fluid Dynamics
FFL = Finished Floor Level
DB = Dry bulb temprature
WB = Wet bulb temprature
RH = Relative Humidity
SET= Standard effective temperature

REFERENCES

Akhdar, A. & Ghoussoub, Y. (2014). Towards Sustainable Design with Cutting Edge CFD Simulations, paper presented at The First International ASHRAE Conference & Exhibition on Efficient Building Design: Materials and HVAC Equipment Technologies, Beirut, 2-3 October.

Awbi, H.B. (1989). Application of computational fluid dynamics in room ventilation. Building and Environment, Vol. 24, No. 1, pp. 73-84.

Bouyer, J., Vinet, J., Delpech, P. and Carre, S. 2007. Thermal comfort assessment in semi-outdoor environments: Application to comfort study in stadia. Journal of Wind Engineering and Industrial Aerodynamics, 95 (9), pp. 963-976.

Chen, Q. (2009). Ventilation performance prediction for buildings: A method overview and recent applications. Building and Environment, Vol. 44, pp. 848-858.

Chen, Q. & Srebric, J. (2000). Application of CFD tools for indoor and outdoor environment designs. International Journal on Architectural Science, Vol. 1, pp. 14-29

Chow, W.K. (1996). Application of computational fluid dynamics in building services engineering. Building and Environment, Vol. 31, pp. 425-36.

Maughan, R., Shirreffs, S., Ozgunen, K., Kurdak, S., Ersoz, G., Binnet, M. and Dvorak, J. (2010). Living, training and playing in the heat: challenges to the football player and strategies for coping with environmental extremes. Scandinavian Journal of Medicine & Science in Sports, 20 (s3), pp. 117-124.

Nws.noaa.gov. (2005). National Weather Service Heat Safety. [online] Available at: http://www.nws.noaa.gov/os/heat/index.shtml.

Patel, C. (2013). Heat Stress Guidelines QGL-CHH-001. [e-book] Qatar Petroleum / RLIC. p. 3. Available through: http://www.raslaffan.qp.qa/irj/go/km/docs/documents/Qatar%20Petroleum/e-RLC/Attachments/Regulations/RLC%20Heat%20Stress%20Guidelines.pdf.

Sawka, M., Leon, L. and Montain, S. (2011). Integrated Physiological Mechanisms of Exercise Performance, Adaptation, and Maladaptation to Heat Stress. [e-book] Comprehensive Physiology. p. 1. Available through: Wiley Online Library http://onlinelibrary.wiley.com/doi/10.1002/cphy.c100082/full.

Szucs, A., Moreau, S. and Allard, F. (2007). Spectators' aerothermal comfort assessment method in stadia. Building and environment, 42(6), pp.2227--2240.

Ucuncu, M., Woolf, D. and Zikri, M. (2010). Thermal Comfort of Spectators in Stadia Built in Hot Climate, paper presented at Adapting to Change: New Thinking on Comfort, Windsor, 9-11 April. London: Network for Comfort and Energy Use in Buildings, p. 8. Available at: http://nceub.org.uk/dokuwiki/lib/exe/fetch.php?media=nceub:uploads:members:w2010:extra_papers:53-01-14-ucuncu.pdf

Effect of PCM Packet Mass and Melting Temperature on Cooling Vest Performance

Mariam Itani

Nesreen Ghaddar, PhD
Member ASHRAE

Kamel Ghali, PhD
Member ASHRAE

Djamel Ouahrani, PhD

Walid Chakroun, PhD
Member ASHRAE

ABSTRACT

Workers in hot environment experience high heat stresses that affect their productivity and working intervals. A cooling vest containing phase change material packets (PCMs) is used as a passive body cooling method to remove body heat for a relatively long duration. The aim of this work is to improve the design of the PCM-vest by making its weight lighter if possible or choosing a suitable melting temperature of the PCMs without jeopardizing the duration of its applicability.

The study uses a validated transient mathematical model of heat and mass transfer through clothing layers containing PCM packets (fabric-PCM model) and integrated with a segmental bio-heat model to perform a parametric study on the effect of the melting temperature of PCM packets and their mass on trunk skin temperature and heat loss. The bio-heat model divides the torso into upper and lower segments to allow the incorporation of non-uniformity in skin temperature of torso as an interacting boundary for the PCM-vest.

A parametric study is performed to investigate the effect of the mass of the PCM packets and PCM melting temperature on the upper and lower torso segments. The cooling duration of the cooling vest and cooling rate are determined and a number of recommendations are developed on desired PCM mass as well as melting temperature for maintaining lowest trunk skin temperature for the longest period in given environmental conditions and activity level. It is found that a lower melting temperature is favored when fast cooling is needed. Moreover, the mass of the PCM affects directly the total duration of melting but not the cooling rate.

Key-words: *Phase Change Material; bio-heat and passive cooling vest modeling; transient heat transfer in fabric and PCM.*

INTRODUCTION

Workers performing activities at high metabolic rates and in hot environmental conditions are subject to thermal strains due to elevated body skin and core temperatures that affect their working performance (Gao et al., 2011). To combat thermal strain and heat stresses, passive cooling systems such as phase change material (PCM) cooling vests, have been used (Ghaddar et al., 2011).

The cooling vests that contain PCMs are of many types and include those made up of ice, paraffin and salt hydrates (Loumala et al., 2012, Mondal, 2008). The ice cooling vests are known for decreasing body core and skin temperatures as well as reducing heat strain, but they could result in erythema, due to their low melting temperature, leading to discomfort (House et al., 2013). On the other hand, cooling vests with PCMs made up of salt hydrates, whether in the form of microcapsules or packets, do not have the disadvantage of the ice cooling vests. In spite of that, the salt PCMs in the form of microcapsules as compared to the salt PCM packets, were not very successful in warm climates, due to their small volume and heat capacity and their inability to regenerate by themselves (Ghali et al., 2004). Thus, many studies have focused on investigating the performance of salt PCM packets in hot environments and considered different melting temperatures and thermo-physical properties of the PCMs (Mondal, 2008).

Those studies that considered the use of salt PCM packets in the cooling vests, have searched for the factors that affect the performance of the cooling vest in order to try and find an optimum design of the vest. One of the factors include the weight and the design of the vest that affect the mobility of the workers and their performance under certain activity levels (Tate et al., 2008, Chou et al., 2008). In the study of Reinertsen et al. (2008) experiments on human subjects considered different positioning of the PCM packets on the upper and lower trunk segments of the human body, while the subjects were wearing well-insulated clothing in a hot environment. The results of the study revealed that positioning PCM packets on both the upper and lower trunk segments caused lower skin temperatures, however, better thermal sensation was observed when only the upper trunk was covered since sweat was allowed to transport through the clothing with lower number of PCM packets.

In another study by Bendkowska et al. (2010), a thermal manikin was used with a constant skin temperature to investigate the cooling effect of small PCM packets inserted on the inner surface of the cooling vest and worn under chemical protective garments. Their results showed that a low initial temperature of the PCMs provided the best performance. Other findings by Gao et al. (2007) showed that the surface temperature of the manikin as well as the PCM melting temperature affect the cooling vest performance. Gao et al. (2010) derived an empirical correlation for the PCM cooling effect and temperature gradient between skin and PCM melting temperature, mass, and covering area using three different melting temperatures. The results of the study of Gao et al. (2010) were validated by experiments done on human subjects and it was found that a stronger cooling effect of the torso and lower mean skin temperature were achieved at lower PCM melting temperature (Gao et al., 2011).

In this study, the aim is to use a validated transient mathematical model of heat and mass transfer through clothing layers containing PCM packets (fabric-PCM model) and integrated with a segmental bio-heat model (Karaki et al., 2013, Hamdan et al., 2016). The integrated model is used to perform a parametric study on the effect of the mass of the PCM packets and melting temperature on the trunk cooling duration and skin temperature.

INTEGRATED BIO-HEAT AND FABRIC-PCM MODEL

In order to investigate the performance of the cooling vest with different PCM thermo-physical properties (mass, melting temperature, area and latent heat of fusion), a fabric-PCM model is needed (Hamdan et al., 2016). The fabric-PCM model is then integrated with a multi-node segmental and transient bioheat model capable of predicting the segmental skin and core temperature of all body segments (Karaki et al., 2013). The integrated model is validated with published experimental data on human subjects and the formulation of the fabric-PCM model can be found in the study of Hamdan et al. (2016).

The cooling vest, as shown in Figure 1(a), covers the upper and lower segments of the torso, where in each segment, the number of PCM packets can be defined to cover a certain area of the torso. Figure 1(b) shows a schematic of the side view of the cooling vest, where the PCM packets are sandwiched between the microclimate and macroclimate air layers, the outer fabric layer is subjected to the ambient conditions, and the inner fabric layer is exposed to the skin surface of the torso segments.

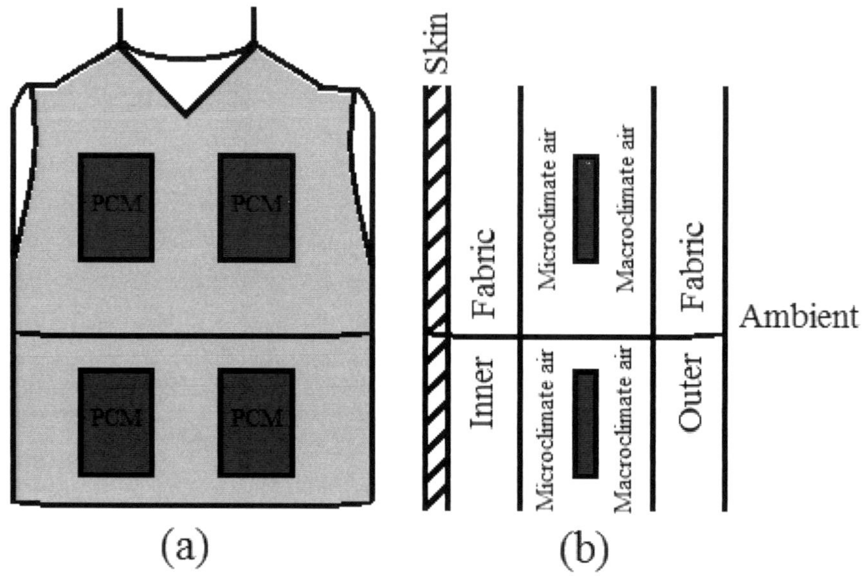

Figure 1: Schematic of the cooling vest showing (a) a front view of the upper and lower torso segments covered with PCM packets and (b) a side view of the cooling vest with the skin layer and the different fabric, PCM, and air layers

The integrated model was used by setting the initial conditions of temperature and pressure of all the different clothing and air layers to be at equilibrium with the environment, and the initial PCM temperatures. In addition to that, the ambient conditions and metabolic rate are set in the bio-heat model. The number of PCM packets and their thermo-physical properties covering each segment of the torso are also inputted to the integrated model. After setting the initial conditions, the time loop starts to calculate the new temperatures, humidity ratios, and vapor pressures of the inner and outer fabric layers as well as the humidity ratios and temperatures of the microclimate and the macroclimate air layers, by calling the fabric-PCM model for each torso segment covered by PCM packets. At each time step, for updating the PCM temperature, it was compared against the melting temperature to decide if the PCM is in the melting phase or not and to find the PCM melted fraction, accordingly. Then, the human body core temperature and the segmental and mean skin temperatures can be found. The flowchart, showing the coupling methodology between the fabric-PCM model and the bio-heat model at each time step, is illustrated in Figure 2.

PARAMETRIC STUDY: EFFECT OF PCM MELTING TEMPERATURE AND VEST WEIGHT

In order to perform a parametric study to investigate the effect of different PCM melting temperatures and mass of PCM packets placed in the cooling vest on the performance of the cooling vest, an integrated fabric-PCM and bio-heat model was used. A metabolic rate of 1 Met and a clothing insulation equivalent to 0.53clo were inputted into the integrated model.

The melting temperature of the PCM packets is a significant parameter to study since it directly affects the microclimate air temperature near the skin, due to their closeness. Three PCM melting temperatures were investigated, 21°C (69.8°F), 24°C (75.2°F) and 26 °C (78.8°F). All PCM packets had the same initial temperature of 20 °C (68°F). The three vests contained the same number of PCM packets; the total cooling area was 0.22m^2 (2.37ft^2), representing

47% of the whole torso area. Therefore, the upper and the lower torso segments had the same number of PCM packets. In addition, a simulation, referred to as the base-case, was done to represent the case when when no PCM packets were used.

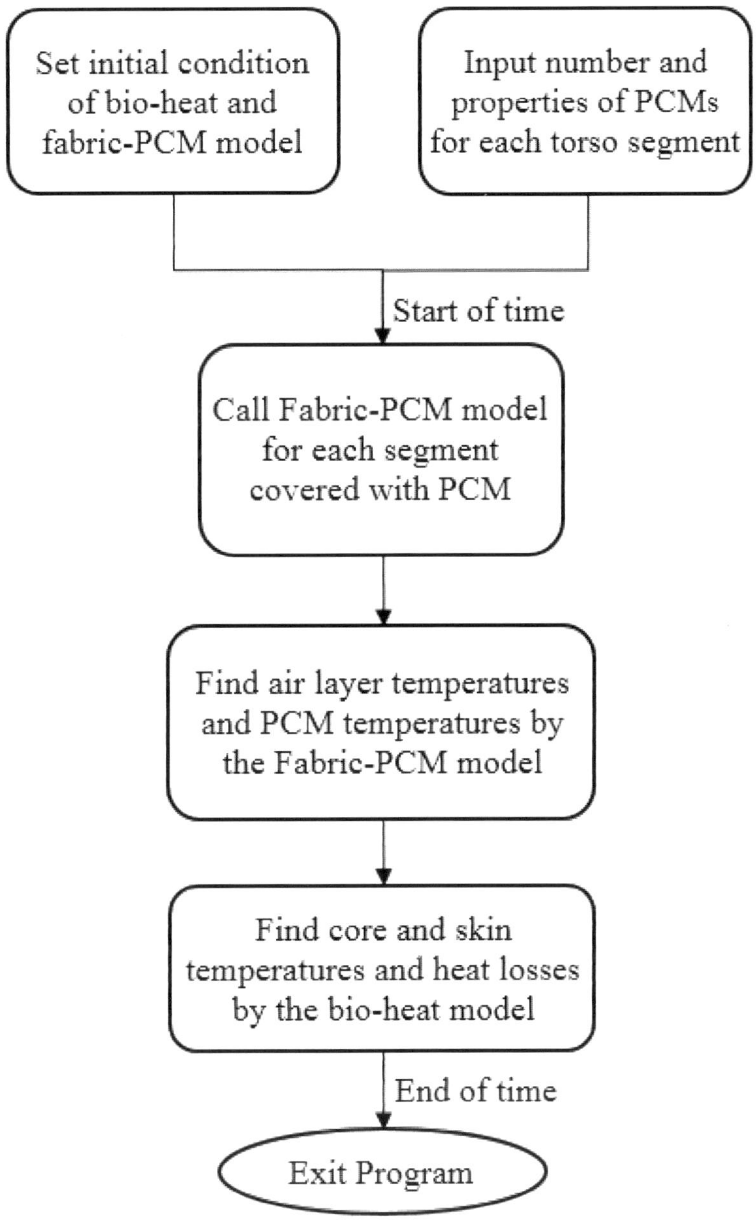

Figure 2: Flowchart showing the coupling methodology between the fabric-PCM and bio-heat models at each time step

The simulations at different melting temperatures were conducted at the same ambient condition of 32°C (89.6°F) temperature and 50% relative humidity. Then a comparison was done between the torso heat losses and torso mean skin temperatures of the different cases.

The mass of the phase change material is also a parameter that affects the performance of the cooling vest significantly. Two cases were simulated at the same total covering area of the PCM packets but with different weights (thicknesses of PCMs). The weight of VestA was 0.88 kg (1.94 lbs) while that of VestB was 1.7 kg (3.75 lbs). The two vests had the same melting temperature of 21°C (69.8°F) and same PCM initial temperature of 20°C (68°F). The PCM packets were covering 23% of the torso area. Then a comparison was done between the cooling duration of the two vests and their cooling rate.

RESULTS AND DISCUSSION

The mean skin temperature of the torso segments is shown in Figure 3, by plotting its variation with time for the three PCM melting temperatures of 21°C (69.8°F), 24°C (75.2°F) and 26 °C (78.8°F) and the base-case. It can be noticed from Figure 3, that the three vests had a sharp decrease in the torso skin temperature during the first few minutes due to the low initial PCM temperature. It can also be noticed that the mean skin torso temperature of the base-case reached a maximum of 36.60°C(97.88°F) and the vest with the lowest PCM melting temperature (Vest21) had the lowest mean skin torso temperature reaching a minimum of 36.05°C (96.89°F), indicating a 0.55°C(32.99°F) decrease in the torso mean skin temperature. However, after some time (92 minutes), the torso skin temperature corresponding to Vest21 started increasing due to the total melting of the PCM packets.

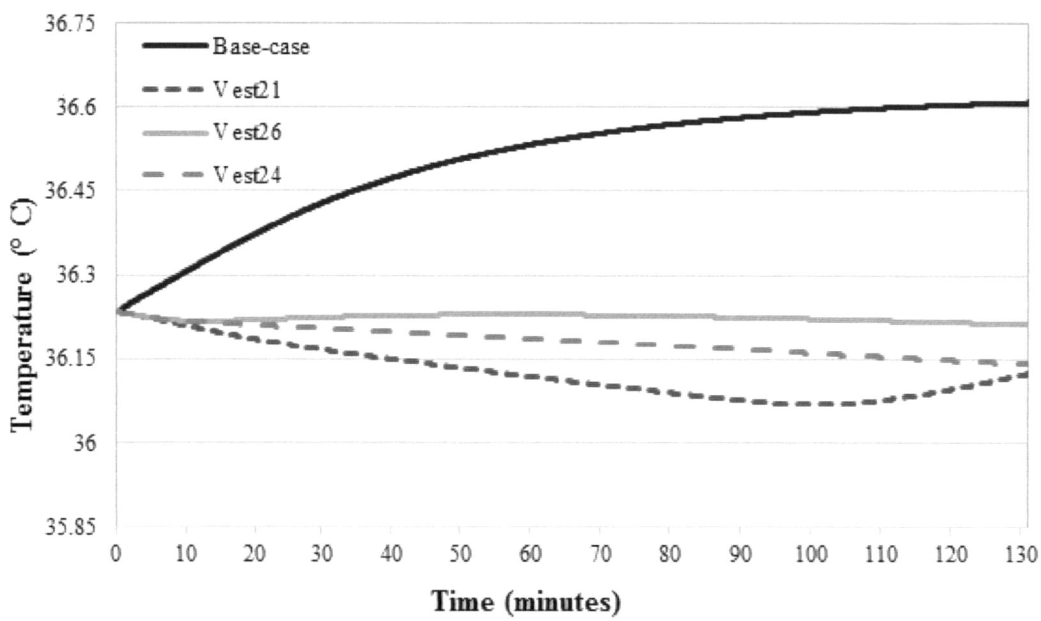

Figure 3: Plots of the variation of the mean skin temperature of the torso with time for the cooling vest with different PCM melting temperatures and a base-case.

Table 1. Start and End of PCM Melting Phase and Average Cooling Rate for the Different PCM Vest Melting Temperature

	Vest21	Vest24	Vest26
PCM melting temperature in °C (°F)	21 (69.8)	24 (75.2)	26 (78.8)
Melting starts after in min	1.7	8.9	15
Duration of melting in min	90	120	180
Average cooling rate in W (BTU/s)	25 (0.024)	21.7 (0.021)	19.5 (0.018)

Table 2. Start and End of PCM Melting Phase and Average Cooling Rate for the Different PCM Vest Weights

	VestA	VestB
PCM vest weight in kg (lbs)	0.88 (1.94)	1.76 (3.75)
Melting starts after in min	1.8	3.65
Duration of melting in min	79.2	162
Average cooling rate in W (BTU/s)	18.42 (0.017)	19.17 (0.018)

A comparison was done between the time when the PCM packets started melting, the duration of PCM melting and the average cooling rate of the three different cooling vests (Vest21, Vest24 and Vest26). The results are shown in Table 1, and it was observed that as the PCM melting temperature increased, the time needed for the PCM to start melting increased and the duration of melting was also extended. In addition, the average cooling rate of the vest with the lowest PCM melting temperature was the largest. It can be inferred from this comparison that for an extended cooling duration, a higher melting temperature is needed, however, for faster cooling, a lower melting temperature is recommended. These findings were analogous to the experimental findings of Gao et al. (2010).

Table 2 shows a comparison between the time when the PCM packets started melting, the duration of PCM melting and the average cooling rate of the two cooling vests with different weights (VestA and VestB). It can be noticed that increasing the weight of the PCM packets (thickness) lead to an extended cooling duration, but had no significant effect on the average cooling rate of the cooling vests.

CONCLUSION

A validated integrated fabric-PCM and segmental bio-heat model was used to predict the temperatures of the air and the fabric layers as well as the temperatures of the salt PCM packets placed inside the cooling vest, at certain ambient conditions and activity level of the human body. The integrated model can also be used to predict the core, segmental and mean skin temperature of the human body as well as the heat losses. A parametric study was performed to examine the effect of different parameters such as the PCM melting temperature and PCM weight in the cooling vest on the performance of the cooling vest. The results showed that when a faster cooling effect is required, a lower PCM melting temperature is needed, however, for an extened cooling duration, a higher PCM melting temperature is desirable. Moreover, it was noticed that the mass of the PCM (thickness) affected the melting duration of the PCM packets and not their average cooling rate.

ACKNOWLEDGMENTS

The authors would like to acknowledge the financial support of the Qatar National Research Funds (a member of the Qatar Foundation), Grant #NPRP 7 - 486 - 2 – 185, State of Qatar.

NOMENCLATURE

PCM = phase change material

REFERENCES

C. Gao, K. Kuklane, I. Holmér, Cooling vests with phase change materials: the effect of melting temperature on heat strain alleviation in an extremely hot environment, European Journal of Applied Physiology 111 (2011) 1207-1216.

N. Ghaddar, K. Ghali, S. Chehaitly, Assessing thermal comfort of active people in transitional spaces in presence of air movement, Energy and Buildings 43 (2011) 2832-2842.

M.J. Loumala, J. Oksa, J.A. Salmi, V. Linnamo, I. Holmer, J. Smolander, B. Dugue, Adding a cooling vest during cycling improves performance in a warm and humid conditions, Journal of Thermal Biology 37 (2012) 47-55.

S. Mondal, Phase change materials for smart textiles-An overview, Applied Thermal Engineering 28 (2008) 1536-1550.

J.R. House, H.C. Lunt, R. Taylor, G. Milligan, J.A. Lyons, C.M. House, The impact of a phase-change cooling vest on heat strain and the effect of different cooling pack melting temperature, Eur J Appl Physiol 113 (2013) 1223-1231.

K. Ghali, N. Ghaddar, J. Harathani, Experimental and numerical investigation of the effect of phase change materials on clothing during periodic ventilation, Textile Res. J. 74(3) (2004) 205-214.

M. Tate, D. Forster, D.E. Mainwaring, Influence of garment design on elite athlete cooling, Sports Technology 1(2-3) (2008) 117-124.

C. Chou, Y. Tochihara, T. Kim, Physiological and subjective responses to cooling devices on firefighting protective clothing, European Journal of Applied Physiology 104 (2) (2008) 369-374.

R.E. Reinertsen, H. Færevik, K. Holbǿ, R. Nesbakken, J. Reitan, A. Røyset, M. Suong Le Thi, Optimizing the performance of phase-change materials in personal protective clothing systems, International Journal of Occupational Safety and Ergonomics (JOSE) 14(1) (2008) 43–53.

W. Bendkowska, M. Kłonowska, K. Kopias, A. Bogdan, Thermal manikin evaluation of PCM cooling vests, Fibres & Textiles in Eastern Europe 18(1) (2010) 70-74.

C. Gao, K. Kuklane, I. Holmér, Cooling effect of a PCM vest on a thermal manikin and on humans exposed to heat, Environmental Ergonomics 12 (4) (2007) 146-149.

C. Gao, K. Kuklane, I. Holmér, Cooling vests with phase change material packs, Ergonomics 53(5) (2010) 716-723.

W. Karaki, N. Ghaddar, K. Ghali, K. Kuklane, I. Holmer, L. Vanggaard, Human thermal response with improved AVA modeling of the digits, International Journal of Thermal Sciences 67 (2013) 41-52.

H. Hamdan, N. Ghaddar, D. Ouahrani, K. Ghali, M. Itani, PCM cooling vest for improving thermal comfort in hot environment, International Journal of Thermal Sciences 102 (2016) 154-167.

Energy Performance of Industrial Steel Buildings in Different Climates

Pascal Brinks, PhD

ABSTRACT

The origin of the industrial steel building industry is in the US and Europe. Here large industrial steel halls have a long tradition and the building design was adapted to the climate in these western regions. With an increasing industrial development of other regions also their demand for industrial buildings raised in the last years. Such emerging markets are often located in very different climates, which should be respected better by the building designers. However, little knowledge in energy efficient building design and missing building regulations in many emerging markets cause the opposite. Industrial steel buildings are often imported by European or American suppliers without any climate adaptation. This leads to a non-optimal energy performance and discomfort in the buildings. In practice such drawback is often compensated by oversized HVAC systems, causing high investments and energy costs and increase the carbon emissions.

For a better assessment of the most important climate sensitive building parameters, extensive building energy simulations were carried out. Such simulations covered the air infiltration using an air flow network model, ground interaction by linking the building simulation to a transient finite difference model of the ground and solar optimization of window and sky light orientation. To gain input data for the simulation of air infiltration, first extensive measurements of the leakages in typical industrial buildings were carried out. For this purpose an air tightness test stand was build and moreover fan pressurization tests in whole buildings were executed. The assessment was carried out for hot climates such as North Africa, for cold regions such as different locations in Russia and for climates with hot summers and cold winters such as Turkey. The simulation results were compared to the energy performance in the origin climate (Central Europe).

The outcomes of this study are concepts for improving the energy performance of industrial steel buildings exported to different countries, which still have a lack of experience and regulations for energy efficient building design. In addition a focus was on design aspects for damage free building constructions regarding building physics. As the climate has a huge impact on the humidity transport in building components this aspect was also crucial to consider for the design of optimized building components.

INTRODUCTION

In the past the research regarding low-energy concepts for buildings and building physics in general was mainly concentrated on residential and office buildings. The industrial building sector usually just aimed for low investment costs and on meeting the minimum requirements of building regulations as discussed in (Brinks 2015). Many countries have even lower requirements on the energy performance of industrial buildings than on the energy performance of other buildings, as e.g. France (Méthode de Calcul, 2012). Thereby the request for high performing industrial buildings was very low, which again caused little research activities in that field. This lack of research was already claimed in (Oschatz et al. 2012, Kuhnhenne 2009, Brinks et al. 2015 a). However, in Europe and the United States of America industrial steel buildings have a long tradition and were adapted to their specific climates, even if their energy performance is often improvable.

With an increasing industrial development of other regions also their demand for industrial buildings raised in the last years. But due to little experience in the construction of large industrial facilities and due to the possibility to ship light steel buildings easily, many industrial halls are exported to these new markets by European and American suppliers. However, the importing countries are often located in very different climate zones, which is often not considered by the western building designers. Little knowledge in energy efficient building design and building physics in the importing countries, as well as missing building regulations, cause that no climate adaptation for the specific region is conducted. Usually just the insulation thickness is adjusted dependent on if the building is exported to a hot or cold climate, but without any calculation or simulation. And the roof is painted in bright colours, when buildings are exported to hot climates. This leads to a non-optimal energy performance, partially also to damages and often to discomfort in such buildings. In practice such drawback is often compensated by oversized HVAC systems, causing high investment and energy costs and increase the carbon emissions. Building regulations, which could prevent these problems are partially missing at all or often not fully developed. In Russia for example the SNIP (СНиП 2003) does not require an energy performance calculation for industrial buildings at all. Many of these "simplified" building regulations just require certain U-values or R-values. Although, it is not reasonable to rate the energy performance just based on the thermal insulation of a building. Important parameters, such as air tightness, ventilation, building cubature or solar orientation of the glazed surfaces are often not considered. To gain better knowledge of the interdependences of industrial steel buildings and different climates, an extensive analysis was undertaken.

Figure 1 Exported industrial halls without significant climate adaptation; left: Russia; right: North Africa

METHOD

The main method for this analysis was building energy simulation. The building energy model "Type 56" of the software TRNSYS was linked to a transient 3D finite difference model of the ground (Type 49). To model the ground in detail was important, as the large floor area of industrial buildings causes an important interaction of the soil and the building. Besides there are big differences in national standards for floor slab insulation, which should be evaluated in this study. Because air infiltration is another important source of energy losses, the building energy simulation was also linked to an air flow network model. However, to simulate air infiltration into a building with such model requires to know the specifics of typical leakages in the analyzed buildings. As only little data about leakages in industrial light steel buildings is available, own data was generated. Therefore an air tightness test stand was built. In this test stand typical leakages between building components were measured. Examples are the eaves and verge connection, the connection between the floor slab and the walls, leakages around glazing elements, skylights etc. and leakages at vapor retarder connections. The measurements and the test stand are described in detail in (Brinks et al. 2015 b). Another aspect was to determine a preferable solar orientation of the glazed surfaces in industrial buildings. To date passive building design is not yet standard for such buildings.

Today the natural lighting in industrial halls is often exclusively ensured by sky lights in the roof. In the building simulation their effect on the energy performance and the discomfort hours in industrial facilities was analyzed. Moreover the climate has a significant impact on the hygro-thermal transport in wall and roof components. As in light steel buildings vapor retarders or vapor barriers are often used, to avoid condensation inside the constructions, it is important to analyze the moisture transport dependent on the climate. Therefore the transient hygro-thermal simulation software WUFI (Zirkelbach et al. 2015) was used.

Climates

For the simulations different climates were used. One maritime, hot and dry climate (Casablanca, Morocco), one cold continental climate (Samara, Russia) and one climate with hot summers and cold winters (Ankara, Turkey) was chosen. These climates represent different countries, which often import industrial steel buildings that were produced in Central Europe. Thus, they were compared to the German reference climate (Potsdam), which represents the climate, buildings were originally designed for. Figure 2 - 4 show the main characteristics of these climates (all values represent the monthly averages).

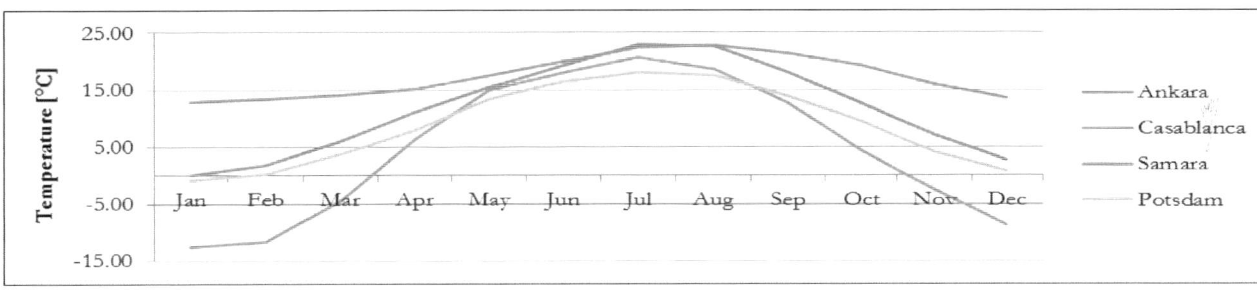

Figure 2 Outside temperature for different analyzed climates

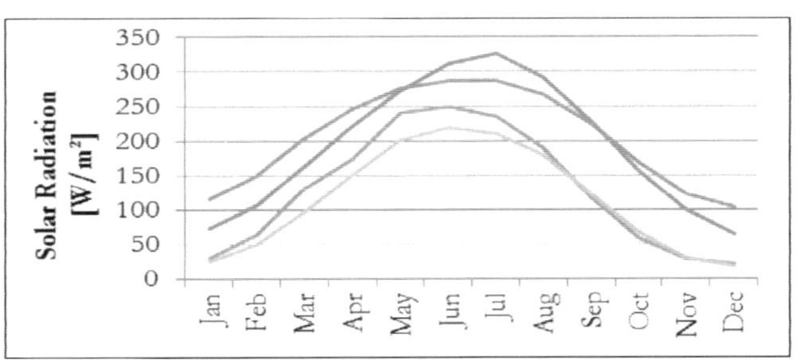

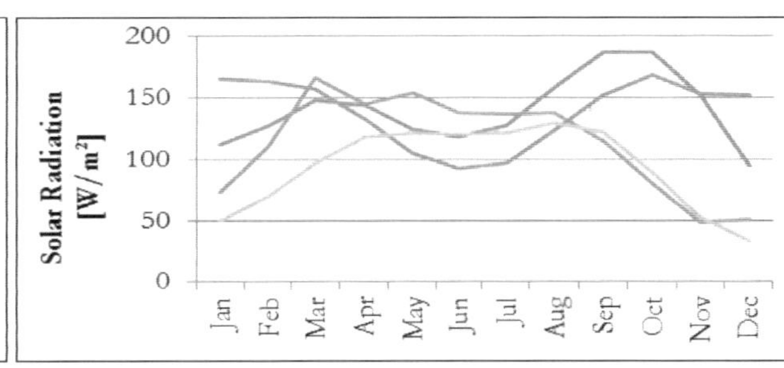

Figure 3 (a) Intension of solar radiation (horizontal) (b) Intension of the solar radiation (south)

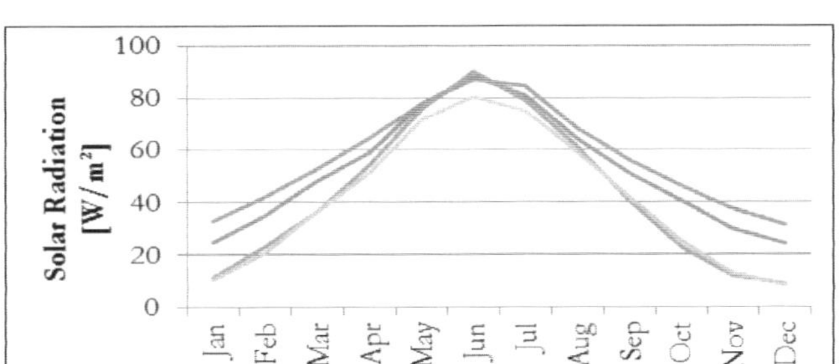

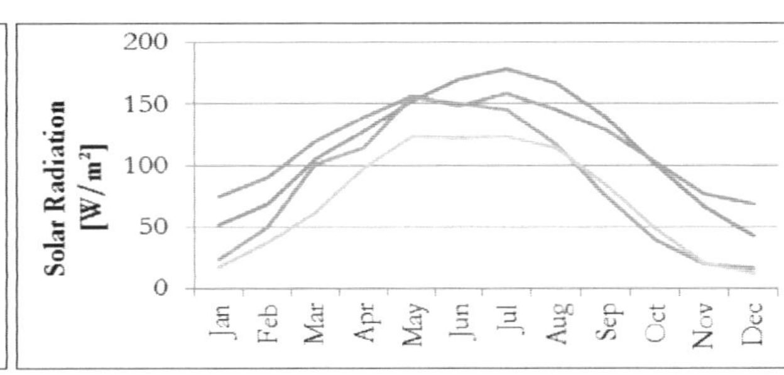

Figure 4 (a) Intension of solar radiation (north) (b) Intension of the solar radiation (east)

RESULTS

Glazing Orientation & Passive Cooling

In Europe warehouses and production buildings often have skylights or dome lights in the roof to have a good natural lighting despite a relatively small glazing area. In exported industrial buildings this common practice of glazed surfaces in the roof is usually not changed, especially, if no specific regulations for summer overheating exist, as in many African countries or in Russia. However, for many climates this can be critical. To highlight the consequences of using a large number of sky lights and to determine an optimal glazing orientation several simulations were carried out. Figure 5 shows the impact of the glazing orientation and the glazing area on the discomfort hours. In this chart the discomfort hours higher than 26 °C are shown. 500 Kh/yr is the common limit for summer overheating in commercial buildings in Germany (DIN 4108-2). The calculations were conducted for a production building with 2000 m² floor area, heat loads by machines of 40 W/m² during the work shift, U-values of wall and roof of 0.20 W/m²K and an air change rate of 1.5 per hour.

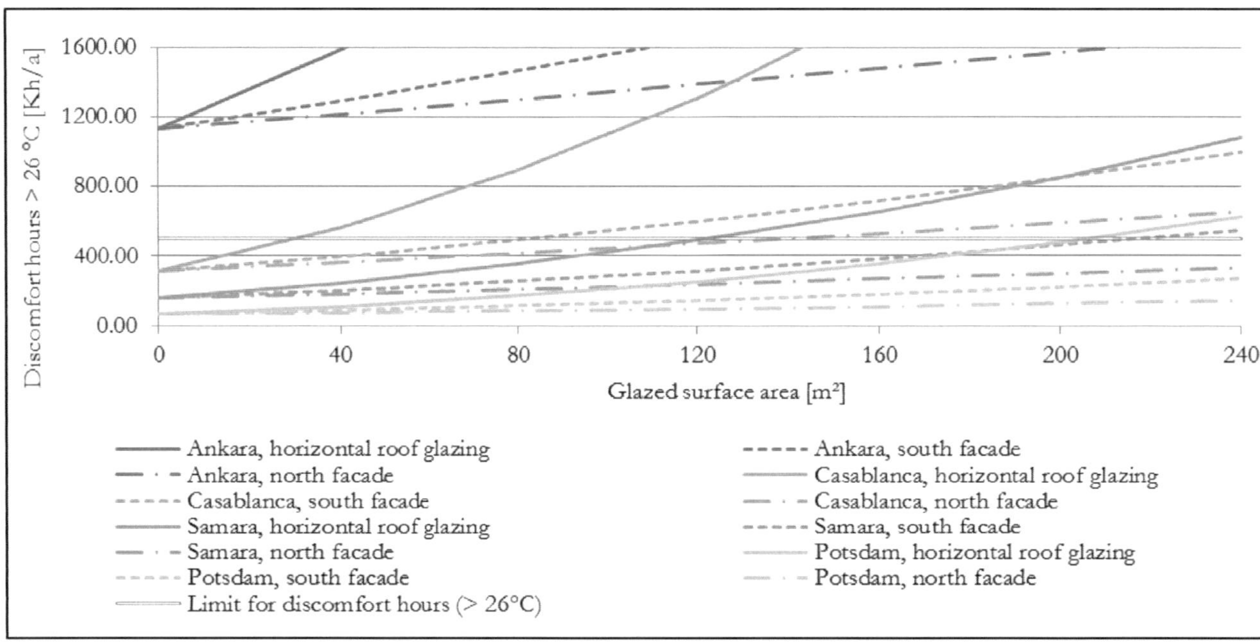

Figure 5 Discomfort hours for a production building in different climates dependent on the glazing area

Figure 5 shows, that in the origin country (Germany, Potsdam climate) horizontal roof glazing areas up to 200 m² (10 % of the floor area) are still acceptable. The limit of discomfort hours is not met. However, for all other countries shown in the graph, this glazing area and orientation is already critical. Even in Samara, Russia, the limit of discomfort hours would have been exceeded. Due to the high solar radiation and the high continental temperatures in summer the production building in Samara would already be overheated. Hence, it would be reasonable to move the glazed surfaces in the roof at least partially to the south façade. Due to the high solar radiation on the south during winter and spring (Fig. 3 b) this would also reduce the space heating demand in Samara. Because the heating period in this area is very long, a larger share of the solar radiation can be used than in Germany.

For Ankara and Casablanca, horizontal glazing in the roof of production buildings should be prevented at any rate. Here the north and south façades should be used for glazed surfaces.

However, meeting the German requirements with a limit of 500 Kh/yr > 26 °C will not be possible in these climates. In addition other measures, such as passive cooling, should be used. Hence, also simulations for a controlled ventilation were conducted. In Figure 6 the discomfort hours above 27 °C were simulated for the same production building and for the same critical climates (Ankara and Casablanca). Here the impact of night ventilation is shown. The ventilation is always turned on, if the inside temperature exceeds the outside temperature. It is visible, that this measure is very effective for both climates. Even with south oriented glazed surfaces in the façade the discomfort hours > 27°C can be reduced to a sufficient level without any active cooling.

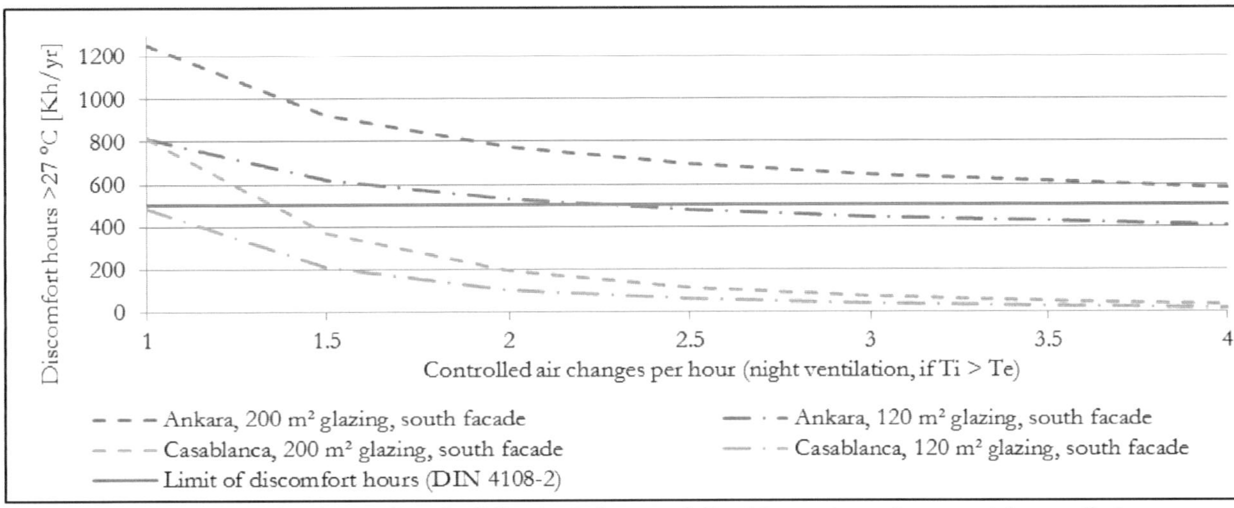

Figure 6 Discomfort hours for a building in Ankara and Casablanca dependent on night ventilation rates

Air Tightness

In the building regulations of European countries usually a minimum level for the air permeability of buildings is defined as a q_{50}-value in m³/m²h. This permeability has to be proven with a fan pressurization test after erecting the building. Otherwise a significantly higher permeability has to be considered in the energy performance calculation. Thus, the building´s air tightness is tested and usually meets the particular requirements. In Germany e.g. the maximum permeability is 4.5 m³/m²h (EnEV 2013).

In many countries, which import industrial buildings such requirements do not exist for industrial buildings, e.g. in Russia (SNIP 2003). Hence, the erection quality regarding air tightness is often much worse. However, due to the cold climate air tightness is very important. The high temperature differences between inside and outside increase the pressure differences at the building envelope, caused by stack effects. The higher pressure differences again cause higher infiltration, even at the same permeability. In the here carried out air flow network simulations the air infiltration into production buildings with the same q_{50}-value were simulated for different climates. Figure 7 (a) shows the average infiltration into a building with a permeability of 3 m³/m²h (at 50 Pascal). In the coldest climate (Samara) the air infiltration is almost three times higher than in the warmest climate (Casablanca). Hence, it is extremely important to increase the air tightness of buildings in cold climates.

Figure 7 (b) shows a typical damage caused by a high air infiltration in cold climate. Via convection warm air is introduced into the façade. The warm air cools down and water condensates. The water comes out at the socket of the façade and freezes.

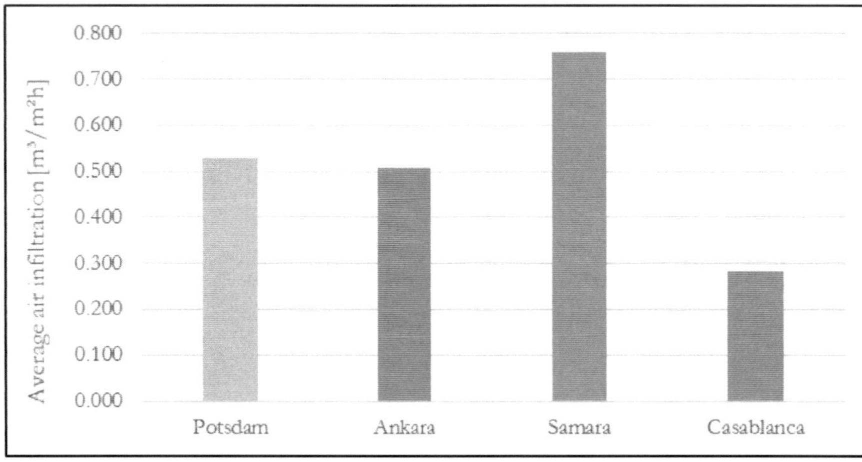

Figure 7 (a) Average infiltration into a building (q_{50} = 3 m³/m²h, A = 2000 m², $T_{i,min}$ = 17 °C) in different climates (b) Frozen condensation water at a façade, caused by convective heat flow in a leaky wall in Russia

For Figure 8 the space heating demand of a production building was simulated, dependent on the q_{50}-value. The calculations show that air tightness is extremely important for heat energy saving in cold climates. An example shows that: Reducing the q_{50}-value from 10 to 7 m³/m²h saves as much energy as reducing the U-value of all walls and the roof from 0.40 to 0.20 W/m²K. However, today the Russian standard for energy use in buildings (SNIP 2003) only requires a certain thermal insulation level, but no specific air permeability of the envelope.

Also for the climate of Ankara the air tightness still has a certain impact on the space heating demand. The cold temperatures in winter cause a high stack effect. Hence, the air infiltration is almost as high as in Germany (Figure 7 (a)).

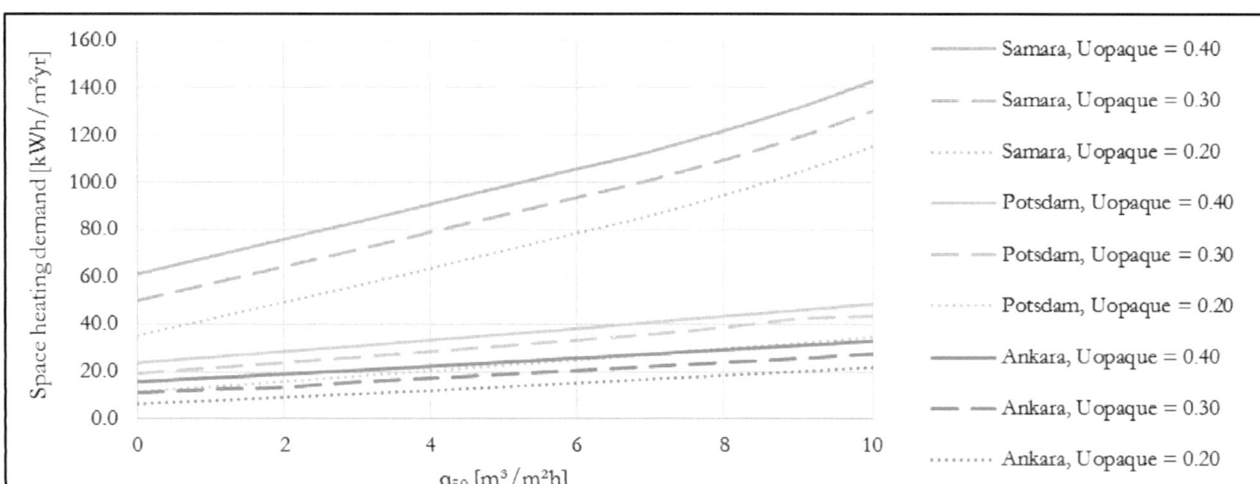

Figure 8 Space heating demand of a production building dependent on its air permeability and U [W/m²K]

Insulation Of Floor Slabs On Grade

The insulation of large floor slabs on grade can have a significant impact on the energy performance of industrial buildings. Warehouses and partially also production buildings are often low heated. In Germany the minimum inside temperatures of permanent working areas must be at least 12 °C. But for non-permanent work areas the temperatures can also be lower. Below 12 °C horizontal floor slab insulation is not required. This is reasonable, as the average ground temperature under large slabs mostly does not fall below the yearly averaged air temperature. For the German reference climate this temperature is 9.5 °C. During summer the ground under large slabs on grade is heated up. Thus, the average soil temperature below the building is even higher. This causes very low energy losses via floor slabs of low heated buildings, even when they are not insulated. Partially even gains from the ground into the building can exist, when the ground temperature exceeds the required minimum inside temperature of the building.

With the 3D finite difference ground model, transient simulations of the ground temperature under industrial buildings were conducted. The simulations were carried out for the climates of Potsdam, Germany and Samara, Russia. In Russia no floor slab insulation is required according to (СНиП 2003). Figure 9 shows the yearly averaged temperature directly under the floor slab of a production building with 2000 m² floor area and 11 °C minimum inside temperature for the first eight years after erection. In Figure 10 the corresponding yearly energy demand for space heating is shown. It is visible that in the first years the ground temperature under the building heats up for both climates. For Potsdam the temperature evens out at approximately 11.3 °C with horizontal insulation and at 12.9 °C without horizontal insulation. Because the average ground temperature in this case is higher than the required minimum inside temperature, gains from the ground to the building exist. To increase these gains it is reasonable to use no horizontal insulation under the floor slab. In Figure 10 it is shown that without an insulation the long term energy demand will be lower than with an insulation. For the cold climate in Samara, a horizontal insulation under the slab would be reasonable, even if it is not required according to Russian legislation. Due to the low average temperatures under the slab, there are higher losses without an insulation. More information about thermal heat transport via slabs on grade in Central European climate is described in (Brinks et al. 2016).

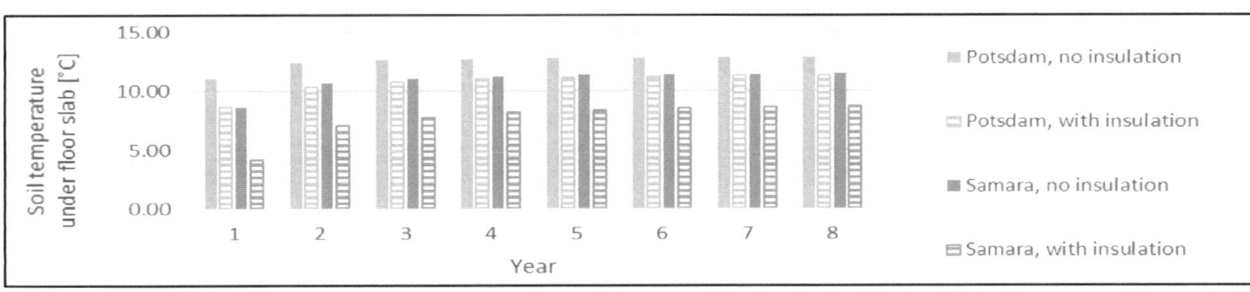

Figure 9 Yearly averaged temperature under a slab on grade of a production building ($T_{i,min}$ = 11 °C)

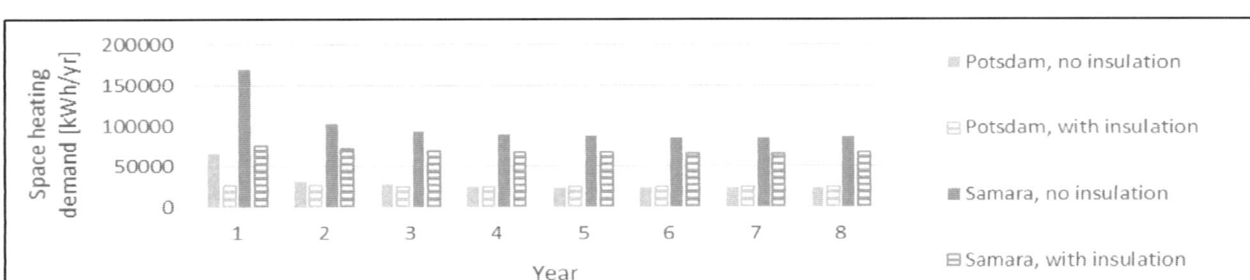

Figure 10 Yearly space heating demand of a production building ($T_{i,min}$ = 11 °C)

CONCLUSION

As a conclusion it can be summarized, that there is need for stricter building regulations for industrial buildings in countries like Russia, that import industrial steel buildings from Europe. However, even if such regulations are missing, also the exporting companies need to ensure a climate adaptation of exported buildings.

For hot climates this means mainly to avoid large glazed surfaces in the flat roofs and to replace them as much as possible by north (and south) oriented façade glazing. Reflective coatings should also be used to reduce overheating. However, the most effective energy cooling measure will be passive cooling by controlled night ventilation.

For cold climates, horizontal insulation under the floor slabs should be used, even for low heated buildings. In addition large energy savings can be realized by large glazed facades with south orientation. The most important measure to reduce the space heating demand will be to increase the air tightness, because air infiltration is even higher in cold climates, due to very high stack effects.

For climates with hot summers and cold winters no horizontal slab insulation should be used, as the ground heats up during summer. This reduces the number of discomfort hours in winter and increases the gains from the ground into the building during winter. Sky lights should be avoided in such buildings and should be replaced by south and partially north oriented façade glazing.

ACKNOWLEDGMENTS

The presented research was carried out with financial support by the Fonds National de la Recherche Luxembourg (FNR) in framework of the PhD project “Concepts for Nearly Zero-Energy Industrial Buildings” at Technical University Kaiserslautern, Germany. We thank the FNR and the TU Kaiserslautern for the support.

REFERENCES

Brinks, P., 2015. Concepts for nearly zero-energy industrial buildings. Dissertation, Technical University Kaiserslautern.

Brinks, P., Kornadt, O., Oly, R., 2015 a. Imprecisions Deficits and Possible Improvements in Current Energy Balance Calculation Methods for Industrial Buildings. Proceedings World Sustainable Energy Days 2015. Wels, Austria.

Brinks, P., Kornadt, O., Oly. R. 2015 b. Air Infiltration Assessment for Industrial Buildings. Energy and Buildings 86: 663–676.

Brinks, P., Kornadt, O., Oly, R., 2016. Development of concepts for cost-optimal nearly zero-energy buildings for the industrial steel building sector. Applied Energy 173: 343-354.

СНиП 23-02-2003 СТРОИТЕЛЬНЫЕ НОРМЫ И ПРАВИЛА РОССИЙСКОЙ ФЕДЕРАЦИИ ТЕПЛОВАЯ ЗАЩИТА ЗДАНИЙ - THERMAL PERFORMANCE OF THE BUILDINGS (Russian building energy requirements, SNIP 2003). 2003.

DIN 4108-2. Thermal protection and energy economy in buildings – Part 2: Minimum requirements to thermal insulation. 2013.

EnEV - Energieeinsparverordnung, Nichtamtliche Lesefassung zu der am 16.10.2013 von der Bundesregierung beschlossenen, noch nicht in Kraft getretenen zweiten Verordnung zur Änderung der Energieeinsparverordnung. EnEV 2014 (German building regulations 2014), 2013.

Kuhnhenne, M. 2009. Energetische Qualität von Gebäudehüllen in Stahl-Sandwichbauweise (Thermal quality of building envelopes in steel sandwich technology). Dissertation, RWTH Aachen.

Méthode de Calcul - ANNEXE A l'arrêté portant approbation de la méthode de calcul Th-BCE 2012 (Regulation for the utilization of the calculation method Th-BCE 2012). 2012.

Oschatz, B., Rosenkranz, J., Mailach, B., Gritzki, R., Kaiser, J., Perschk, A., Rösler, M., Seifert, J. and Otto, F. 2012. Gesamtanalyse Energieeffizienz von Hallengebäuden (Global analysis of the energy efficiency of hall buildings).

Zirkelbach, D., Schmidt, T., Kehrer, M, Künzel, H.M. WUFI® Pro – Manual, Fraunhofer Institute for Building Physics

New Concept of Power Generation from TEG Using the Exhaust Airflow of All-Air HVAC Systems and the Condenser Hot Air

Mohamad Ramadan, PhD. **Khaled Chahine, PhD.** **Mahmoud Khaled, PhD.**

ABSTRACT

The present work proposes a new design for electric power generation using a thermoelectric power generator (TEG). It recovers waste energy of the HVAC condenser and reuses the exhaust air flow for thermo-electric power generation. The air of the condenser is considered as heat source, whereas the exhaust air flow is utilized as cooler. The advantage of this concept is that it allows recovering waste energy of HVAC systems to produce green energy through TEG. To proceed, a mathematical model is first developed and a parametric analysis is then conducted to evaluate the power generated with the new concept when the air velocity of the exhaust airflow and that of the condenser vary. It is shown that for a space cooling load of 100 kW, a 40x40 cm² flat plate is capable to generate 90 W of electrical power.

INTRODUCTION

Energy management involves several methods to improve energy performance, of which waste heat recovery is one. Waste heat is abundantly and freely available in industries as well as residential areas. In this context, thermoelectric generators present an excellent approach for converting waste heat into electrical power. Such generators are silent, scalable, with no moving parts and can generate electricity whenever they undergo a temperature gradient. A significant source of waste heat that is almost ubiquitous is a Heating, Ventilating and Air Conditioning HVAC system. Indeed, these systems, despite being an essential need of modern life, dissipate large amounts of waste energy, especially through the condenser. The exhaust air flow of All-Air HVAC systems is also a source of waste energy and thus may be utilized as a cooling fluid in many applications. In this context, the present work suggests a new design that permits to couple the two energy domains described above: power generation from TEG and HVAC. It particularly uses the exhaust airflow of All-Air HVAC systems to cool one surface of a TEG, whereas its second surface is heated by the hot airflow leaving the condenser unit. This permits to induce a temperature gradient across the thickness of the TEG and generate electric power. First, one-dimensional steady-state solution of the heat diffusion equation is considered with convective boundary conditions at the two TEG surfaces. A parametric analysis is then conducted to evaluate the power generated with the new concept when the air velocity of the exhaust airflow and that of the condenser vary.

The remaining of the paper is organized as follows: section 2 presents the parametric analysis for different real thermal configurations a TEG must encounter, section 3 shows the thermal modeling of the system as well as the corresponding results, and finally section 4 draws the conclusions of the work.

Authors are faculty members at the Mechanical Department of the School of Engineering of the Lebanese International University

PARAMETRIC ANALYSIS AND RECOMMENDATIONS

This section concerns a parametric analysis of the TEG thermal behavior subjected to several conditions. The aim is to find recommendations as to obtaining significant temperature gradient through the TEG module. Although a temperature gradient is the trigger for a TEG to generate power, practically this gradient is not exactly nor necessarily obtained by fixing the temperatures at the two TEG surfaces. Moreover, fixing the temperatures at the two surfaces of the TEG requires external means requiring power. The most common boundary conditions that are encountered in practice will be exposed and treated below. For each case, the temperature gradient between the hot and cold surfaces of the TEG will be evaluated at steady state and parametrically analyzed. Figure 1 shows schematics of the different configurations that will be tested.

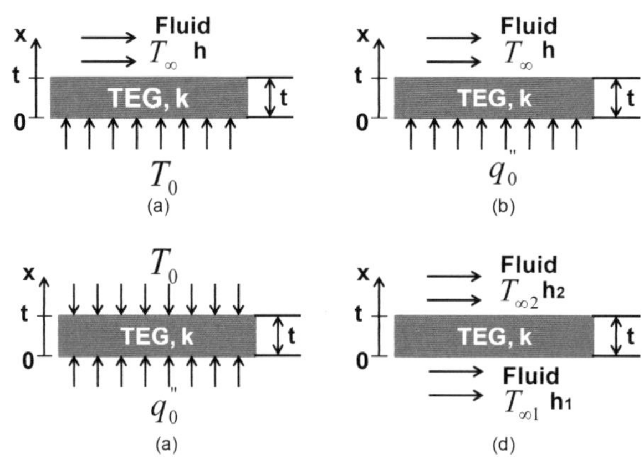

Figure 1 Schematics of (a) configuration 1, (b) configuration 2, (c) configuration 3, and (d) configuration 4.

In configuration one (Figure 1-a), the TEG of thickness t and thermal conductivity k is subjected to a constant temperature T_0 at its hot side and simultaneously cooled by air having a convective coefficient h and temperature T_∞. In configuration two (Figure 1-b), the TEG is subjected to a constant heat flux q''_0 at its hot side and simultaneously cooled by air having a convective coefficient h and temperature T_∞. In configuration three (Figure 1-c), the TEG is heated at its hot side with a constant heat flux q''_0 and simultaneously subjected to a constant temperature T_0 at its cold side. In configuration 4 (Figure 1-d), the TEG is heated at one side with air having a temperature $T_{\infty 1}$ and convective heat transfer coefficient h_1 while being simultaneously cooled with air having a temperature $T_{\infty 2}$ and convective heat transfer coefficient h_2 at the other side.

For the different conditions, the TEG is considered as a plane wall and the conduction through it as one-dimensional and steady. Then, the heat diffusion equation is reduced to:

$$\frac{d^2T}{dx^2} = 0 \tag{1}$$

The temperature distribution and the temperature gradient in all configurations can both be obtained by solving the differential equation of the heat diffusion equation above, but will vary according to the boundary conditions. In all of the above configurations, integrating the reduced form of the heat diffusion equation with respect to x yields the below linear variation dependent on two constants A and B function of the boundary conditions:

$$T(x) = Ax + B \tag{2}$$

Below, the boundary conditions, temperature distributions and temperature differences corresponding to the four configurations are exposed.

Configuration 1

$$T(0,\theta)=T_0 \tag{3-a}$$

$$-k\left.\frac{dT}{dx}\right|_{x=L}=h[T(t,\theta)-T_\infty] \tag{3-b}$$

$$T(x)=\frac{h(T_\infty-T)_0}{K+ht}x+T_0 \tag{3-c}$$

$$\Delta T=\frac{ht(T_0-T_\infty)}{k+ht} \tag{3-d}$$

Configuration 2

$$-k\left.\frac{dT}{dx}\right|_{x=0}=q_0'' \tag{4-a}$$

$$-k\left.\frac{dT}{dx}\right|_{x=L}=h[T(t,\theta)-T_\infty] \tag{4-b}$$

$$T(x)=-\frac{q_0''}{k}x+T_\infty+\frac{ht+k}{hk}q_0'' \tag{4-c}$$

$$\Delta T=\frac{q_0''t}{k} \tag{4-d}$$

Configuration 3

$$-k\left.\frac{dT}{dx}\right|_{x=0}=q_0'' \tag{5-a}$$

$$T(t,\theta)=T_0 \tag{5-b}$$

$$T(x)=-\frac{q_0''}{k}x+T_0+\frac{q_0''}{k}t \tag{5-c}$$

$$\Delta T=\frac{q_0''t}{k} \tag{5-d}$$

Configuration 4

$$-k\left.\frac{dT}{dx}\right|_{x=0}=h_1(T_{\infty 1}-T(0)) \tag{6-a}$$

$$-k\left.\frac{dT}{dx}\right|_{x=L}=h_2(T(t)-T_{\infty 2}) \tag{6-b}$$

$$T(x)=\frac{h_1h_2(T_{\infty 2}-T_{\infty 1})}{k(h_1+h_2)+h_1h_2t}x+T_{\infty 1}+\frac{kh_2(T_{\infty 2}-T_{\infty 1})}{k(h_1+h_2)+h_1h_2t} \tag{6-c}$$

$$\Delta T=\frac{h_1h_2(T_{\infty 1}-T_{\infty 2})}{k(h_1+h_2)+h_1h_2t}t \tag{6-d}$$

A parametric analysis based on the equations derived above is conducted to draw the main recommendations towards increasing the temperature difference across the TEG plate. These recommendations are summarized in Table 1.

Table 1. Summary of recommendations.

CONFIGURATION	DETAILS	RECOMMENDATIONS
1	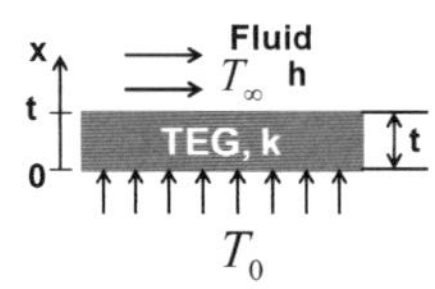	- The thickness of the TEG, the temperature prescribed at the hot surface of the TEG, and the convective coefficient should be high; - The fluid temperature at the cold side and the thermal conductivity of the TEG should be low.
2	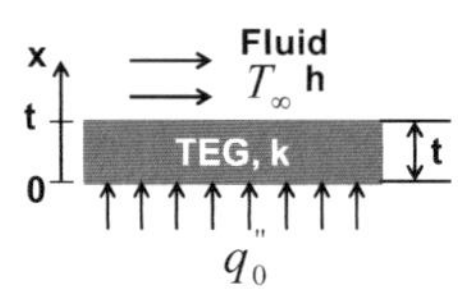	- The thickness of the TEG and the heat flux prescribed at the hot surface of the TEG should be high; - The thermal conductivity of the TEG should be low.
3	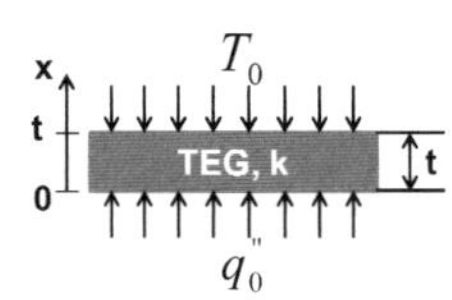	- The thickness of the TEG and the heat flux prescribed at the hot surface of the TEG should be high; - The thermal conductivity of the TEG should be low.
4	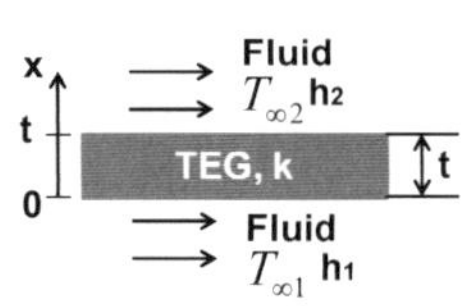	- The thickness of the TEG, the temperature of the hot fluid, the convective coefficients of the hot and cold sides should be high; - The thermal conductivity of the TEG and the temperature of the cold fluid should be low.

As conclusion, the most promising configuration is configuration 4 provided that the designer selects the appropriate couple of convective heat transfer coefficients since it has the most promising magnitude orders and corresponds at the same time to real configurations that can be encountered in engineering practice. All the configurations can be more promising if the thermal conductivity of the TEG can be lowered and its thickness increased which is related to the enhancement of the design of the TEG itself. Configurations 2 and 3 can be also promising if the applications involve high and extreme heat fluxes. Configuration 1 can also be more promising for applications where high temperatures are present. The following section will be devoted to the implementation of configuration 4 in a real case and its corresponding thermal behaviors.

NEW CONCEPT AND CALCULATIONS

The study presented in the previous section has shown that configuration 4 presents the higher gradient of temperature for the same conditions. As an application, a system that uses the hot condenr air and the relatively cold exhaust air flow of All-Air HVAC systems is suggested and investigated. The suggested system (Figure 2) consists in using a flat plate composed of several TEG modules in series with its upper surface exposed to the condenser hot airflow and its lower surface to the exhaust air of the HVAC system. Each of the TEG modules will then have its upper surface heated by the condenser airflow and its lower surface cooled by the exhaust airflow, which will induce a temperature gradient and then a voltage (electrical power).

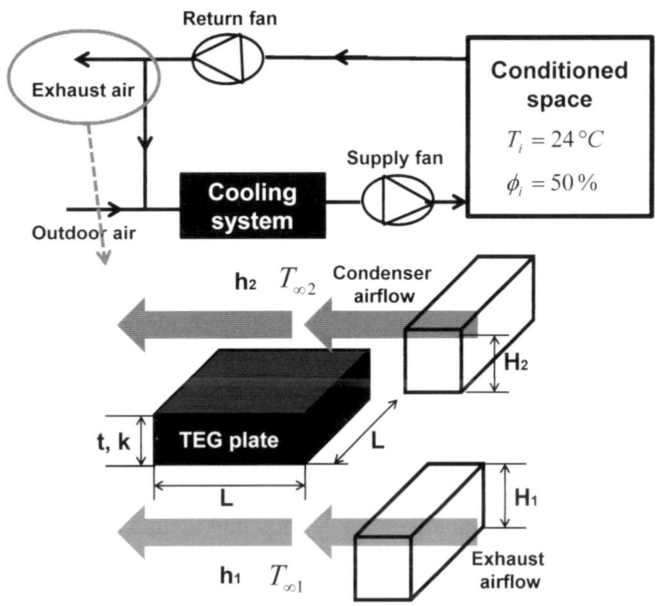

Figure 2 Schematic of the suggested concept of power generation.

The mass flow rate of air exhausted from the system is written as follows:

$$\dot{m}_e = E\dot{m}_s \quad (7)$$

Where E is the fraction of exhausted air from supply air that depends on the type of conditioned space (apartment, industry, hospital, etc.) and $\dot{m}_s$ is the mass flow rate of air supplied to the conditioned space calculated from the following relation:

$$\dot{m}_s = \frac{\dot{Q}_{cooling}}{i_r - i_s} \quad (8)$$

Where $\dot{Q}_{cooling}$ is the cooling load of the conditioned space, i_r the specific enthalpy of air in the room, and i_s the enthalpy of supplied air. According to ASHRAE standards [9], the indoor temperature and relative humidity are respectively 24 ^{0}C and 50 % which corresponds to an enthalpy of 48 kJ/kg. Then, with a typical value of the sensible heat factor of 0.8 and for a supply temperature of 13 ^{0}C, the enthalpy of supply air is 32 kJ/kg. Considering now the exhausted air flow below the TEG plate, the infinite velocity is caluclted as follows:

$$u_{\infty 1} = \frac{4\dot{m}_e}{\rho\pi D_{h1}^2} \quad (9)$$

Where ρ is the air density and D_{h1} the hydraulic diameter if the duct where the exhaust airflow exits calculated as follows:

$$D_{h1} = \frac{4S}{P} = \frac{2(H_1 L)}{H_1 + L} \quad (10)$$

Then, the Reynolds number at the edgle of the bottom surface is calculated from the following relation:

$$\mathrm{Re}_{L1} = \frac{\rho u_{\infty 1} L}{\mu} = \frac{4\dot{m}_e L}{\pi D_{h1}^2 \mu} \tag{11}$$

Where μ is the dynamic viscosity of air. Now to calculate the heat transfer coefficient h_1 of air at the lower surface of the TEG plate, the flow regime should be checked. If the Reynolds number Re_{L1} is lower than 5x10^5, the flow is laminar all over the plate and then the convective transfer coefficient can be calculated from the following correlation [10]:

$$Nu_1 = \frac{h_1 L}{k} = 0.664\,\mathrm{Re}_{L1}^{0.5}\,\mathrm{Pr}^{1/3} \tag{12}$$

Where Nu_1 is the Nusselt number, k the thermal conductivity, and Pr the Prandtl number. If the Reynolds number Re_{L1} is greater than 5x10^5, the flow is mixed (laminar then turbulent) over the plate and then the convective transfer coefficient can be calculated from the following correlation [10]:

$$Nu_1 = \frac{h_1 L}{k} = \left(0.037\,\mathrm{Re}_{L1}^{4/5} - 871\right)\mathrm{Pr}^{1/3} \tag{13}$$

Considering now the condenser air flow above the TEG plate, the infinite velocity is caluclted as follows:

$$u_{\infty 2} = \frac{4\dot{m}_C}{\rho \pi D_{h2}^2} \tag{14}$$

Where $\dot{m}_C$ is the mass flow rate of air traversing the condenser equal to 1 kg/s for cooling loads less than 100 kW, 2 kg/s for cooling loads between 100 and 200 kW, 3 kg/s for cooling loads between 200 and 300 kW, 4 kg/s for cooling loads between 300 and 400 kW, 5 kg/s for cooling loads between 400 and 500 kW. D_{h2} is the hydraulic diameter of the duct where the condenser airflow exits calculated as follows:

$$D_{h2} = \frac{4S}{P} = \frac{2(H_2 L)}{H_2 + L} \tag{15}$$

Then, the Reynolds number at the edge of the upper surface is calculated from the following relation:

$$\mathrm{Re}_{L2} = \frac{\rho u_{\infty 2} L}{\mu} = \frac{4\dot{m}_C L}{\pi D_{h2}^2 \mu} \tag{16}$$

Now to calculate the heat transfer coefficient h_2 of air at the lower surface of the TEG plate, the flow regime should be checked. If the Reynolds number Re_{L2} is lower than 5x10^5, the flow is laminar all over the plate and then the convective transfer coefficient can be calculated from the following correlation [10]:

$$Nu_2 = \frac{h_2 L}{k} = 0.664\,\mathrm{Re}_{L2}^{0.5}\,\mathrm{Pr}^{1/3} \tag{17}$$

If the Reynolds number Re_L is greater than 5x10[5], the flow is mixed (laminar then turbulent) over the plate and then the convective transfer coefficient can be calculated from the following correlation [10]:

$$Nu_2 = \frac{h_2 L}{k} = \left(0.037\,\mathrm{Re}_{L2}^{4/5} - 871\right)\mathrm{Pr}^{1/3} \tag{18}$$

On the other hand, the rate of heat extracted by the condenser is written as:

$$\dot{Q}_C = \dot{Q}_{cooling}\left(1 + \frac{1}{COP}\right) \tag{19}$$

Where COP is the coefficient of performance of the refrigeration cycle of the HVAC system under consideration and set equal to 3 in this study. Then, the air temperature downstream of the condenser (upstream of the upper surface of the TEG plate) $T_{\infty 2}$ is equal to:

$$T_{\infty 2} = T_a + \frac{\dot{Q}_C}{\dot{m}_C C_p} \tag{20}$$

Where T_a is the ambient temperature and C_p is the specific heat of air.

Then substituting the different parameters obtained above $(h_1, T_{\infty 1}, h_2, T_o)$ in equation (6-d), the temperature difference across each TEG module is finally:

$$\Delta T = \frac{h_1 h_2 (T_o - T_{\infty 1})}{k(h_1 + h_2) + h_1 h_2 t}\, t \tag{21}$$

Figure 3 shows the variation of the temperature difference across each TEG module as well as the total power obtained by the suggested generator in function of the space cooling load. For this set of calculations, the room temperature is fixed at 24 ^{0}C, the fraction of exhausted from supply air E is set equal to 0.4, the height of the exhaust duct H_1 is taken equal to 0.1 m, the height of the of the condenser airflow exit duct H_2 is 0.1 m, the length and width of the bottom plate composed of TEG modules are equal to $L = 0.4\,m$, the ambient temperature T_a is set equal to 30 ^{0}C, and the thickness and conductivity of each TEG module are respectively 0.12 m and 0.3 W/m.K [11-12]. As shown in Figure 3, the temperature difference across each TEG module (Figure 3-a) and the power generated by the assembly (Figure 3-b) increase significantly with the cooling load at low cooling loads and then varies slightly around given values at high cooling loads. As illustration when the cooling load increases from 20 to 500 kW, the temperature difference across each TEG module increases from 29.6 to 134.4 ^{0}C and the power generated by the assembly increases from 19.9 to 90.3 W for the first 100 kW of increase then vary slightly around 121 ^{0}C and 81 W for the following 400 kw of increase. Indeed when the cooling load increases from 20 to 100 kW at constant condenser air flow rate of 1 kg/s, the temperature of air leaving the condenser and heating the upper surface of the TEG plate increases significantly which increases significantly the temperature difference across each TEG module and the power generated by the TEG plate. For each condenser flow rate higher than 1 kg/s and corresponding to cooling loads higher than 100 kW, the same trends are observed. However when the cooling load increases from the range of 0-100 kW to the range of 100-200 kW or higher, the air temperature increase through the passage by the condenser decreases with respect to the case of lower flow rate corresponding to 0-100 kW. The decrease in the temperature difference between the hot and cold fluid sides $(T_{\infty 2} - T_{\infty 1})$ of equation 21 is domiant on the increase of the temperature difference caused by the increase of the convective heat transfer coefificen h_2 related to the hot side.

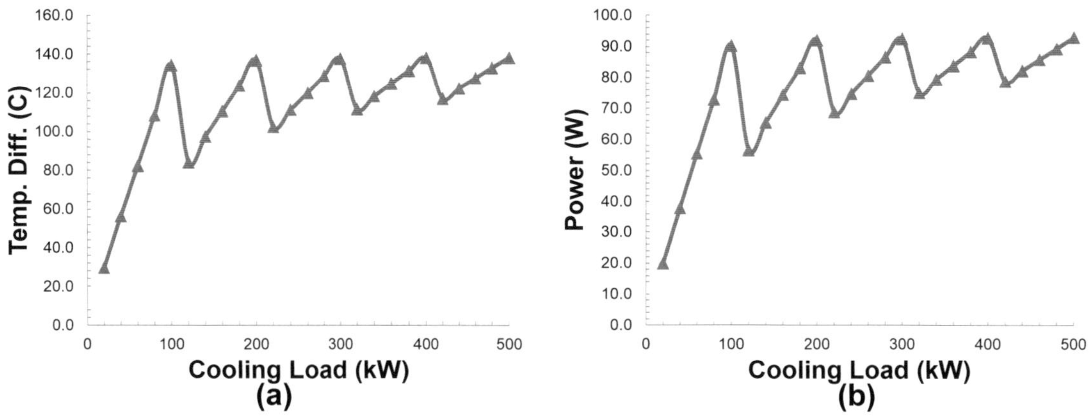

Figure 3 Variation of (a) temperature difference across each TEG module and (b) power generated by the assembly of TEG modules in function of the cooling load.

CONCLUSION

In this manuscript, a new concept of power generation from TEGs using the exhaust airflow of all-air HVAC systems and the condenser hot airflow is presented. An appropriate thermal modeling of the system is exposed as well. It was shown that for a space cooling load of 100 kW, a 40x40 cm^2 flat plate is capable to generate 90 W of electrical power.

REFERENCES

Ramadan M, Khaled. M, El Hage H. 2015. Using Speed Bump for Power Generation –Experimental Study. Energy Procedia 75, 867–872.

Khaled. M, Ramadan. M, El Hage H. 2015. Parametric Analysis of Heat Recovery from Exhaust Gases of Generators. Energy Procedia 75, 3295–3300.

Ramadan, M., Gad El Rab, M. and Khaled, M. 2015. Parametric analysis of air–water heat recovery concept applied to HVAC systems: Effect of mass flow rates. Case Studies in Thermal Engineering, 6, 61-68.

Khaled. M, Ramadan. M, Chahine. K, Assi. A. 2015. Prototype implementation and experimental analysis of water heating using recovered waste heat of chimneys. Case Studies in Thermal Engineering, 5,127-133.

Johnson I. and Choate W. T. 2008. Waste Heat Recovery: Technology and Opportunities in U.S. Industry. BCS, Incorporated, March.

Bell L. E. 2008. Cooling, Heating, Generating Power, and Recovering Waste Heat with Thermoelectric Systems. Science 321, 1457.

Vining C. B. 2009. An Inconvenient Truth about Thermoelectrics. Nature Materials 8, 83.

Nagengast, B. 1999. Early Twentieth Century Air-Conditioning Engineering. ASHRAE Journal, March.

McQuiston, F.C., Parker, J.D., and Spitler, J.D. 2005. Heating, Ventilating and Air Conditioning: Analysis and Design. 6th Ed. John Wiley & Sons.

Incorpera, F.P. and DeWitt, D.P. 2007. Fundamentals of heat and mass transfer. Sixth Edition. John Wiley & Sons.

Spices P., Pollak M., and Mateu L. 2015. Handbook of Energy Harvesting Power Supplies and Applications edited by Peter. CRC Press Taylor and Francis Group.

Synder G.J. and Toberer E.S. 2008. Complex thermoelectric materials, Nature Materials 7, 108-116.

Heating Fresh Air by Hot Exhaust Air of HVAC Systems

Ali Khraibany **Ali Najdi** **Amjad Makki**

Mostafa Gad El Rab, PhD. **Mohamad Ramadan, PhD.** **Mahmoud Khaled, PhD.**

ABSTRACT

The present work concerns heating fresh air by the hot exhaust air of all–air heating ventilating and air conditioning HVAC systems. To proceed, a prototype is implemented and a parametric analysis is carried out to test the effect of temperatures and mass flow rates on the performance of the heat recovery system. Experiments are performed under typical conditions simulating the HVAC operation in cold climates. It was shown that powers up to 110 W can be economized in cold climates and for low air flow rates of around 0.1 kg/s and 0.005 kg/s for respectively the hot exhaust air and the fresh cold air. Extrpolation shows that economized power can exceed 1 kW for a room of heating load 4 kW if all of the supply air is exhausted.

INTRODUCTION

The actual world tendency is towards the reduction of fuel consumption and carbon dioxide emission . This is mainly accomplished nowadays by using renewable energy and energy management systems [1-4]. Particularly, heat (energy recovery) is one of the main axes in the energy management development [5-6]. There are many applications of heat recovery such as internal combustion engines, chimneys, shower water and Heating Ventilating and Air Conditioning HVAC [5-6]. The principle of heat recovery is to use the waste heat released from a system in another heating or non-heating application. In fact waste heat occurs in many systems in particular the industrial applications [7-8].

On the other hand, Heating, Ventilating and Air Conditioning HVAC systems has changed from being luxury to essential need for people and thus involve many energy components which need to be more and more managed. In this context, the present work suggest a new design that permits to couple between the two energy axes described above: energy recovery and HVAC. It particularly use the hot echausted air from a condirtioned space in cold climates to heat/preheat cold fresh outdoor air.

There are three main categories of HVAC systems: All-Water, Water-Air, and All-Air systems. An All-Air system has acquired this name in HVAC since temperature and humidity are controlled by supplying only air to the conditioned space. Figure 1 shows a schematic of the operational mode of this category of HVAC system [9].

The assembly of components shown in Figure 1 and known as "Air handling unit" has as objective to remove energy from or add energy to airstreams before their supply to conditioned spaces. It particularly permits to heat, cool, humidify, dehumidify, clean, and distribute air to the various conditioned spaces in a zone or zones. Also, outdoor air can be admitted and room air can be exhausted through the air handling system. In general and in order to adopt compromises between comfort and economy, a fraction of the return air from the conditioned spaces will be

Authors are students and faculty members at the Mechanical Department of the School of Engineering of the Lebanese International University; Dr. Mostafa Gad El Rab is also affiliated to the Mechanical Power Engineering Department of the Minoufiya University, Shebin El-Kom, Egypt

exhausted and instead an amount of outdoor air will be mixed. This permits to have continuously fresh air entering the spaces and to reduce the amount of energy required to condition the supply air required completely from the outdoor air.

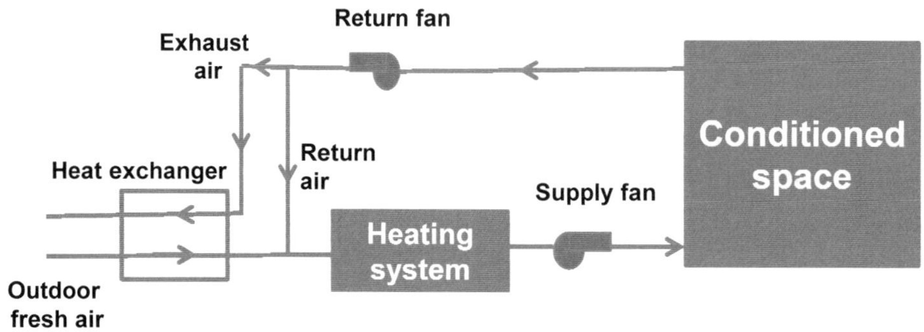

Figure 1 Schematic of all-air HVAC system and the integrated heat recovery system.

The heat recovery concept suggested in this paper consists on pre-heating the cool fresh outdoor air by the relatively hot exhausted air in cold climates. This is accomplished by a cross-flow air-to-air compact heat exchanger having as airstreams the outdoor air and exhaust air flows, as seen in Figure 1.

The remaining of the paper is organized as follows: section 2 presents the experimental setup and configurations, section 3 shows the results and analysis, and finally section 4 draws the conclusions of the work.

EXPERIMENTAL SETUP AND CONFIGURATIONS

To achieve the concept of the heat recovery system and to study its performance, a prototype is implemented. It consists mainly of a room (the conditioned space) simulated by a wood box of 6.7 m3 (Figure 2), an air conditioning split unit of 5.6 kW (18 000 Btu/hr) capacity simulating the hot air supply of the central system to the room, and the ducting system forming the heat recovery system. The heat recovery system corresponds to an insulated metallic duct including an air to air cross flow tubes-and-fins heat exchanger and a variable speed fan of flow simulating the exhaust air of the HVAC system. The flow in the exchanger simulates the oudoor fresh air flow.

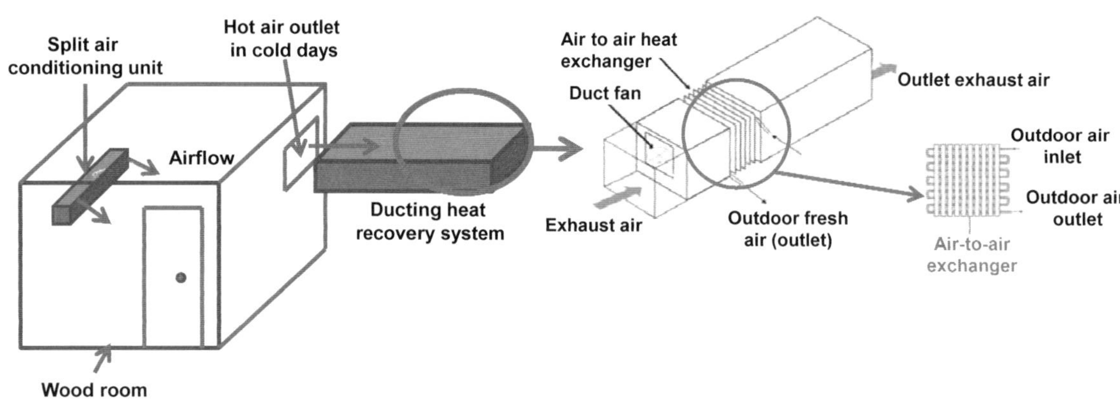

Figure 2 Schematic of the implemented prototype.

To test the performance of the system, it is important to measure the temperature and velocity (flow rate) at different locations of the system. For this purpose, 5 digital type-K thermocouples with 0.5 ^{0}C error are used to read continuously the temperature in the wood room, at the inlet and outlet of the exchanger, and in the duct upstream and downstream of the exchanger. Two velocity anemometers are used to measure the air velocity at different locations in the outlet sections of the exchanger and different locations in a plane downstream of the exchanger and parallel to it. On the other hand, a compressor is used to simulate the outdoor cold weather (outdoor air flow).

In the experimental setup, there are two main flow rates that need to be measured: the air flow passing through the exchanger and the air flow passing in the exchanger geometry. After measuring the velocity at different locations on the outlet section of the exchanger and on a plane downstream of the exchanger, the flow rates passing in the exchanger $\dot{m}_{exchanger}$ and through it $\dot{m}_{duct}$ can be calculated as follows:

$$\dot{m}_{exchanger} = \rho_a A_{out,exchanger} V_{av,out} \tag{1}$$

$$\dot{m}_{duct} = \rho_a A_{duct} V_{av,duct} \tag{2}$$

Where ρ_a is the air density evaluated at an average temperature of 30 ^{0}C, $A_{out,exchanger}$ is the cross-sectional area of the exchanger outlet, A_{duct} is the cross-sectional area of the duct where the exchanger is integrated (which is equal to the area of a plane upstream or downstream of the exchanger and parallel to it), $V_{av,out}$ is the average velocity at the exchanger outlet, and $V_{av,duct}$ is the average velocity on a plane downstream of the exchanger.

On the other hand and from the temperature readings, the heat transfer rates in both sides can be calculated as follows. The heat gained by the air passing in the exchanger can be calculated from the following relation:

$$\dot{Q}_{exchanger} = \dot{m}_{exchanger} C_{p,a} \left(T_{a,outlet} - T_{a,inlet}\right) \tag{3}$$

Where $C_{p,a}$ is the specific heat of air evaluated at an average temperature of 30 0C, $T_{a,outlet}$ is the air temperature at the outlet of the exchanger, and $T_{a,inlet}$ is the air temperature at the inlet of the exchanger.

The heat lost by the air passing through the exchanger can be calculated as follows:

$$\dot{Q}_{duct} = \dot{m}_{duct} C_{p,a} \left(T_{a,downstream} - T_{a,upstream}\right) \tag{4}$$

Where $T_{a,downstream}$ and $T_{a,upstream}$ are the temperatures respectively downstream and upstream of the exchanger.

Two main tests are performed by simulating cold day: one with constant duct flow rate and various exchanger flow rates while the second at constant exchanger flow rate and various duct flow rates. Note that in these two tests, low temperatures of air entering the exchanger are achieved by using the compressor.

In each experiment, the velocity at the outlet of the exchanger and on a plane downstream of the exchanger is measured at many locations in order to calculate the mass flow rates in the exchanger and passing through the exchanger using equations (1) and (2). For each couple of flow rates, four readings of temperature are recorded: at the inlet and outlet of the exchanger and upstream and downstream of the exchanger. These readings permit to calculate the heat rate gained in the exchanger and heat rate lost by the air stream traversing the exchanger. The next section will be devoted to the results of the different tests as well as the corresponding analysis.

RESULTS AND ANALYSIS

By simulating the cold days by circulating air from a compressor in the exchanger, a first set of experiments are carried out for a fixed heat exchanger cold air flow rate of $\dot{m}_{exchanger} = 0.0022\, kg/s$ and exchanger inlet temperature of 5 ^{0}C. The room temperature is set at 22 °C in the space control of the air conditioning unit. The hot air flow rate in the duct is varied from 0.05 to 0.2 kg/s. Figure 3 shows the variation of the heat rate gained by cold air in the exchanger and the heat rate lost by hot air in the duct (simulating the exhaust air) in function of the hot mass flow rate (flow rate of air in the duct, the exhaust flow rate).

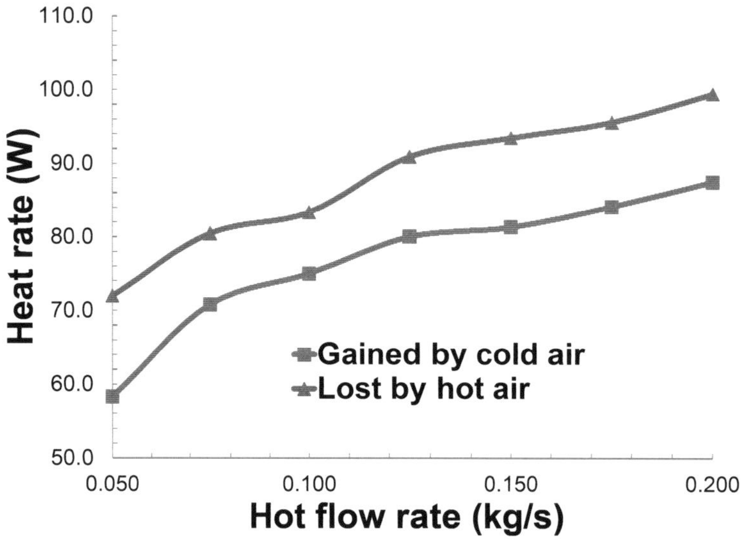

Figure 3 Variation of the heat rate gained by cold air and lost by hot air in function of the hot air mass flow rate.

It can be shown that both the heat rate gained by the air in the exchanger and the heat rate lost by the air in the duct increases with the mass flow rate in the duct. The heat lost by air passing through the exchanger is much greater than the heat gained by air passing in the exchanger. As illustration when the duct mass flow rate is increased from 0.05 to 0.2 kg/s, the heat gained by air in the exchanger increases from 58.3 to 87.5 W while the heat lost by air passing through the exchanger increases from 72 to 99.4 W, let an average efiiciency of the exchanger of 87.1 %.

A second set of experiments are performed for simulation of the cold days situations by fixing the hot air flow rate in the duct at $\dot{m}_{duct} = 0.111\,kg/s$. The room temperature is set at 22 °C in the control of the air conditioning unit. The temperature at the inlet of the exchanger is fixed at 9 ^{0}C and the cold air flow rate in the exchanger is varied from 0.0022 to 0.005 kg/s. Figure 4 shows the variation of the heat rate gained by the air in the exchanger and the heat rate lost by air in the duct in function of the exchanger air flow rate (cold flow rate).

It can be shown that both the heat rate gained by cold air in the exchanger and the heat rate lost by hot air in the duct increases with the cold air mass flow rate in the exchanger. The heat lost by air passing through the exchanger is much greater than the heat gained by air passing in the exchanger. As illustration when the the exchanger cold air flow rate is increased from 0.0022 to 0.005 kg/s, the heat gained by air in the exchanger increases from 75 to 114.3 W while the heat lost by air passing through the exchanger increases from 90.4 to 127 W, let an average efficiency of the exhcnager of 89.1 %.

Now considering a room of 4 kW heating load, indoor conditions of 22 ^{0}C temperature and 50 % reltive humidity, and air supplied at 45 ^{0}C with a sensible heat factor SHF of 0.8. In this case, the enthaplies at indoor conditions and supply conditions are respectively 44 kJ/kg and 74.5 kJ/kg. Then, the mass flow rate of air supplied is 4/(74.5-44) = 0.131 kg/s. Assuming that 20 % are exhausted from the supply mass flow rate, this corresponds then to 0.0262 kg/s in the exchanger representing fresh air and 0.0262 kg/s of air in the duct representing the exhaust hot air. Then assuming an inlet temperature of 9 ^{0}C and extrapolating values of Figure 4 using rational functions developed by Khaled et al. [10], heat rate up to 240 W can be obtained. On the other hand, if all of the supply air is exhausted than heat rate up to 1.5 kW can be obtained.

Considering an apartment of area equivalent to 6 rooms as that presented above, the appartemnt will have an average heating load of 24 kW. Then, 1.44 kW and 9 kW can be recovered respectively if 20 % and 100 % of air are exhausted. Considering the system is running at an average of 12 hours per day, the energy consumption reductions per day will be respectively 17.28 and 108 kWh respectively for the two cases. Then with an average kWh rate in Lebanon of 0.13 $/kWh, the daily running cost reductions will be respectively 2.24 $/day and 14.04 $/day.

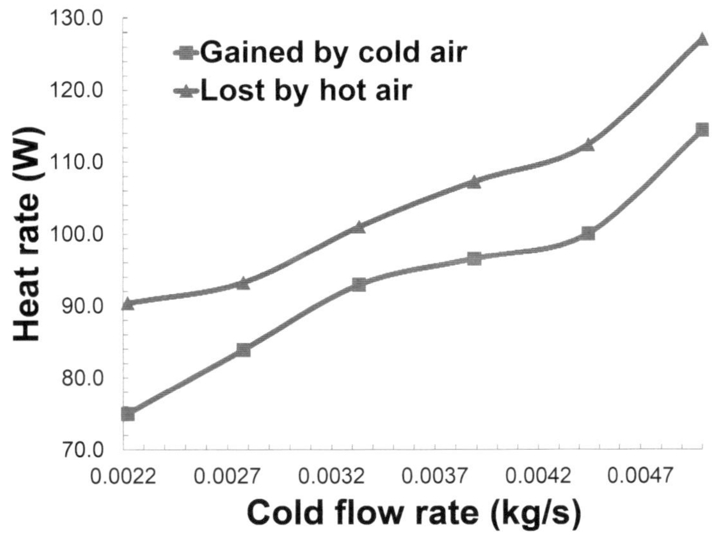

Figure 4 Variation of the heat rate gained by cold air and lost by hot air in function of the cold air mass flow rate.

CONCLUSION

The present work concerned the presentation of the concept, design, prototype implementation and parametric analysis of a system permitting to heat/preheat fresh air using the relatively hot exhaust air of all–air HVAC systems. To proceed, a prototype was implemented and a parametric analysis was carried out to test the effect of mass flow rates on the performance of the heat recovery system. Tests were performed for conditions simulating the HVAC operation in cold climates. An air conditioning unit of 5.3 kW was used in the prototype to generate hot air in the room.

It was shown that powers up to 110 W can be economized in hot climates and for low air flow rates of around 0.1 kg/s and 0.005 kg/s for respectively the duct and the condenser flows. Extrapolation calculations for higher air flow rates show that economized powers can exceed 1 kW.

It was shown also that both the performance and the efficiency of the heat recovery system proposed increase with both the condenser cold and the duct hot flow rates.

REFERENCES

Joselin Herbert, G.M., Iniyan, S., Sreevalsan, E., and Rajapandian, S. 2007. A review of wind energy technologies. Renewable and Sustainable Energy Review, 11:1117-1145.

Khaled, M., Harambat, F., and Peerhossaini, H. 2011. Analytical and empirical determination of thermal performance of louvered heat exchanger – Effects of air flow statistics. International Journal of Heat and Mass Transfer, 54: 356-365.

Lopez, I., Andreu, J., Ceballos, S., Martinez de Alegria, I., and Kortabarria, I. 2013. Review of wave energy technologies and the necessary power-equipment. Renewable and Sustainable Energy Review, 27:413-434.

Mardiana-Idayu, A., and Riffat, S.B. 2012. Review on heat recovery technologies for building applications. Renewable and Sustainable Energy Review, 16:1241-1255.

McQuiston, F.C., Parker, J.D., and Spitler, J.D. 2005. Heating, Ventilating and Air Conditioning: Analysis and Design. 6th Ed. John Wiley & Sons.

O'Connor, D., Calautit, J.K., and Hughes, R. 2016. A review of heat recovery technology for passive ventilation applications. Renewable and Sustainable Energy Reviews, 54: 1481-1493.

Panwar, N.L., Kaushik, S.C., and Kothari, S. 2011. Role of renewable energy sources in environmental protection: A review. Renewable and Sustainable Energy Review, 15:1753-1766.

Shi, L.S., Yit Lin Chew, M. 2012. A review on sustainable design of renewable energy systems. Renewable and Sustainable Energy Review, 16:192-207.

Singh, D.V., and Pederson, E. 2016. A review of waste heat recovery technologies for maritime applications, 111: 315-328.

Solangi, K.H., Islam, M.R, Saidur, R., Rahim, N.A., and Fayaz, H. 2011. A review on global solar energy policy. Renewable and Sustainable Energy Review, 15:2149-2163.

Improving the Efficiency of Photovoltaic Panels Through the Use of Phase Change Materials

Amin A. Wahab, P.E
Student Member

Fadl Moukalled, Ph.D

Marwan Darwish, Ph.D

ABSTRACT

Enhancing the efficiency of renewable energy technologies has been the focus of numerous research projects in the last decades. One such area of interest has been improving the efficiency of photovoltaic (PV) panels. While PV panels use the most abundant and sustainable energy sources in the planet, they have quite a low efficiency in terms of energy conversion. Partly this low efficiency is due to the increase in surface temperature that occurs during the operation of PV panels. A small number of proven techniques have been developed to decrease the surface temperature. These techniques fall into two groups denoted by active cooling techniques and passive cooling techniques, respectively. Active cooling methods include forced ventilation or hydraulic cooling that requires maintenance and additional costs. On the other hand, the passive cooling methods include natural ventilation, and phase change (PCM) materials. This latter technique has shown promises in harnessing the nocturnal cooling effect with its inherent latent energy storage capacity. The aim of this project is to develop a rigorous numerical approach for the application of PCM to the passive cooling of PV panels. To this end, numerical models that simulate PV/PCM system are developed using the "ANSYS Fluent" software package. The validation of these models is done by comparison with experimental data obtained from the literature; consequently, optimum configurations that allow for largest decrease in PV surface temperature are evaluated.

INTRODUCTION

Under ordinary insolation, commercial photovoltaic cell can only convert 10 – 22 % of the solar radiation it receives into electricity depending on the PV cell price (Twidell and Weir 2006). The predicted efficiency of a PV cell at standard test condition (STC) or at standard operating condition is usually higher than the actual efficiency which is greatly affected by reflection losses and solar cell temperature; while STC takes place at 1000 W/m² irradiance and 25°C cell temperature, normal operating cell temperature (NOCT) rating is done under 800 W/m² irradiance and at fixed air temperature of 20°C. For crystalline silicon solar cells, a reduction in voltage and efficiency up to 0.4 ~ 0.5 %/°K can be noticed due to increased cell temperatures (Krauter and Hanitsch 1996). Experiment done by Radziemska and Klugmann (2002) for increasing the solar cell temperature in the range of 22°C to 70°C showed that the output power deterioration of the PV module due to temperature elevation can reach 0.65 %/°K. Consequently, many cooling methods have been investigated to reduce excessive heating of the PV cell and achieve higher operating efficiency of the PV module. Among these methods, integrating latent heat storage systems (LHTS) with the PV cells seems promising as they proved to be more reliable than sensible heat storage systems in applications with small temperature variations because of their high storage capacity and nearly isothermal performance (Lamberg and Sire´n 2003); LHTS materials such as phase change material (PCM) can store 5-14 times more heat per unit volume than sensible heat storage materials such as rock and masonry (Sharma, Tyagi, Chen, and Buddhi 2009). Upon changing phase from solid to liquid or vice versa, PCMs can absorb or discharge large amount of energy in the form of latent heat at a constant phase transition temperature. Thus, using PCM material can be very convenient to absorb excessive heat from a PV module which would be later discharged to the surrounding air during night when no electric power is extracted from the module.

An internally finned PV/PCM system was capable of maintaining the temperature of the PV cell at 29°C for about two hours in an experiment conducted under 750 W/m^2 insolation and 23°C ambient temperature (Huang, Eames, and Norton 2006a). Increasing fin spacing reduces convection and thermal stratification inside the PV/PCM system and allows for a longer time for the complete melting of the PCM (Huang et al. 2011b). Hasan, McCormack, Huang, and Norton (2010) experimentally reported that a PCM with lower thermal conductivity and melting point can better regulate a PV module with lower operating temperature while a higher operating temperature PV module performs better with a PCM that have higher thermal conductivity and melting point.

Huang et al. (2004) were the first to report an experimentally validated numerical model simulating the cooling effect of PCM in a PV system where a temperature of 36.4°C was maintained for 80 minutes under 1000 W/m^2 insolation. This numerical model was later developed by Huang et al. (2006b) to become three dimensional; a good agreement was found in the predictions of the two models. Numerical modeling of two different algorithms using FEMLAB showed greater accuracy in Enthalpy method compared to heat capacity method regarding narrow phase change temperature range of 2°C (Lamberg, Lehtiniemi, and Henell 2004). The effect of combining different phase change materials in a PV/PCM system was numerically investigated by Huang (2011a) where the RT27-RT21 pair resulted in the highest decrease in temperature. A parametric study on a photovoltaic solar-thermal system that contains layers of PCM was conducted by Malvi, Dixon-Hardy, and Crook (2011); simulations showed that any enhancement in the PV performance would be on the expense of the solar-thermal system efficiency. Numerical simulations of a PV device integrated with a microencapsulated PCM layer revealed an improvement in efficiency by 0.13 % and 0.42 % for the summer and winter seasons respectively (Ho, Tanuwijava, and Lai 2012). Hendricks and van Sark (2013) used MATLAB to calculate the energy output gains of a PV/PCM system at distant locations and for different PCM types. Two experiments incorporating particle image velocimetry (PIV) were used to validate a mathematical model that was solved numerically using finite element method; the panel's temperature was sustained under 40°C for 80 minutes (Biwole, Eclache, and Kuznik 2013). Ciulla, Piacentino, and Cardona (2013) proposed a simplified one dimensional numerical model of a PV/PCM module; the model was validated experimentally resulting in a small discrepancy between the calculated and measured PV temperature not exceeding 5 %. Park, Kim, and Leigh (2014) assessed the performance of a PV/PCM module under realistic outdoor climatic conditions; the experiments were conducted at Incheon, South Korea in May and June for different PCM thicknesses and yielded a maximum drop in PV temperature by 5 °C compared to the conventional PV module. This experimental data was used to validate a simulation method that solves for the module temperature and energy generation using TRNSYS; the resulting PV temperature simulation results didn't deviate more than 5.4 % from the corresponding experimental results. The validated simulation method was then used to carry out a parametric study on the PV/PCM module regarding the installation direction of the module and the melting temperature of the PCM. Outdoors examination of a PV/PCM system was conducted experimentally by Hasan et al. (2015) in two different climates in Ireland and Pakistan using two types of PCMs; numerical simulations were performed under the same conditions of the experiments and the difference between the two results was no more than 6.3 %. Both types of PCMs performed better regarding limiting PV temperature rise at Pakistan where hot and stable climate dominates than in Ireland that is overwhelmed by variant weather.

As shown in the literature, most of the PV/PCM modules investigated experimentally or numerically were set indoors under controlled conditions; whereas, few experiments and simulations were conducted outdoors under actual weather conditions. In this study, "FLUENT" software was used to assess the performance of PV/PCM module both indoors and outdoors. The validation of these simulations was achieved using experimental data from Hasan et al. (2010) and Park et al. (2014). Afterwards, modifications are proposed for the validated PV/PCM modules to further enhance their performance compared to a conventional PV cell.

MATHEMATICAL MODEL (GOVERNING EQUATIONS)

Below are the governing equations used in the different regions across all the simulations conducted in the study.

$$\frac{\partial}{\partial t}(\rho) + \nabla.(\rho \boldsymbol{v}) = 0 \quad (1)$$

$$\frac{\partial}{\partial t}(\rho \boldsymbol{v}) + \nabla.(\rho \boldsymbol{v}\boldsymbol{v}) = \nabla.\tau - \nabla \mathrm{p} + \rho \boldsymbol{g} \quad (2)$$

$$\frac{\partial}{\partial t}(\rho c_p T) + \nabla.(\rho \boldsymbol{v} c_p T) = \nabla.(K \nabla T) \quad (3)$$

$$\frac{\partial}{\partial t}(\rho H) + \nabla.(\rho \boldsymbol{v} H) = \nabla.(K \nabla T) \quad (4)$$

$$H = h + \Delta H \quad (5)$$

$$h = h_{ref} + \int_{T_{ref}}^{T} c_p dT \cong c_p T \quad (6)$$

$$\frac{\partial}{\partial t}(\rho c_p T) + \nabla.(\rho \boldsymbol{v} c_p T) = \nabla.(K \nabla T) - \left(\frac{\partial}{\partial t}(\rho \Delta H) + \nabla.(\rho \boldsymbol{v} \Delta H)\right) \quad (7)$$

$$\frac{\partial}{\partial t}(\rho c_p T) = \nabla.(K \nabla T) - \frac{\partial}{\partial t}(\rho \Delta H) \quad (8)$$

$$\Delta H = \alpha L \quad (9)$$

In the phase change material region, Enthalpy method is commonly used to solve diffusion phase change problems. Fixed grid solutions can be obtained because conditions on temperature, velocity, and heat removal that are required at the phase change front will not be needed when the enthalpy approach is used (Voller, Cross, and Markatos 1987; ANSYS 2011). In the Phase change material region, only the conservation of mass equation (Equation 1) and energy conservation equation (Equation 8) will be applied, whereas the equation of conservation of momentum (Equation 2) will be neglected because weak natural convective currents exist in the PCM material which is usually confined in microcapsules. Equation 8 is derived from Equation 4 which represents the enthalpy form of the energy conservation equation. Equation 5 and Equation 6 represent the equations of the total and sensible enthalpy respectively. Equation 7 resulted from substituting Equation 5 and Equation 6 into Equation 4. Equation 7 is reduced into Equation 8 by neglecting the convective heat transfer terms because the PCM material experiences weak natural convective currents. As the enthalpy-porosity method will be used to model the melting/solidification process, Equation 9 represents the relation between latent enthalpy and PCM latent energy through the liquid mass fraction α which varies between 0 and 1.

In the outdoor air region, Equation 1, Equation 2, and Equation 3 will be used, which represent the conservation equations of mass, momentum, and energy respectively. Equation 3 is the energy conservation equation in temperature form.

In other regions such as PV cell, glass, Perspex, aluminum container, and concrete wall, only conservation of mass and energy will be considered; thus, only Equation 1 and Equation 3 are solved.

Solution

To solve this problem and obtain the temperature distribution on the PV cell surface, the Finite Volume Method

(FVM) is used. In this numerical approach, the modeled geometry is first discretized into small control volumes constituting the mesh. Then, the systems of partial differential equations that express conservation of mass, momentum, energy, etc. are discretized into a system of algebraic equations in every control volume. This method uses the Divergence theorem, known also as Green's theorem, to change volume integrals into surface integrals (Hackbusch 1985; Hackbusch 1994; Murthy and Mathur 2002).

In this study, the geometric discretization is done via "ANSYS Meshing Software", while solutions are generated using the CFD "FLUENT" software package.

INDOORS ASSESSMENT OF PV/PCM MODULE

Hasan et al. (2010) conducted a set of experiments on PV modules indoors in a controlled environment; two of these experiments were chosen to validate the CFD simulations of PV/PCM systems using "FLUENT". Thermo-physical properties of the materials used in "FLUENT" models as listed in Table 1 are typical to those used in the experiments by Hasan et al. (2010).

Table 1: Thermo-physical Properties of the Materials Used in the Simulations

	Paraffin Wax RT20	Aluminum	Glass	Perspex	Polycrystalline Silicon	Concrete
Density Solid, (Kg/m^3)	880	2719	3000	1190	2330	2240
Density Liquid, (Kg/m^3)	770	–	–	–	–	–
Specific Heat Capacity (J/Kg.°K)	2100	871	500	1470	677	921
Thermal Conductivity (W/m.°K)	0.2	237	1	0.189 - 0.25	148	1.2978
Solidus Temperature (°K)	294.38	–	–	–	–	–
Liquidus Temperature (°K)	298.88	–	–	–	–	–
Latent Heat (J/Kg)	140300	–	–	–	–	–
Dynamic Viscosity (Kg/m.s)	0.205	–	–	–	–	–

Validation of the "FLUENT" simulation of PV module indoors

The data gathered from the experiment conducted by Hasan et al. (2010) under 750 W/m^2 insolation for measuring the PV front surface temperature was used to validate the "FLUENT" radiation model. In the experiment, the PV module constituted of polycrystalline silicon PV cells with dimensions of 10 cm * 10 cm * 0.05 cm that were embraced between two sheets of transparent Perspex of 3 mm thickness. A 3D model of the PV module was drawn and meshed in "ANSYS" with the same dimensions as in the experiment. The boundary conditions applied in the "FLUENT" case were similar to those used in the experiment. All the sides of the model were set to "wall" type. The free stream temperature of air around the model was set to 293.15 K; heat transfer coefficient of air at the front side of the model was set to 12.5 W/m^2.K; however, it was set to 10 W/m^2.K at the top, bottom, lateral, and back sides; the emissivity and temperature of the external radiation source were set to 0.135 and 559.92 K respectively. The "FLUENT" case temperature and velocity were initialized to 283 K and (0, 0, 0.0001) m/s respectively. Using first order upwind schemes and 10 seconds time step size, "FLUENT" simulated a temperature distribution of the PV front surface temperature that is very close to that in the experiment as shown in Figure 1.

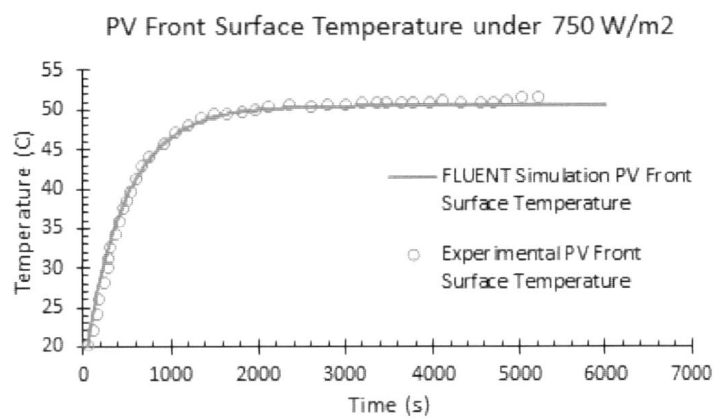

Figure 1: PV front surface temperature simulated in "FLUENT" and measured experimentally by Hasan et al. (2010) under 750 W/m2.

Validation of the "FLUENT" simulation of PV/PCM module indoors

The PV front surface temperature measured by Hasan et al. (2010) in an experiment incorporating PV/PCM module was used to validate simultaneously the radiation and solidification/melting models in "FLUENT". An aluminum enclosure with dimensions of 10 cm * 10 cm * 6 cm and thickness of 0.5 cm was filled with paraffin wax (RT20) whose thermos-physical properties are shown in Table 1. The PV module previously used was attached to the aluminum enclosure and the PV/PCM module was placed under 750 W/m^2 insolation. A typical 3D model of the module was reproduced and meshed in "ANSYS" to be simulated in "FLUENT" under similar boundary conditions. All the sides of the model were set to "wall" type. The free stream temperature of air around the model was set to 293.15 K; heat transfer coefficient of air at the top, bottom, lateral, front, and back sides of the model was set to 5 W/m^2.K; the emissivity and temperature of the external radiation source were set to 0.135 and 559.92 K respectively. The "FLUENT" case temperature and velocity were initialized to 283 K and (0, 0, 0.001) m/s respectively. First order upwind schemes and 30 seconds time step size were employed to solve the "FLUENT" case. The maximum discrepancy between the simulation and experimental results regarding PV front surface temperature doesn't exceed 2°C as shown in Figure 2.

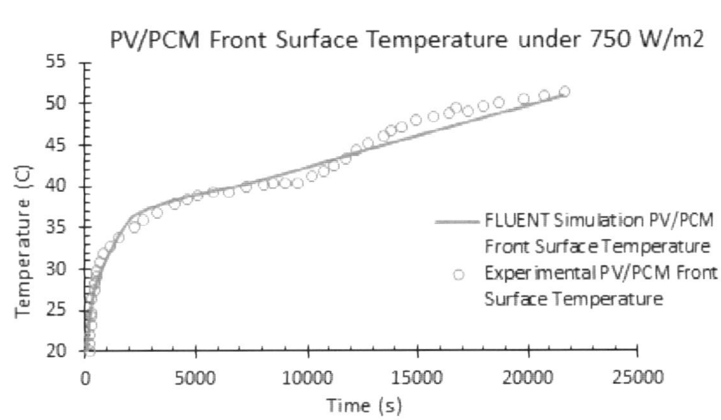

Figure 2: PV/PCM front surface temperature simulated in "FLUENT" and measured experimentally by Hasan et al. (2010) under 750 W/m2.

Evaluation of the performance of a PV/PCM module placed on a wall containing PCM using "FLUENT"

After the PV/PCM module has been validated in "FLUENT" against experimental results, an extension of this 3D model is proposed to further improve the temperature drop in the PV cell. The PV/PCM module was placed on a concrete wall of 10 cm thickness and containing in the middle of it paraffin wax (RT20) 5 cm in depth. The extended 3D model was simulated in 'FLUENT" with the same boundary conditions, initial conditions, schemes, and time step size that were used with the original 3D model. The thermo-physical properties of concrete are listed in Table 1. As shown in Figure 3 the front surface temperature of the PV cell dropped by more than 6°C after 6 hours compared to the original model.

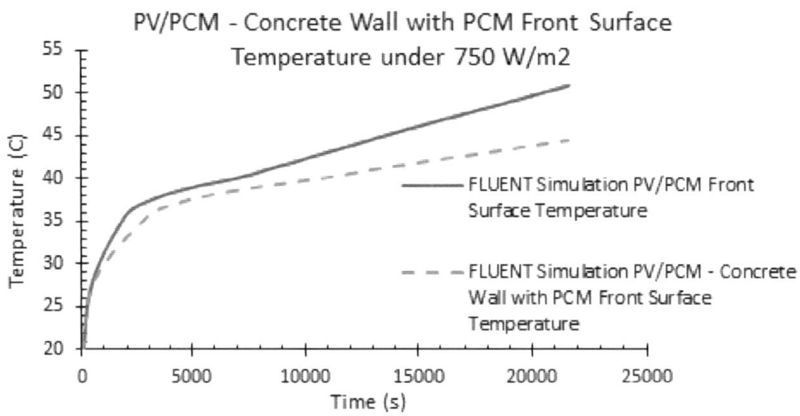

Figure 3: Comparison between the front surface temperature of PV/PCM module and PV/PCM - Concrete Wall with PCM module simulated in "FLUENT" " under 750 W/m2.

OUTDOORS ASSESSMENT OF PV/PCM MODULE

Data from two experiments conducted by Park et al. (2015) was employed to validate the usage of the solar radiation model in conjunction with solidification/melting model in "FLUENT" under realistic weather conditions. The experiments lasted for 24 hours and were accomplished on 15 June 2012 at Incheon, South Korea (37.38 degrees North, 126.65 degrees East). The thermos-physical properties of the materials used in the simulation are listed in Table 1 and conform to the materials used in the experiments.

Validation of the "FLUENT" simulation of PV module outdoors under solar radiation

Park et al. (2015) used two layers consisting of a glass cover and a PV cell to model the PV module; both layers have dimensions of 35 cm * 28 cm. The PV module was placed in a vertical position facing South and the temperature of its front surface was recorded for 24 hours starting from mid night. "ANSYS" was used to draw and mesh an identical 3D model of the PV module. The solar radiation model was activated in "FLUENT" with settings (day, time zone, longitude, and latitude) typical to those in the experiment. A convective thermal boundary condition was applied to the top, bottom, lateral, and back sides of the model which were set to "wall" type such that the free stream temperature and heat transfer coefficient were set to 293.15 K and 5 $W/m^2.K$ respectively. An air zone was meshed and coupled to the front side of the model to allow the application of the solar load at the front side of the model whose absorptivity and transmissivity were set to 0.05 and 0.76 respectively. While the top, lateral, and front sides of the air zone were set to "velocity inlet" type with (-0.27, 0, 0) m/s velocity and 293.15 K temperature, the bottom side of the air zone was set to "wall" type with 0.6 absorptivity and 293.15 K temperature. Only the bottom side of the air zone and the front side of the model were allowed to participate in the "Solar Ray Tracing". The "FLUENT" case temperature and velocity were initialized to 293.15 K and (0.27, 0, 0) m/s respectively. First order upwind schemes and 300 seconds time step size were used to obtain the temperature distribution on the front surface of the 3D model for 24 hours. Figure 4 shows the temperature of the front surface that was obtained via experiment and simulation by Park et al (2015) compared to that obtained using "FLUENT" simulation. The "FLUENT" simulation and experimental temperature data exhibit a close match as the maximum difference between the two doesn't exceed 2.5°K.

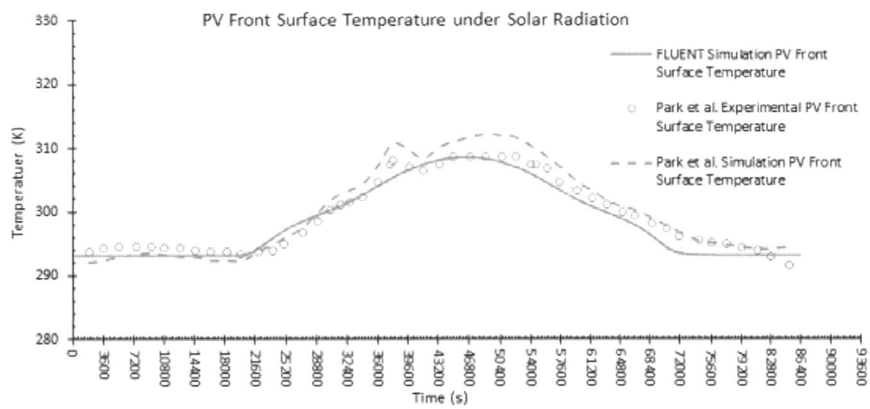

Figure 4: PV front surface temperature simulated in "FLUENT" and measured experimentally and simulated by Park et al. (2015) under solar radiation

Validation of the "FLUENT" simulation of PV/PCM module outdoors under solar radiation

Validating "FLUENT" solar radiation model used simultaneously with solidification/melting model for a PV/PCM module is currently in progress against experimental data from Park et al. (2015)

FUTURE WORK

We plan to complete the validation of the PV/PCM module under solar radiation in "FLUENT" and investigate an extension of the validated model and to conduct a parametric study regarding the position of the module under solar radiation and the PCM thickness to optimize the design of a PV/PCM system in terms of achieving the highest drop in PV surface temperature and eventually acquire the highest efficiency.

NOMENCLATURE

ρ = density (Kg/m^3)

v = velocity vector (m/s)

T = temperature of the control volume (°K)

K = thermal conductivity (W/m. °K)

c_p = specific heat (J/Kg. °K)

H = total enthalpy (J/Kg)

h = sensible enthalpy (J/Kg)

ΔH = latent enthalpy (J/Kg)

L = latent heat of PCM (J/Kg)

α = liquid fraction in a control volume in PCM region
T_m = melting temperature of PCM (°K)
τ = surface stresses (Pa)
p = Pressure (Pa)

REFERENCES

ANSYS, Inc. 2011. ANSYS FLUENT Theory Guide 14.5. www.arc.vt.edu/ansys_help/flu_th/flu_th.html

Biwole, P.H., Eclache, P., and F. Kuznik. 2013. Phase-change materials to improve solar panel's performance. Energy and Buildings 62: 59–67.

Hackbusch, W. 1985. Multi-Grid Methods and Applications. Berlin: Springer Inc.

Hackbusch, W. 1994. Iterative Solution of Large Sparse Systems of Equations, New York: Springer-Verlag Inc.

Hasan, A., McCormack, S.J., Huang, M.J., and B. Norton. 2010. Evaluation of phase change materials for thermal regulation enhancement of building integrated photovoltaics. Solar Energy 84: 1601-12.

Hasan, A., McCormack, S.J., Huang, M.J., Sarwar, J., and B. Norton. 2015. Increased photovoltaic performance through temperature regulation by phase change materials: Materials comparison in different climates. Solar Energy 115: 264-76.

Hendricks, J. H. C., and W. G. J. H. M. van Sark. 2013. Annual performance enhancement of building integrated photovoltaic modules by applying phase change materials. Progress in Photovoltaics: Research and Applications 21: 620-30.

Ho, C.J., Tanuwijava, A.O., and C.M. Lai. 2012. Thermal and electrical performance of a BIPV integrated with a microencapsulated phase change material layer. Energy and Buildings 50: 331-8.

Huang, M.J. 2011a. The effect of using two PCMs on the thermal regulation performance of BIPV systems. Solar Energy Materials and Solar Cells 95: 957-63.

Huang, M.J., Eames, P.C., and B. Norton. 2004. Thermal regulation of building-integrated photovoltaics using phase change materials. International Journal of Heat and Mass Transfer 47: 2715-33.

Huang, M.J., Eames, P.C., and B. Norton. 2006a. Phase change materials for limiting temperature rise in building integrated photovoltaics. Solar Energy 80: 1121-30.

Huang, M.J., Eames, P.C., and B. Norton. 2006b. Comparison of a small-scale 3D PCM thermal control model with a validated 2D PCM thermal control model. Solar Energy Materials and Solar Cells 90: 1961-72.

Huang, M.J., Eames, P.C., Norton, B., and N.J. Hewitt. 2011b. Natural convection in an internally finned phase change material heat sink for the thermal management of photovoltaics. Solar Energy Materials and Solar Cells 95: 1598-603.

Krauter, S., and R. Hanitsch. 1996. Actual optical and thermal performance of PV-modules. Solar Energy Materials and Solar Cells 41-42: 557-74.

Lamberg, P., and K. Sire´n. 2003. Analytical model for melting in a semi-infinite PCM storage with an internal fin. Heat and Mass Transfer 39: 167-76.

Lamberg, P., Lehtiniemi, R., and A.M. Henell. 2004. Numerical and experimental investigation of melting and freezing processes in phase change material storage. International Journal of Thermal Sciences 43: 277-87.

Lo Brano, V., Ciulla, G., Piacentino, A., and F. Cardona. 2013. On the Efficacy of PCM to Shave Peak Temperature of Crystalline Photovoltaic Panels: An FDM Model and Field Validation. Energies 6: 6188-210.

Malvi, C.S., Dixon-Hardy, D.W., and R. Crook. 2011. Energy balance model of combined photovoltaic solar-thermal system incorporating phase change material. Solar Energy 85: 1440-46.

Murthy, J.Y., and S.R. Mathur. 2002. Numerical Methods in Heat, Mass, and Momentum Transfer. engineering.purdue.edu/ME608/webpage/main.pdf

Radziemska, E., and E. Klugmann. 2002. Thermally affected parameters of the current–voltage characteristics of silicon photocell. Energy Conversion and Management 43: 1889-1900.

Sharma, A., Tyagi, V.V., Chen, C.R., and D. Buddhi. 2009. Review on thermal energy storage with phase change materials and applications. Renewable and Sustainable Energy Reviews 13: 318-45.

Twidell, J., and T. Weir. 2006. Renewable Energy Resources, 2nd ed. London: Taylor & Francis Group.

Voller, V.R., Cross, M., and N.C. Markatos. 1987. An enthalpy method for convection/diffusion phase change. International Journal for Numerical Methods in Engineering 24: 271-84.

Park, J., Kim, T., and S.B. Leigh. 2014. Application of a phase-change material to improve the electrical performance of vertical-building-added photovoltaics considering the annual weather conditions. Solar Energy 105: 561-74.

Improving Ventilation Effectiveness and Thermal Comfort Using Chair Fans to Aid Displacement Ventilation

Walid Abou Hweij

Nesreen Ghaddar, PhD
Member ASHRAE

Kamel Aboughali, PhD
Member ASHRAE

ABSTRACT

The aim of this work was to model and simulate the performance of displacement ventilation system (DV) assisted by chair fans (CF) in providing thermal comfort (TC) and good indoor air quality (IAQ) for an office occupant at lower energy consumption. An office space was modeled using commercial computational fluid dynamics (CFD) software. The CFD model was coupled to a segmental bio-heat model to predict simulated occupant skin temperatures, which were used for evaluation of local and overall thermal comfort and sensation of simulated occupant. In addition, the CFD model predicted the IAQ by evaluating the ventilation effectiveness based on concentration fields of CO_2 at the breathing zone of occupant. The use of CF produced better IAQ and TC compared to DV system mainatained at same conditions. The novel system reached 20.84% savings in energy consumption over a standalone DV system providing same performance.

INTRODUCTION

Nowadays, green and environmental technologies have been dedicated great interest as an attempt of providing acceptable TC and good IAQ for occupants in different residential and open space buildings. DV system is gaining great popularity in different countries e.g. Germany, England, Japan, and USA (Mundt 1995) for its ability in providing good IAQ, as reported by Xing et al. (2001), while establishing acceptable thermal comfort, as reported by Lin et al. (2005). However, DV system might provide discomfort thermal environment as a result of draught at the feet, as reported by Melikov et al. (2005). Hence, to enhance the performance of DV system in providing TC, it is proposed in the literature to use fans connected to occupant's seat.

Watanabe et al. (2009) suggested the use of chair fans located under the seat and behind the backrest. Their results revealed that the occupants' whole body thermal sensation (WBTS) was close to slightly warm based on the WBTS range at high space temperature of 30°C (86°F) where the WBTS was reported as warm without the use of chair fans. However, the occupants' local thermal sensation (LTS) was not efficient as that of the WBTS as the head, the arms, and the hands were not affected by this innovation leading to another research study. Thus, Sun et al. (2012) proposed the use of a chair aided by four computer fans placed at the corners of the chair. Their results showed that the use of chair fans improved the convection airflow around the occupants. In addition, they suggested the appropriate fan speeds that can accommodate with various room temperatures while providing thermal comfort for occupants. On the other hand, Habchi et al. (2015) investigated the performance of DV system aided by chair fans, placed at the seating level of an occupant modeled as heated cylinder, in reducing particles inhalation. The results

Walid Abou Hweij is a graduate research assistant. **Nesreen Ghaddar** is a professor and Qatar Chair for Energy Studies in the Department of Mechanical Enegineering, American University of Beirut, Beirut, Lebanon. **Kamel Ghali** is a professor and Chairperson in the Department of Mechanical Enegineering, American University of Beiruit, Beirut, Lebanon.

showed that the inhalation effectiveness was improved using the novel system compared to a standalone DV system. On the other hand, Habachi et al. (2015) had investigated TC using predicted mean vote (PMV) model and recorded TC for occupants revealing acceptable TC levels. Although this system had focused on the effect of chair fans on contaminated particles transport, the effect of chair fans on gas phase contamination (CO_2) had not been studied. Moreover, the use of PMV is not recommended in non-uniform and asymmetrical environmentts as the method was based on uniform and symmetrical ones. In addition, previous studies done by Watanabe et al. (2009) and Sun et al. (2012) were performed on subjects that required a lot of precauations prior to any experiment. Moreover, these studies didn't introduce the effect of using chair fans on IAQ.

Thus, the aim of this study is to compare the performance of a standalone DV system with a DV system aided by chair fans (DV + CF) computationally using a commercial CFD solver. The assessment of IAQ will be investigated using the CFD model and TC will be evaluated using validated bio-heat model. In addition, energy consumption will be evaluated by comparing the performance of the proposed system (DV + CF) with standalone DV system providing same thermal comfort and IAQ.

METHODOLOGY

The objective of the proposed (DV + CF) system is to enhance the convective plumes created by occupants to provide good IAQ and TC while reducing energy consumption.

The interaction between the air jets issued by fans upon blowing air from floor level upwards and the rising thermal plumes result in complex interactions. Thus, the coupling of the DV system and CF must be accurately predicted to investigate the effect of this system on the behavior of the upward convective plumes and its ability in providing clean environment. Airflow pattern and distribution of CO_2 tracer gas had been well predicted using CFD modeling (Makhoul et al. 2013 a). Thus, a 3-dimensional CFD model using the commercial CFD solver, ANSYS Fluent is used to investigate the effectiveness of the proposed (DV + CF) system (see Figure 1a). This model predicts the flow, temperature and CO_2 concentration fields in a typical office space. In order to access TC of occupants, a validated bio-heat model suggested by Salloum et al. (2007) and the comfort and sensation models provided by Zang et al. (2010) will be integrated with CFD code to predict the human segmental thermal conditions and segmental and overall thermal sensation and comfort (Makhoul et al. 2013 a).

Airflow Modeling

Accurate prediction is based on proper modeling of flow physics including turbulence models, buoyancy effects, and boundary layers near the surfaces (Makhoul et al. 2013 b). Proper controlling of the grid resolution at some surfaces is crucial to capture perfectly the shear-layer entrainment process and the different thermal plumes as well as the fluid/thermal boundary layers around the occupant (Makhoul et al. 2013 b). In order to accurately resolve the boundary layer, set of inflation layers were selected to refine the mesh near the surfaces. This refinement was based on the selection of first layer thickness of the boundary layer. The first layer thickness was calculated according to the equations provided by LEAP (Leading Engineering Application Providers). The obtained value will be checked such that the y+ value fall in the range of 0.8 and 4 in order to capture the boundary layer development. Accordingly, a tetrahedral unstructured grid with different element size for the boundary faces was established. Grid Independence test was performed to select the appropriate mesh size with the lowest computational cost and acceptable accuracy. The selection would be based on a maximum relative error between the current and the previous grid size to be less than 5%. The results of the grid independence tests are summarized in Table 1. It can be noticed that mesh 5 produces results that are 4.4% different compared to that of mesh 4; hence, mesh 4 could be selected where it provides the acceptable computational cost and accuracy. Thus, a face size of 0.015 m (0.05 ft) and 0.025 m (0.08 ft) was assigned for the manikin and the walls respectively. Inflation boundary layers were created around the manikin and walls such that a first layer thickness of 0.0015 m (0.005 ft), a growth rate of 1.2, and a total of 4 layers were assigned for the walls and a first layer thickness of 0.0012 m (0.004 ft), a growth rate of 1.1, and a total of 3 layers were assigned for the manikin. These inflation layers resulted in y+ value between 0.8 and 4. These mesh components resulted in a total number of 3,516,946 elements and 876,576 grid nodes for the geometrical domain that will be considered in this study (see Figure 1b).

Table 1. Grid Independence Testing using Five Mesh Case

Grid Type	Face Sizing [m], [ft] Manikin/Walls	Number of Elements	Number of Grid Nodes	Maximum Relative Difference in Predicted Values of Temperature, Velocity, and Concentrations with the Previous Mesh Values %
Mesh 1 (Intial case)	0.02/0.08 0.07/0.26	766,144	184,707	-
Mesh 2	0.02/0.05 0.07/0.16	1,217,543	301,264	43.52
Mesh 3	0.015/0.03 0.05/0.1	2,559,155	638,040	10.32
Mesh 4	0.015/0.025 0.05/0.08	3,516,946	876,576	5.5
Mesh 5	0.015/0.02 0.05/0.07	5,652,622	1,384,638	4.4

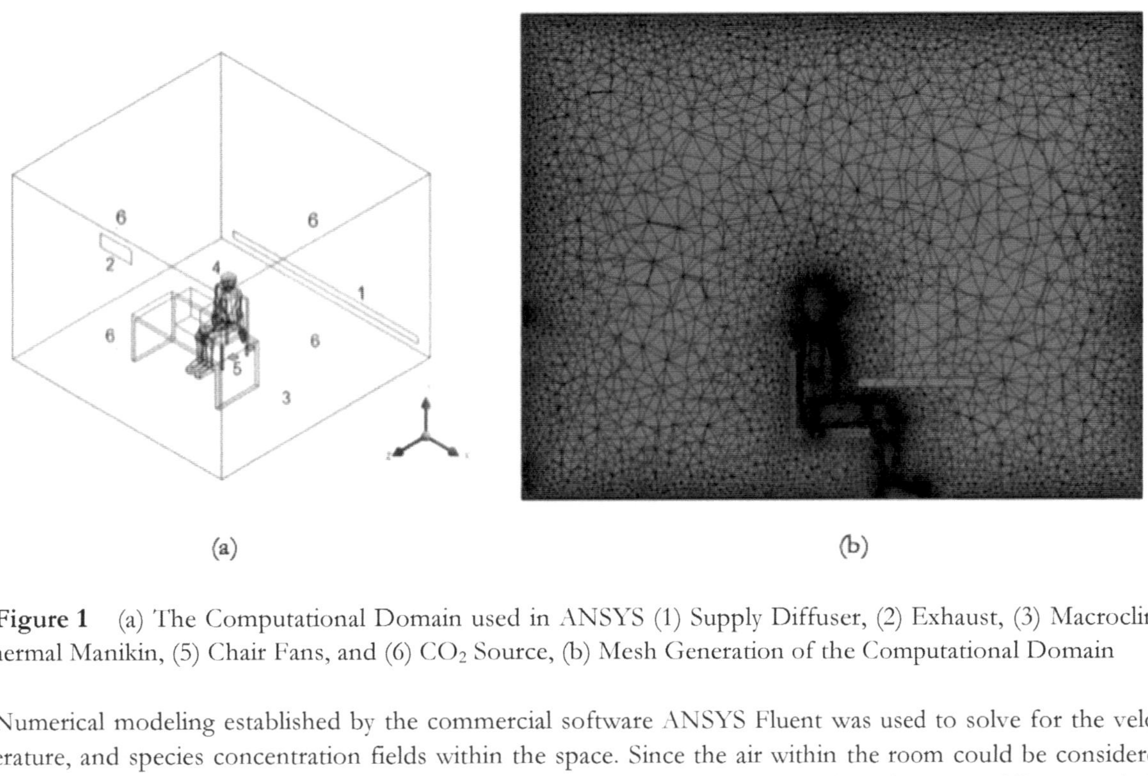

(a) (b)

Figure 1 (a) The Computational Domain used in ANSYS (1) Supply Diffuser, (2) Exhaust, (3) Macroclimate, (4) Thermal Manikin, (5) Chair Fans, and (6) CO_2 Source, (b) Mesh Generation of the Computational Domain

Numerical modeling established by the commercial software ANSYS Fluent was used to solve for the velocity, temperature, and species concentration fields within the space. Since the air within the room could be considered as continuum fluid, the flow could be evaluated using the Eulerian approach (Habchi et al 2015). Different turbulence models were available in the commercial solver, ANSYS Fluent in which it was used to solve the Navier-Stoke equations. Thus, the standard k-ε model was selected based on a study done on investigating the effect of turbulence models in detecting air temperature distribution above a seated thermal manikin (Zelensky et al. 2013). This turbulence model was selected along with enhanced wall treatment option to accurately predict the flow fields that involve separation and recirculation. In addition, the boussinesq approximation was used in the density modeling to account for buoyancy effects. On the other hand, the assessment of the IAQ and the concentration of CO_2 were created by enabling the species transport equation. Moreover, the momentum, k, ε , energy and even the species transport differential equations were discretized using the second order upwind scheme. The "Standard" staggered scheme was used for the pressure where the latter was coupled with velocity fields using the "SIMPLE" algorithm.

The numerical convergence criterion was based on assigning scaled residual of minimum value of 10^{-5}, monitoring the actual values at certain points, and having a net heat flux of the computational domain below 1% of the total heat gain (Makhoul et al. 2013 a).

Bio-heat Coupling

The prediction of occupant segmental skin temperature was computationally expensive using the CFD simulation. Besides, TC and thermal sensation (TS) could not be predicted using the CFD model. Hence, a bio-heat model proposed by Salloum et al. (2007) along with a thermal comfort and sensation models proposed by Zang et al. (2010) would be coupled with CFD code. The bio-heat model, which divides the body into 11 segments (head, chest, back, abdomen, U-arm, L-arm, hands, buttocks, thighs, calves and feet), was used to simulate the physiological thermal responses of each body segment to accurately predict TC based on the TC model established in a high transient thermal environment. Figure 2 presents the coupling procedure where it was repeated (typically 4 iterations) until convergence was achieved. Hence, upon achieving convergence, the corresponding skin temperature was integrated with Zhang's model to predict occupant' s TC and TS.

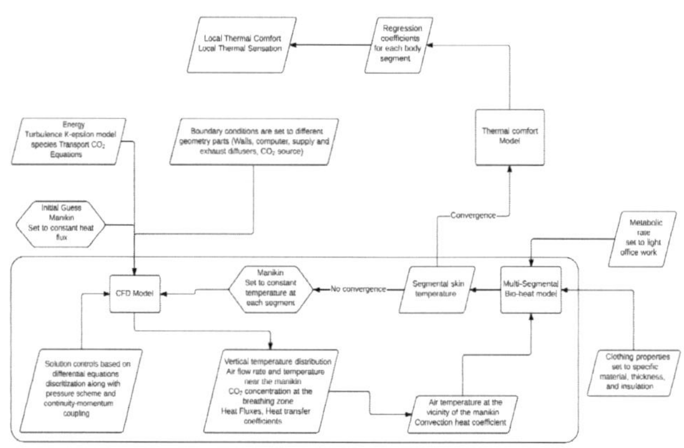

Figure 2 Flow Chart Rpresenting the Integration of the CFD Model with the Bio-Heat Model

Air Quality Modeling

IAQ, which could be defined as the level of freshness inside the room due to mixing between the supplied fresh air and the stagnant contaminated room air, was described in terms of the ventilation effectiveness where the latter was assessed using CO_2 as a tracer gas. Therefore, the higher the ventilation effectiveness the better the air quality would be. The supply grill (supply diffuser) supplies fresh air with 0.4% by volume or 4000 parts per million (ppm) of CO_2 which mimics the real nature of fresh air.

In order to represent a room with contaminated air, CO_2 tracer gas would be uniformly distributed in the room by placing 4 identical sources at 1 m (3.28 ft) from the floor each producing a flow rate of 0.0083 L/s (0.0003 ft^3/s), thus a total flow rate of 0.0332 L/s (0.0012 ft^3/s) would be established.

In addition, the room was designed in a way such that the amount of fresh air supplied was equal to that exhausted through the exhaust, thus preventing the build-up of CO_2 within the space. Therefore, the ventilation effectiveness or the Air Quality Index (AQI) was evaluated as:

$$\varepsilon_v = \frac{C_E - C_B}{C_E - C_{Fr}} \quad (1)$$

Where ε_v is the ventilation effectiveness, C_E is the CO_2 concentration at the exhaust in ppm, C_B is the CO_2 concentration at the breathing zone in ppm, and C_{Fr} is the CO_2 concentration at the supplied air in ppm. It is noted that the concentration of inhaled air is measured as a volume average concentration in the spherical volume of 20 mm (0.066 ft) diameter placed at 25 mm (0.082 ft) from the manikin's nose (Makhoul et al. 2013 a).

Boundary Conditions

Proper CFD modeling was based on the setting of boundary conditions for airflow, energy and CO_2 concentration. Concerning the airflow modeling, the supply diffuser velocity along with the turbulence intensity and hydraulic diameter were specified based on the available equations provided by ANSYS online manual. The exhaust was modeled as a pressure outlet with zero gauge pressure. As for the chair fans, the "fan" boundary condition was assigned for the inlet of chair-mounted fans with a predefined pressure drop in an attempt to find the desired flow rate whereas the outlet was assigned an "interior" boundary condition (Makhoul et al. 2013 a). On the other hand, constant heat fluxes were selected to model thermal surfaces. Concerning the air quality modeling, CO_2 source was assigned recommended values for turbulence intensity, hydraulic diameter and source temperature.

CASE STUDY

A typical office of dimensions 3.4 m x 3.4 m x 2.6 m (11.15 ft x 11.15 ft x 8.53 ft) ventilated by a DV system was investigated (see Figure 1a). A base case (without chair fans) and novel system (with chair fans) were investigated. In both cases, the room temperature was set to 25°C (77°F); the supply diffuser produced 47.5 L/s (1.68 ft3/s) fresh air at 21°C (69.8°F); and each CO_2 source produced 0.0083 L/s (0.0003 ft^3/s) at room temperature. The walls heat flux and lighting loads resulted in a total of 300 W (1023 Btu/hr); a desktop computer of 93 W (317.13 Btu/hr) and an occupant of 70 W (238.85 Btu/hr) under sedentary activity resulted in a total load of 40 W/m^2 (12.7 Btu/hr.ft2) that represenets a typical load removed by a DV system as reported by Yuan et al. (1998) and Vivian et al. (2002). The novel system was equipped with fans mounted on occupant's chair. The configuration of chair-mounted fans is shown in Figure 3.

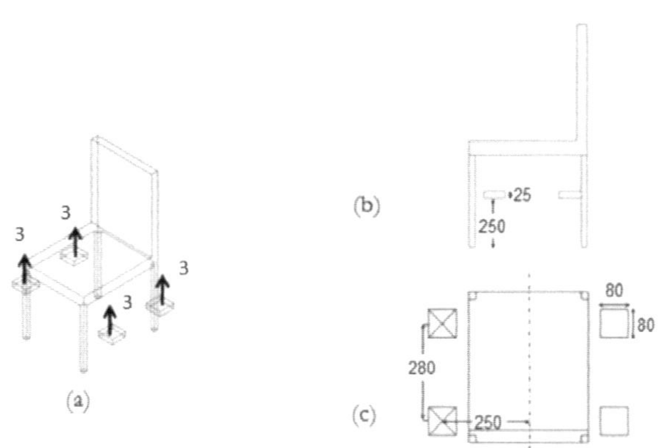

Figure 3 Chair Mounted Fans (a) Isometric View (3 L/s (0.105 ft^3/s) representing fan flow rate), (b) Side View, (c) Top View. All dimensions on the drawing in millimeter (1mm ~ 3.28 * 10^{-3} ft)

RESULTS AND DISCUSSION

The performance of the proposed system (DV + CF) is compared to a standalone DV system uing five parameters, i.e. local thermal sensation (LTS), local thermal comfort (LTC), whole body thermal sensation (WBTS), whole body thermal comfort (WBTC), and venitlation effectiffeness (ε_v). The results are shown in Table 2.

Table 2. LTS, LTC, WBTS, WBTC and (ε_v) for base case and proposed System

Fans Flow rate [L/s], [ft³/s]	Parameter	Head	Chest	Back	Abdomen	Buttocks	U-arm	L-arm	Hands	Thighs	Calves	Feet	Overall	ε_v %
0/0	LTS	2.09	0.32	1.03	-0.36	-0.04	-0.1	-0.22	-0.46	-0.13	-1.05	-0.95	0.82	49
	LTC	0.15	1.13	0.85	1.32	2.05	0.4	0.61	0.19	1.12	1.19	1.08	0.8	
12/0.42	LTS	0.67	0.11	0.89	-0.42	0.02	-0.32	-0.41	-0.83	-0.14	-1.07	-0.97	0.17	55
	LTC	1.79	1.82	1.36	2.05	1.9	1.51	1.8	0.31	1.94	1.87	1.6	1.24	

Table 2 differentiates between a standalone DV system and (DV + CF) system. Notice that a drop in LTS was recorded. This indicated a drop in segmental skin temperatures. The upper body parts were the most affected ones as a result of fans operation. Hence, a significant decrease in WBTS from 0.82 (case without fans) to 0.17 (case with fans) was recorded.

On the other hand, a decrease in LTS and WBTS didn't always reveal acceptable TC. Therefore, LTC was investigated for whole body segments. It is noticed that a remarkable improvement was shown. Hence, a decrease in LTS was accompanied by an increase in LTC. In addition, an increase in WBTC from 0.8 (case without fans) to 1.25 (case with fans) was recorded. This enhancement refered to the ability of fans in sucking cold air from floor level and blowing it upwards; thus, enhancing the convective air at the proximity of the occupant.

On the other side, providing acceptable TC for an occupant didn't reveal an acceptable environment unless good IAQ was fullfiled. Hence, the concentration of CO_2 at the breathing zone was investigated. An enehancement of 12.24% was recorded. This was expected as the fans sucked the fresh air from floor level and blew it upwards; thus, reducing the concentration of CO_2 at the breathing zone.

Energy Consumption

A novel system will not be a subject of use unless it provides low energy consumption. Thus, the energy consumed by this novel system (DV + CF) is compared to a standalone DV system and the results are reported as a percentange of energy savings. This comparison is performed on the basis of providing approximately same WBTC and same CO_2 concentration at the breathing zone. A standalone DV system supplying a flow rate of 54 L/s (1.91 ft³/s) is compared to (DV + CF) system supplying a flow rate of 47.5 L/s (1.68 ft³/s) aided by chair fans supplying a total flow rate of 12 L/s (0.42 ft³/s). Both systems supply 100% fresh air at a temperature of 21°C (69.8°F) and 50% relative humidity (RH). The above systems provide a WBTC of 1.24 and maintain a value of 658 ppm of CO_2 at the breathing zone.

Based on what is stated above, a chiller with COP 3.5 is selected. Fresh air at 32°C (89.6°F) and 70% RH supplied by a fan is cooled and dehumidified by exchanging heat with a cooling coil provided by the chiller. The heat sources in the room heat up the supplied air where it is extracted out by an exhaust fan. The electrical power consumed by the chiller is evaluated and it is added to the electrical power consumed by the fans. The supply and exhaust fans each consumes 310 W (1057.1 Btu/hr) for the novel system and 455.5 W (1553.255 Btu/hr) for the standalone DV system. The chair-mounted fans consume 25 W (85.25 Btu/hr).

Upon evaluation of energy consumption, it was noted that the novel system results in 20.84% energy savings compared to a standalone DV sytem. This refers to the use of chair fans where they provide the acceptable TC and IAQ locally without providing a global comfort in all the space that ends up to be useless as it is the case with a standalone DV sytem with high flow rate.

CONCLUSION

The performance of DV system aided by chair-mounted fans has been investigated using CFD solver. The performance is evaluated on the basis of TC, IAQ, and low energy consumption. This novel system provides an acceptable enhancement in TC and IAQ compared to a DV system held at the same conditions. An energy saving of 20.84% is recorded when the novel system is compared to a standalone DV system providing same performance. However, TC and IAQ could be affected upon variation of fans flow rate or height relative to the floor. In addition, the location of CO_2 source may affect the contaminat concentration at the breathing zone mainly if it is located near the floor level or above the occupied zone. These parameters are subjects of interest to be investigated.

NOMENCLATURE

AQI = Air Quality Index
C = Carbon dioxide concentration
CF = Chair Fans
CFD = Computational Fluid Dynamics
DV = Displacement Ventilation
IAQ = Indoor air Quality
LTC = local thermal comfort
LTS = local thermal sensation
MV = Mixed Ventilation
PMV = Predicted mean Vote
PPD = percentage of people dissatisfied
PPM = parts per million
RH = Relative humidity
TC = Thermal Comfort
TS = Thermal Sensation
WBTS = whole body thermal comfort
WBTS = whole body thermal sensation
Y^+ = dimensionless wall thickness

Greek Symbols:

ε = Effectiveness

Subscripts

B = breathing
E = Exhaust
Fr = fresh air
F = fan
v = Ventilation

REFERENCES

ANSYS Software: ANSYS Inc http://www.ansys.com/.
Habchi, C., Ghali, K., Ghaddar, N., & Shihadeh, A. 2015. Chair fan-enhanced displacement ventilation for high IAQ: Effects on particle inhalation and stratification height. Building and Environment , 84:68-79.

Lin, Z., Chow, T. T., Fong, K. F., Wang, Q., & Li, Y. 2005. Comparison of performance of displacement and mixing ventilations,partI: Thermal comfort. International Journal of Refregeration , 28(2):276-287.

Makhoul, A., Ghali, K., & Ghaddar, N. 2013 a. Desk fans for the control of the convection flow around occupants using ceiling mounted personalized ventilation. Building and environmnet , 59: 336-348.

Makhoul, A., Ghali, K., & Ghaddar, N. 2013 b. Low-mixing coaxial nozzle for effective personalized ventilation. Indoor and Built Environment, 1420326X13508967.

Melikov, A., Pitchurov, G., Naydenov, K., & Langkilde, G. 2005. Field study on occupant comfort and the office thermal environment in rooms with displacement ventilation. *Indoor air*, *15*(3): 205-214.

Mundt, E. (1995). Displacement ventilation systems-Convection flows and temperature gradients. Building and Environment , 30(1): 129-133.

Providers, L. (. (n.d.). Retrieved from www.computationalfluiddynamics.com.au

Salloum, M., Ghaddar, N., & Ghali, K. 2007. A new transient bioheat model of the human body and its integration to clothing models. International Journal of Thermal Sciences , 46(4): 371-384.

Sun, W., Cheong, K. W. D., & Melikov, A. K. 2012. Subjective study of thermal acceptability of novel enhanced displacemnet ventilation system and implication of occupants' personal control. Building and Environment , 57:49-57.

Vivian, L., Rohini, B., Michelle, M., Edward, V., & Mike, M. 2002. Energy Savings Potential of Flexible and Adaptive HVAC Distribution Systems for Office Buildings, ARTI21-CR Research Project 605-30030.

Watanabe, S., Shimomura, T., & Miyazaki, H. 2009. Thermal evaluation of a chair with fans as an individually controlled system. Building and Environmnet , 44(7): 1392-1398.

Xing, H., Hatton, A., & Awbi, H. B. 2001. A study of air quality in the breathing zone in aroom with displacement ventilation. Building and Environment , 36(7): 809-820.

Yuan, X., Chen, Q., & Glicksman, L. R. (1998). A critical review of displacement ventilation. *ASHRAE transactions*, *104*, 78.

Zelenský, P., Barták, M., Hensen, J. L. M., & Vavřička, R. 2013. Influence of turbulence model on thermal plume in indoor air flow simulation. In Proceedings of the 11th REHVA World Congress „Energy Efficient, Smart and Healthy Buildings "–Clima 2013.Prague,Crzech: Energy efficient, smart and healthy buildings.

Zhang, H., Arens, E., Huizenga, C., & Han, T. 2010. Thermal sensation and comfort models for non-uniform and transient environments:Part I:Local sensation of individual body parts. Building and environments , 45(2):380-388.

Zhang, H., Arens, E., Huizenga, C., & Han, T. 2010. Thermal sensation and comfort models for non-uniform and transient environments,part II:Local comfort of individual body parts. Building and Environmnet , 45(2): 389-398.

Zhang, H., Arens, E., Huizenga, C., & Han, T. 2010. Thermal sensation and comfort models for non-uniform and transient environments,part III: Whole body sensation and comfort. Building and environment , 45(2):399-410.

Chair Fans or Desk Fans to Aid Performance of Ceiling Personalized Ventilation for Reduced Disease Transmission Between Occupants?

Carine Habchi, PhD

Kamel Abou Ghali, PhD
Member ASHRAE

Nesreen Ghaddar, PhD
Member ASHRAE

ABSTRACT

Over recent years, the occupancy density in offices has been increasing due to urban and energy efficiency needs which challenged maintaining good indoor air quality (IAQ). Higher occupancy density increases the possibility of disease transmission. Cross-contamination can take place via particle spread in the space and inhalation by exposed occupants or by deposition on surfaces at their vicinity being susceptible to re-suspension by human contact. The ventilation configuration plays a major role in particle distribution. Therefore, a proper design of the ventilation system should be selected in crowded office spaces to insure acceptable IAQ. Localized ventilation consists of providing each occupant locally with its comfort and breathable air quality needs. Ceiling personalized ventilation (CPV) is a recent type of localized ventilation compromising between ease of integration, and reduction of the energy consumption. By proper control of the airflow interactions in the human body microclimate, the breathable air quality can be largely improved. This control is possible by the introduction of fans at the proximity of the occupant assisting the CPV system which suppress the rising occupant thermal plumes and improve the CPV performance. In this work, the performance of CPV was investigated when aided with i) desk fans and ii) chair mounted fans. The CPV with desk and chair fans were compared in terms of reducing disease transmission between occupants. A validated computational fluid dynamics (CFD) was used to simulate the flow and concentration fields in a two-station office space. A parametric study was conducted to assess the effect of the fan type, fan flow rate, and distance between occupants on the probability of cross-contamination. The ventilation configuration was optimized for reduced cross-infection between occupants with efficient use of the space. Chair fans ensured better performance in terms of reducing disease transmission between occupants compared to desk fans. The optimal CF flow rate per occupant was found to be 10 L/s.

INTRODUCTION

Human generated particles during expiratory activities contribute to air pollution constituting a real threat to the human health (Mendell et al. 2002, Schneider et al. 2003). As the occupancy time is the highest for indoor environments (Sexton et al. 2004), public health studies underlined the strong relation between bad indoor air quality (IAQ) and variable diseases as asthma and lung cancer, etc… (Brook et al. 2004, Englert 2004). Particularly, offices in which occupants spend large amount of time constitute critical indoor spaces especially with the increase of occupant density. Cross-contamination between occupants occurs mainly by direct inhalation and contact with contaminated surfaces (Alvin et al. 2000, Chen et al. 2012). Therefore, in order to enhance the IAQ, not only contaminant concentration in the inhaled air should be decreased but also particles deposition at the vicinity of the susceptible

Carine Habchi is a graduate PhD student in the Department of Mechanical Engineering, American University of Beirut, Beirut, Lebanon. **Kamel Ghali** is a professor and chair in the Department of Mechanical Engineering, American University of Beirut, Beirut, Lebanon. **Nesreen Ghaddar** is a professor in the Department of Mechanical Engineering, American University of Beirut, Beirut, Lebanon and the corresponding author, email: karinehabchy2733@gmail.com occupants. In fact, viruses and bacteria carried by particles can deposit on surfaces where they survive for a long period of time (Vasickova et al. 2010) increasing the possibility of their transfer by contact to the exposed occupants (Rusin et al. 2002, Weber et al. 2010).

Occupancy density in offices is increasing over the years (Bedford et al. 2013) making cross-infection between occupants a serious issue with decreased work-stations separating distance. The widely used ventilation system is mixing ventilation (MV) in which high distance should be kept between the occupants for decreased probability of cross-contamination challenging the use of space efficiently. For this reason, new ventilation configurations emerged to insure good IAQ with efficient space usage as the airflow pattern largely affects particles behavior and spread within the space. Ceiling personalized ventilation (CPV) is an innovative ventilation method practical to use since it can be retrofitted to MV systems without additional installation cost. Yang et al. (2010) was the first to propose this system consisting of a single nozzle personalized ventilator delivering fresh air at the center of a peripheral diffuser supplying recirculated air. Then the performance of this system was enhanced by Makhoul et al. (2013) using coaxial personalized ventilation jets to lengthen the potential core region of the fresh air jet for more efficient delivery to the breathing level. Nevertheless, the coaxial system requires additional ducting and difficulty of controlling two jets at equal velocities. Therefore, there was a need for more practical CPV designs.

The CPV jet delivery is opposed by the rising occupant plumes. Hence, suppressing the convective plumes lead to an enhancement of the CPV performance. In order to control the human plumes, a practical design was based on assisting *CPV* by fans mounted on desk or chair to suck the thermal plumes (El-Fil et al. 2015, Makhoul et al. 2013). In both cases, the suppression of plumes emanating from the occupant facilitated the penetration of the fresh air jet towards the breathing with less mixing with the surrounding air and provided individual occupant control of its microclimate environmental conditions.

In this work, the performance of CPV in terms of reducing direct and indirect cross-contamination was investigated when assisted by desk fans (DF) versus chair fans (CF). For this purpose, a computational fluid dynamics (CFD) model computing particle behavior under variable ventilation configurations was developed. A parametric study was conducted to assess the effect of the fan type, fan flow rate, and distance between occupants on the probability of cross-contamination. The CPV system operation was optimized to insure high IAQ for reduced separating distance between work-stations.

CFD MODEL

To accurately simulate the ventilated space, the CFD model should capture the variable physics affecting the air flow behavior involving turbulence, buoyancy effects, and formation of surfaces boundary layers. The velocity, thermal, and concentration fields were computed using the commercial software ANSYS Fluent. The mesh was generated with unstructured tetrahedrons elements. In order to resolve for boundary layers of surfaces and refine the mesh, surface inflation and face sizing were performed. For final mesh selection, a grid independence test was performed. The criterion for numerical convergence was to attain scaled residuals below 10^{-5}.

Room air, treated as continuous fluid, was simulated by the Eulerian approach (Chen et al. 2006). ANSYS Fluent solves for the Navier-Stokes equations by discretization after transforming them to algebraic equations. Fluent software offers variable discretization schemes and turbulence modeling options (ANSYS Software). The second-order upwind scheme was used for discretization of the momentum, kinetic energy and dissipation rate equations. The PRESTO! scheme was adopted to solve for the pressure equation and the SIMPLE algorithm was selected to couple between the pressure and velocity fields (ANSYS Software). For turbulence modeling, the Realizable k-ε model was adopted for its accuracy in modeling recirculation of air and predicting round jets spread rate (ANSYS Software). Besides, for turbulence modeling better adaptation near surfaces, enhanced wall treatment was performed allowing to switch, depending on the grid size, between the enhanced wall function and the two-layer model (ANSYS Software). For buoyancy effects modeling, the incompressible ideal gas law was adopted (ANSYS Software).

In order to model particle distribution within the space, the Lagrangian technique was used. This approach consist on tracking particles motion through the trajectory method based on Newton's second law incorporating the variable forces acting on the particle (Zhang et al. 2007, Zhu et al. 2006). The discrete random walk model (DRW) was selected to account for the effect of local turbulence intensities on particle trajectories. DRW models velocity fluctuations' effect on particle paths using a stochastic approach (Wang et al. 2012) requiring that the trajectories number is high enough to assure statistically stable results (Makhoul et al. 2013, Zhang et al. 2007). In order to correlate within each computational cell between particle concentration and trajectories the particle source in-cell (PSI-C) scheme was adopted (Zhang et al., 2006). A stable solution for the concentration field was reached when 200,000 particle trajectories were used.

Appropriate boundary conditions are needed for accurate CFD simulations of the thermal, velocity, and concentration fields. For thermal field modeling, surfaces' heat fluxes were specified. For velocity field computation, the velocity direction, magnitude, and turbulent intensity were determined at inlets and zero gage pressure outlet condition was selected at exhausts. Furthermore, for fans' jets modeling, the fan boundary condition was selected at fan inlets and the pressure drop through the fan was specified (Makhoul et al., 2013). For concentration field prediction, particle generation characteristics of sedentary breathing were specified at the breathing level of the infected occupant (Li et al., 2011). In order to model particle deposition, the trap treatment of the Lagrangian technique was selected at walls. At the outlet, the escape treatment was adopted to end the particle trajectory. Re-suspension of particles was neglected because threshold velocities for particle detachment are much higher than indoor airflow velocities (Braaten, 1994).

EVALUATION OF CROSS-CONTAMINATION

Infected people generate through exhalation contaminated droplets carrying viruses or/and bacteria that might infect healthy occupants by two main paths (Habchi et al., 2016). The first infection path is direct via inhalation (Alvin et al., 2000), while the second path is indirect taking place through contact of contaminated surfaces (Chen et al., 2012). In order to investigate cross-contamination in both direct and indirect paths, two indexes were assessed. The first one is the intake fraction (*IF*) to study direct cross-contamination between occupants and is given by:

$$IF = \frac{Particle\ concentration\ at\ breathing\ level\ of\ healthy\ person}{Particle\ generation\ concentration} \quad (1)$$

The IF parameter represents the ratio of contaminants concentration inhaled by the susceptible occupant to the one released by the infected occupant (Dygert and Dang 2010, Gao and Niu 2006). The second index is the particle deposited fraction (*PDF*) at the proximity of the exposed occupant defined as the ratio of particles' number deposited at the vicinity of the exposed person to the particles' number generated by the infected person:

$$PDF = \frac{Number\ of\ particles\ deposited\ at\ the\ vicinity\ of\ the\ exposed\ person}{Number\ of\ particles\ generated\ by\ the\ infected\ person} \quad (2)$$

CASE STUDY

The purpose of this work is to come up with a practical ceiling installed localized system for crowded office spaces. Therefore, the performance of different CPV systems combined with angled diffusers' flows and assisted by either chair or desk fans was considered. The reference case was the conventional MV with a large distance of 4m separating the occupants. The same total supply air flow rate of 90 L/s was considered for both mixing and personalized ventilation to be comparable. The CPV system studied (Fig. 1) is composed of a single nozzle (directly above the breathing level of the occupant) delivering fresh air at the center of a peripheral ceiling diffuser supplying recirculating air at an angle of 45° which was found to be optimal in creating a localized flow around the occupant (Makhoul et al., 2013). The combined nozzle jet and canopy are expected to reduce significantly direct and indirect cross-contamination. CF and DF assistance were compared to optimize the fan operation. As one of the main concerns of this study is to reduce cross-contamination with minimal distance between the workstations the performance of the studied air-conditioning configurations was assessed for two separating distances between the occupants (Sd1=2m; Sd2=4m).

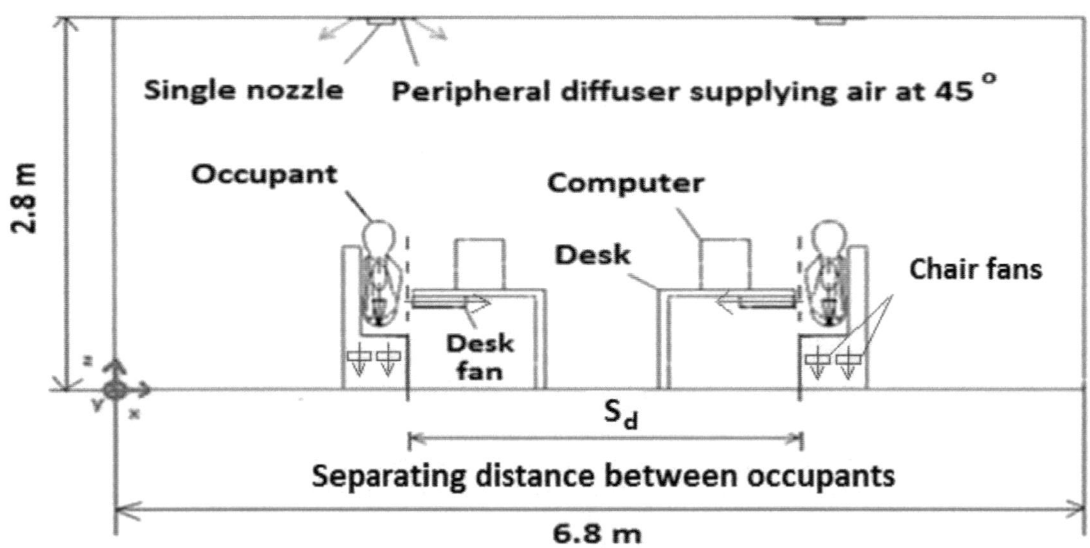

Figure 1 2D representation of the studied two-station office at the symmetry plane.

A full-scale computational domain representing a typical two station-office space of dimensions (3.4 m x 6.8m x2.6 m) was considered (Fig 1). The occupants were simulated by two seated multi-segment manikin characterized each by a total surface area of 1.78 m^2, a height of 1.2 m and generating a typical sensible heat flux of 70 W for sedentary activity (ASHRAE, 2009). Two ceiling mounted nozzles integrated with slot diffusers constitute the air supply inlets while four exhausts were considered at the corners. Each work-station was equipped by a computer of 93W load (ANSI/ASHRAE, 2013) placed on the desk in front of the manikin (Fig 1). In addition, typical lighting load of 10 W/m^2 (ANSI/ASHRAE, 2013) was imposed on the ceiling, while the remaining load was distributed equally as heat flux through the walls such that the total room load was 60 W/m^2 of floor area. Desks and chairs were equipped with fans in order to assess their performance in enhancing the IAQ (Fig. 1). The optimized ventilation configuration should reduce cross-infection for the critical inhalable range (particles of diameter lower than 10 μm) because they constitute infection threat to the exposed person (Habchi et al., 2014). For this reason, the performance of the different ventilation systems was examined for two extreme particle diameters at 1 μm representing the fine mode and 10 μm representing the coarse mode with a density of 1000 Kg/m^3 for the particulate matter.

RESULTS AND DISCUSSIONS

Different ceiling supplied air systems were simulated in order to come-up with the optimal configuration enhancing the IAQ by minimizing direct and indirect cross-contamination with efficient use of the space. In order to be comparable the total supply flow rate for the different configurations studied was equal to 85 L/s.

The intake fraction (IF) and particle deposited fraction (PDF) were determined according to equations (1-2) for the different configurations for the fine and coarse modes of particles. Results are summarized in tables 1-2 representing respectively the IF and PDF for the different ventilation configurations studied.

Table 1. Ventilation configuration effect on 1 and 10 μm intake fraction (IF) at the breathing level of the exposed person

HVAC configuration	Separating Distance	IF of 1 μm ($x10^{-4}$)	IF of 10 μm ($x10^{-4}$)
MV	Sd1=4m	1.71	0.61
MV	Sd2=2m	4.02	1.65
CPV	Sd1=4m	1.33	0.51
CPV	Sd2=2m	2.23	0.97
CPV+CF(5 L/s)	Sd1=4m	0.61	0.31
CPV+CF(5 L/s)	Sd2=2m	1.03	0.57
CPV+DF(5 L/s)	Sd1=4m	0.77	0.36
CPV+DF(5 L/s)	Sd2=2m	1.27	0.65
CPV+CF(10 L/s)	Sd1=4m	0.38	0.17
CPV+CF(10 L/s)	Sd2=2m	0.79	0.38
CPV+DF(10 L/s)	Sd1=4m	0.49	0.22
CPV+DF(10 L/s)	Sd2=2m	0.91	0.45
CPV+CF(15 L/s)	Sd1=4m	0.73	0.33
CPV+CF(15 L/s)	Sd2=2m	1.35	0.71
CPV+DF(15 L/s)	Sd1=4m	0.89	0.38
CPV+DF(15 L/s)	Sd2=2m	1.62	0.79

Table 2. Ventilation configuration effect on 1 and 10 μm particle deposition fraction (PDF) at the proximity of the exposed person

HVAC configuration	Separating Distance	PDF of 1 μm ($x10^{-3}$)	PDF of 10 μm ($x10^{-3}$)
MV	Sd1=4m	4.55	2.31
MV	Sd2=2m	8.28	3.79
CPV	Sd1=4m	4.05	1.89
CPV	Sd2=2m	5.41	2.65
CPV+CF(5 L/s)	Sd1=4m	2.79	1.55
CPV+CF(5 L/s)	Sd2=2m	3.33	1.92
CPV+DF(5 L/s)	Sd1=4m	3.18	1.71
CPV+DF(5 L/s)	Sd2=2m	3.89	2.21
CPV+CF(10 L/s)	Sd1=4m	2.37	1.38
CPV+CF(10 L/s)	Sd2=2m	2.82	1.62
CPV+DF(10 L/s)	Sd1=4m	2.68	1.61
CPV+DF(10 L/s)	Sd2=2m	3.21	1.89
CPV+CF(15 L/s)	Sd1=4m	3.53	1.53
CPV+CF(15 L/s)	Sd2=2m	4.42	1.94
CPV+DF(15 L/s)	Sd1=4m	3.83	1.75
CPV+DF(15 L/s)	Sd2=2m	4.87	2.32

The separating distance between the workstations and particle diameter largely affect the risk of cross-contamination (Tables 1-2). As the distance between the workstations increases, a lower number of particles reach the exposed station leading to a decrease of IF and PDF (Tables 1-2). On the other hand, for a fixed distance between the occupants, as the diameter increases, the performance of the downward CPV system in terms of reducing cross-infection is enhanced because it is aided by the gravitational settling effect supporting the downward transport of generated particles increasing floor deposition, which constituted a particles' sink, and reducing contaminants diffusion between the workstations resulting in significant decrease of the IF and PDF compared to the fine particle mode (Tables 1, 2).

In order to enhance more the efficiency of the CPV compared to MV system it was assisted by desk/chair fans. Different fans flow rates per occupant were investigated (0, 5, 10 and 15 L/s). Concerning the IF, it was shown that the performance of desk and chair fans was comparable for the different scenarios studied but always lower IF were reached with CF versus DF. The fans in the chair or desk position allow the flow to reach lower levels compared to standalone single jet. Furthermore, the presence of desk/chair fan reduces the transport of particles between the two stations by increasing the momentum of the jet reaching the breathing level of the infected person and providing more protection to the exposed person. As the fan flow rate increases the momentum of the CPV jet reaching the breathing level increases hence increasing the jet protection but at the same time particles accumulated between the diffuser canopy and the jet are sucked downward which tend to accumulate more particles at the breathing level. These two opposing effects explain the presence of an optimal fan flow rate of 10 L/s for both DF and CF (Table 1). Figure 2 illustrates the 1μm particle concentration field in the exposed person station at the symmetry plane for MV, standalone CPV, and CPV assisted by DF and CF operated at the optimal flow rate showing the significant enhancement induced by the fans operation.

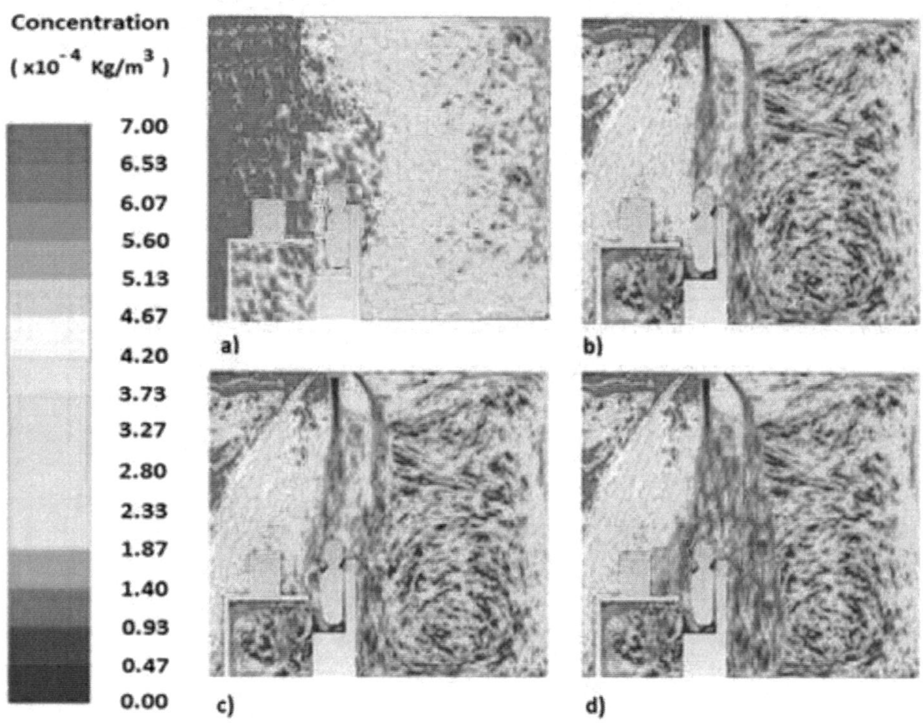

Figure 2 1μm particle concentration field in the exposed person station at the symmetry plane for: a) MV; b) standalone CPV; c) CPV+DF (10 L/s); d) CPV+CF (10 L/s)

From the presented results in Table 2, it is clear that up to a flow rate of 10 L/s per fan, the presence of a desk or chair fan decrease the number of particles deposited at the vicinity of the exposed occupants compared to MV and standalone CPV configurations. In fact, as the fan flow rate increases particles deposited at some surfaces at the vicinity of the exposed occupant decreases while the number deposited at some surfaces at the proximity of the fan (e.g. chair in case of CF, desk, computer in case of DF) increases which lead to the presence of an optimal fan flow rate of 10 L/s for which the PDF is minimized. Generally, for the same flow rate the PDF reached with CF is lower than with DF due to the relative position of the surfaces at the vicinity of the occupant to the fan.

In fact, the CF jet path is relatively farer from the different surfaces at the proximity of the occupant compared to the DF jet path.

CONCLUSION

CPV system permitted supplying fresh air locally to the occupant explaining the lower concentration of particles within the space and higher protection of the exposed occupant compared to MV.

The presence of both chair and desk fans decreased the intake fraction at the breathing level of the exposed occupant and particle deposited fraction at its vicinity compared to the standalone CPV. The relative position of the CPV jet and fan affect the fan performance. With proper selection of the fan flow rate (10 L/s), the introduction of CF to the CPV system was shown to be the more efficient in terms of reducing direct and indirect-contamination with efficient use of the space.

The current work does not consider high momentum respiratory activities (coughing, sneezing) that affect particle behavior leading to increased probability of cross-contamination due to higher particle penetration of the CPV jet. Furthermore, the shift of occupant from the design position might affect significantly the system performance which is worth of future studies.

ACKNOWLEDGMENTS

The authors would like to thank the Lebanese National Council for Scientific Research (CNRS) for their financial support. The support of the University Research Board (URB) at the American University of Beirut is also acknowledged as well as the Shammas PhD Fellowship.

NOMENCLATURE

CF = chair fan

CFD= computational fluid dynamics

CPV= ceiling personalized ventilation

DF = desk fan

DRW = discrete random walk model

IAQ = indoor air quality

MV = mixing ventilation

PDF= particle deposited fraction

REFERENCES

Alvin L. and Nazaroff W., Modeling indoor particle deposition from turbulent flow onto smooth surfaces, J. Aerosol Sci. 2000, (31), No. 4, 463-476.

ANSI/ASHRAE Standard 90.1: Energy Standard for Buildings Except Low-Rise Residential Buildings. Atlanta, American Society of Heating, Air-Conditioning and Refrigeration Engineers, Inc., 2013.

ANSYS Software: ANSYS Inc. http://www.ansys.com/

ASHRAE: ASHRAE Handbook-Fundamentals. Atlanta, American Society of Heating Air-conditioning and Refrigeration Engineers, Inc., 2009.

Bedford M, Harris R, King A, and Hawkeswood A. Occupier density study. British council for offices (BCO) 2013.

Braaten DA. Wind tunnel experiments of large particle reentrainment-deposition and development of large particle scaling parameters. Aerosol Science and Technology 1994; 21(2):157-169.

Brook D, Franklin B, Cascio W, Hong Y, Howard G, Lipsett M, Luepker R, Mittleman M, Samet J, Smith C, Tager I. Air pollution and cardiovascular disease A statement for healthcare professionals from the expert panel on population and prevention science of the American Heart Association. Circulation 2004; 109: 2655-2671.

Chen Q., Ph.D.; James M., Ph.D., Jitendra G., Ph.D.; Byron J., Sagnik M., Stephane P., and John S., Infectious Disease Transmission in Airliner Cabins, Report No. RITE-ACER-CoE-2012-01.

Chen FZ, Yu S, Lai A. Modeling particle distribution and deposition in indoor environments with a new drift–flux modelAtmospheric Environment 2006; 40: 357-367.

Dygert, RK, & Dang TQ. Mitigation of cross-contamination in an aircraft cabin via localized exhaust. Building and Environment 2010; 45(9): 2015-2026.

El-Fil B, Ghali K, Ghaddar N. Optimizing performance of ceiling mounted personalized ventilation system assisted by chair fans: assessment of thermal comfort and indoor air quality. Master thesis. American University of Beirut; 2015.

Englert N. Fine particles and human health—a review of epidemiological studies. Toxicology letters 2004; 149: 235-242.

Gao N, Niu J. Transient CFD simulation of the respiration process and inter-person exposure assessment. Building and Environment 2006; 41(9): 1214-1222.

Habchi C, Ghali K, Ghaddar N. A simplified mathematical model for predicting cross contamination in displacement ventilation air-conditioned spaces. Journal of Aerosol Science 2014; 76:72–86.

Habchi C, Ghali K, Ghaddar N, Chakroun W, Alotaibi S. Ceiling Personalized Ventilation Combined with Desk Fans for Reduced Direct and Indirect Cross-Contamination and Efficient use of an Office Space. Energy Conversion and Management 2016; 111: 158-73.

Li X, Niu J, Gao N. Spatial distribution of human respiratory droplet residuals and exposure risk for the co-occupant under different ventilation methods. HVAC&R Research 2011; 17: 432-445.

Makhoul A, Ghali K, Ghaddar N. Low-mixing coaxial nozzle for effective personalized ventilation, Indoor and Built Environment 2013; 22(3): 508-519.

Makoul A, Ghali K, Ghaddar N. Desk fans for the control of the convection flow around occupants using ceiling mounted personalized ventilation. Building and Environment 2013; 59: 336-348.

Makhoul A, Ghali K, Ghaddar N. Investigation of Particle Transport in Offices Equipped with Ceiling-Mounted Personalized Ventilators. Building and Environment 2013; 63: 97-107.

Mendell J, Fisk J, Petersen R, Hines J, Dong M, Faulkner D, Deddens A, Ruder M, Sullivan D, Boeniger F. Indoor particles and symptoms among office workers: results from a double-blind cross-over study. Epidemiology 2002; 13: 296–304.

Rusin P, Maxwell S, Gerba C. Comparative surface-to-hand and fingertip-to-mouth transfer efficiency of gram-bacteria, gram-negative bacteria, and phage. Journal ofpositive Applied Microbiology 2002; 93(4): 585-592.

Schneider T, Sundell J, Bischof W, Bohgard M, Cherrie W, Clausen A, Dreborg S, Kildesø J, Kjærgaard K, Løvik M, Pasanen P, Skyberg K, „EUROPART' Airborne particles in the indoor environment. A European interdisciplinary review of scientific evidence on associations between exposure to particles in buildings and health effects. Indoor Air 2003; 13: 38–48.

Sexton K, Ramachandran G, Pratt C, Mongin J, Stock H, and Morandi T, Comparison of personal, indoor, and outdoor exposures to hazardous air pollutants in three urban communities. Environ. Sci. Technol. 2004; 38: 423-430.

Vasickova P, Pavlik I, Verani M, Carducci A. Issues concerning survival of viruses on surfaces. Food and Environmental Virology 2010; 2(1): 24-34.

Wang M, Lin CH, Chen Q. Advanced turbulence models for predicting particle transport in enclosed environments. Building and Environment 2012; 47: 40-49.

Weber DJ, Rutala WA, Miller MB, Huslage K, Sickbert-Bennett E. Role of hospital surfaces in the transmission of emerging health care-associated pathogens: norovirus, Clostridium difficile, and Acinetobacter species. American journal of infection control 2010; 38(5): S25-S33.

Yang B, Sekhar C, Melikov AK. Ceiling mounted personalized ventilation system in hot and humid climate - An energy analysis. Energy and Buildings 2010; 42: 2304-2308.

Zhang Z, Chen Q. Comparison of the Eulerian and Lagrangian methods for predicting particle transport in enclosed spaces. Atmospheric Environment 2007; 41: 5236–5248.

Zhang Z, Chen Q. Experimental measurements and numerical simulations of particle transport and distribution in ventilated rooms. Atmospheric Environment 2006; 40(18): 3396–3408.

Zhu S, Kato S, Yang JH. Study on transport characteristics of saliva droplets produced by coughing in a calm indoorenvironment. Building and Environment 2006; 41(12): 1691-702.

Feasibility of Using Natural Ventilation for Indoor Thermal Comfort in Non-Residential Application in Warm Climate

Ghina R. Annan
ASHRAE Member

Balsam A. Nehme

ABSTRACT

Due to large cooling energy demands for buildings located in warm climates and in line with international principles of energy efficiency for non-residential building design, the implementation and optimization of a natural ventilation strategy can assist in minimizing cooling and fan energy annual consumption.The objective of this study is to demonstrate that significant savings in energy consumption can be achieved by further reducing the need for mechanical ventilation and relying on the usage of external windows and skylight openings to cool the building spaces by means of natural ventilation without compromising the occupants' thermal comfort. The study focuses on the benefits of using the outdoor climate to condition non-residential buildings based on the building façade, structure and orientation as well as the existing occupancy, lighting and equipment load profiles within building spaces while taking several factors into consideration. The study focuses on two distinct scenarios for natural ventilation; it assesses the impact of single-sided ventilation (Scenario 1) and cross ventilation (Scenario 2) based on CIBSE design guide for "Natural Ventilation in Non-domestic Buildings".

INTRODUCTION

In this paper, an energy efficient, naturally ventilated building, is proven to provide the required levels of thermal comfort and acceptable indoor air quality if designed under specific seasonal conditions for external dry bulb temperatures, relative humidity and available wind speed and directions. The feasibility study allows for a wider range of temperature and humidity levels within occupied spaces as it proves that occupants of such naturally ventilated areas are often more tolerant of fluctuations in the indoor climate. Accordingly, and since the study would legitimately allow for resetting the thermal comfort range for higher set points for temperature and relative humidity, it will thus result in reduced annual energy consumption.

Previous studies using Beirut typical construction materials and Beirut occupants' behavior in regulating the opening and closing of the windows established in a conducted survey, is indicative of a large potential for energy savings using natural ventilation instead of air-conditioning in Beirut area (Annan 2014).

The facades' opening areas enabling natural ventilation during adequate comfortable external climatic conditions are optimized through calculations onto a typical non-residential application in Algeria's warm climate. This study is conducted in an effort to advise the building's architectural and structural schemes of any potential possible modification, through an integrated design process. This allows for optimal design of building natural ventilation and ensures improvements in thermal comfort and air quality for larger periods of time and most importantly guarantees managing a building with lower energy consumption.

METHODOLOGY

An energy efficient naturally ventilated building will provide the required levels of thermal comfort and acceptable indoor air quality under specific seasonal conditions. As such; a preliminary study analyzing the feasibility of natural ventilation within a convention center in Algeria has been conducted.

The first step in designing a natural ventilation system is to determine the airflow quantities necessary to satisfy the required indoor air quality (IAQ) and thermal comfort criteria. The ventilation system must therefore be designed in such a way to deliver the required quantity of air to the right places, at the right time.

Fundamentally, four important sources of heat gain are present; people, lighting, equipment, and solar load are essentially considered in the analysis of natural ventilation. As well as assessing the magnitude of those loads, it is also important to note that peak values for lighting, people, and equipment loads are reflected in the study, while mean values for solar load are considered assuming the building is shaded. With respect to radiation gain, it has been neglected in small spaces with low ceiling, but has been accounted for in large spaces with large floor-to-ceiling height due to perceptible temperature difference.

It is assumed that the flow rates through openings have been quantified in magnitude and direction using Algeria Wind rose supplied by Ecotect under specific wind and temperature conditions as shown in figure 1.

Before carrying out the calculations, isolated rooms and spaces are identified, parts of the building treated as single space are determined, required data including opening positions, discharge coefficients, relevant wind and temperature are determined. Subsequently, window openings are sized and tested for practicality so as to complement the proposed facade area, such that the required flows are delivered avoiding draughts. While natural ventilation is highly recommended, especially in moderate climatic conditions, several factors are to be taken into consideration as listed below:

1. Indoor temperature during naturally ventilated mode is recommended not to exceed a range of 27-28°C to maintain thermal comfort for space occupants.
2. Indoor relative humidity during naturally ventilated mode is recommended not to exceed a value of 60% to maintain thermal comfort for space occupants.
3. As such, it is recommended to consider natural ventilation for an outside temperature range of 18-24°C and a relative humidity less than 50% outdoors.

The ventilation system is made up of a number of key components, ventilation inlet(s), supply point(s), flow path, exhaust path and ventilation outlet. Consideration must be given to the route the air will take through the building. For instance, it may be beneficial to draw cooler air from a shaded side of the building in order to maximize the cooling potential.

The design allowing sizing for openings for natural ventilation across the convention center has been based on CIBSE design guide for "Natural Ventilation in Non-domestic Buildings", such that:

Single-sided ventilation relies on opening(s) on one side of the ventilated enclosure; this single opening will allow both air intake and exhaust. It is closely approximated that this type of ventilation can be applied within spaces approaching 12 m in depth, if the windows have sufficiently large openable areas. Single sided ventilation is used in most of the peripheral occupied spaces of the convention center with single exposure. Calculations of required areas to be opened within the vertically glazed facade to allow for efficient natural ventilation comply with section 4.3.1 of the CIBSE guide for natural ventilation.

Wind driven cross-ventilation occurs through opposite openings across the space or through oppositely opened doors. Cross-ventilation is effective for a depth equivalent to up to 5 times the floor-to-ceiling height (CIBSE 2005). Calculations of required areas to be opened within the vertically glazed facade to allow for efficient natural ventilation are compliant with section 4.3.1 of the CIBSE guide for natural ventilation. The calculations do account for the potential opening of room doors on the opposite side.

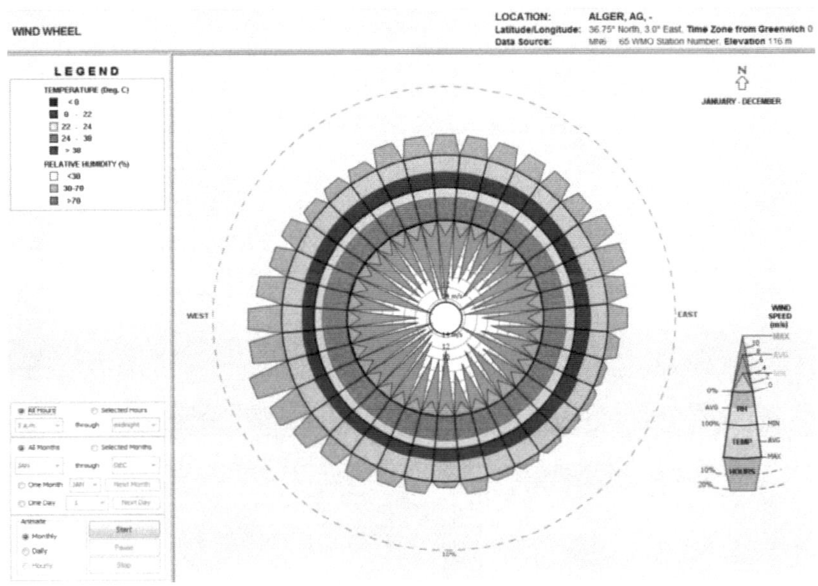

Figure 1 Algeria Wind rose supplied by Ecotect

Methodology

CIBSE design guide describes four common cases, cases 1, 2, and 3 are examples of single sided natural ventilation and case 4 is an example of cross flow studied below. Case 3 investigates a single-sided ventilation case with a single opening, driven by wind alone as per figure 2. Case 4 examines a cross flow ventilation case with a multiple openings, driven by wind alone as per figure 3. The openings are assumed identical.

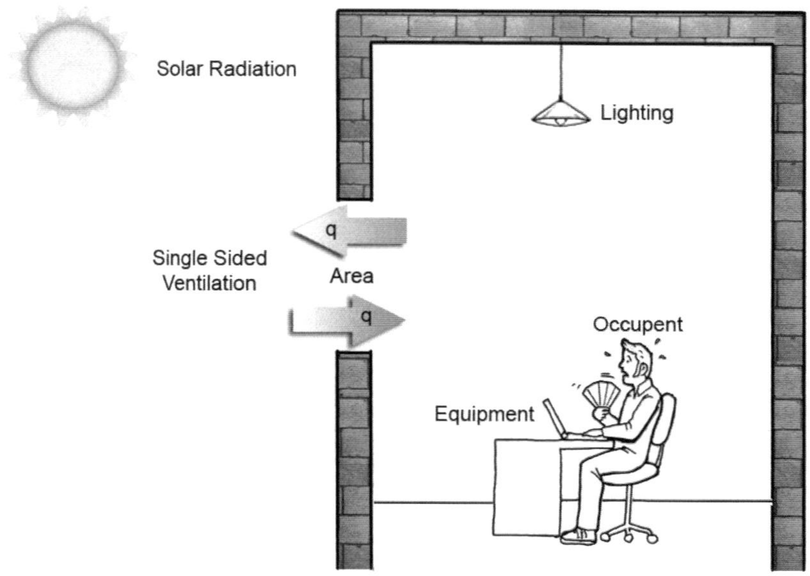

Figure 2 Single sided ventilation with single opening, wind – driven

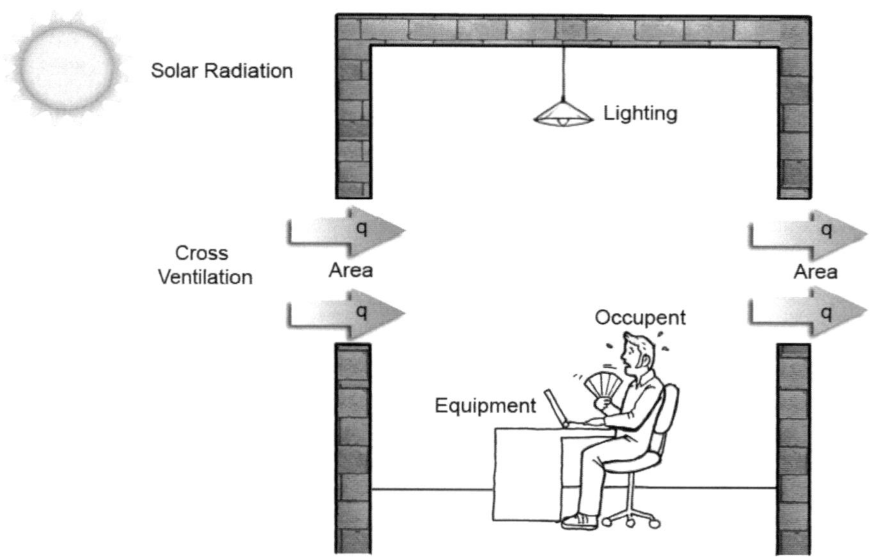

Figure 3 Cross flow ventilation, wind – driven

First, the exposed rooms of the convention center are located, and for each room the following paramters are identified:

1. Room area (m^2)
2. Total façade area (m^2)
3. Total number of occupant

Next, the total given loads comprising of the solar, lighting, people and equipment loads are summed up for each room. Subsequently, the required flow rate is computed based on equation 1:

$$Q_{total} = Q_P + Q_L + Q_E + Qs \quad (1)$$

Subsequently, the required flow rate is computed based on Equation 2:

$$V = Q_{total} / (5.16 * 1000) \quad (2)$$

Finally, the CIBSE spreadsheet is used to compute the required window area for Cases 3 and 4 representing single sided ventilation through an open window by wind alone and cross-flow ventilation by wind alone respectively. In order to compute the area of all single sided ventilation spaces pertaining to Case 3, the computed flow rate, wind speed at building height and flow coefficients are inserted in the CIBSE spreadsheet. By referring to Algeria wind rose supplied by Ecotect, the wind speed at building height is assumed to be 3 m/s. According to CIBSE, the value of the flow coefficient ranges from 0.01 to 0.05 when the wind velocity is measured at the same height of the building. Since the value of the flow coefficient decreases as a result of turbulent diffusion, it is assumed to be 0.05, neglecting any diffusion turbulence.

Similarly, in order to calculate the area of all single sided ventilation spaces representing Case 4, the computed flow rate, wind speed at building height, in addition to the discharge coefficients, and wind pressure cofficients are inserted in the CIBSE spreadsheet. While the openings of cross flow ventilation are assumed identical, the discharge coefficient is assumed to be 0.61 (CIBSE section 4.3.1). Furthermore, the wind pressure coefficents for both windward side and leeward side are assumed 0.5 and – 0.2 respectively, such that the wind pressure coefficients have been extracted from environmental wind tunnel building physical models (CIBSE section 4.2). This will further generate various wind pressure coefficients as listed in Section 4.4.1. The values assumed in our calculations are in line with the CISBE on-line calculator tool (www.cibse.org/venttool).

Case 3 - Single sided ventilation through an open window by wind alone	
Required flowrate (m3/s)	0.770
Wind speed at building height (m/s)	3.0
Flow coefficient	0.05
Required area	5.133

Figure 4 Sample CIBSE spreadsheet calculator – Case 3

Case 4 - Cross-flow ventilation by wind alone	
Required flowrate (m3/s)	0.770
Wind speed at building height (m/s)	3.0
Discharge coefficient	0.61
Wind pressure coefficients	
windward side	0.50
leeward side	-0.20
Required area	0.711

Figure 5 Sample CIBSE spreadsheet calculator - Case 4

RESULTS

Results in table 1 show that the percent openable areas for single sided naturally ventilated spaces with all rooms doors kept closed at all times, (i.e.: Case 3) vary from 3.86 — 17.14 %. To reduce % opening areas in those spaces, natural ventilation has been calculated while accounting for room doors being open (i.e.: Case 4); thus the percent openable decreases to 0.29— 12.86 %.

The required openable areas for naturally ventilated spaces within the palace during adequate comfortable external climatic conditions are displayed in Table 1.

Table 1. Required Openable Area (m^2) for Single Sided and Cross Ventilation

Floor	Single Sided (Case 3)	Cross Ventilation (Case 4)	Façade Area (m2)	Glazing percentage case 3 (%)	Glazing percentage case 4 (%)
Ground	2.2	0.15	42	5.24	0.36
	2	0.14	38	5.26	0.37
	2	0.14	38	5.26	0.37
	2.8	0.2	60	4.67	0.33
	1.4	0.1	34	4.12	0.29
	2.07	0.15	39	5.31	0.38
	1.07	0.08	12	8.92	0.67
	0.8	0.06	16	5.00	0.38
	0.6	0.04	14	4.29	0.29
	1.27	0.09	19.5	6.51	0.46
	2.93	0.21	55	5.33	0.38
	3.4	0.24	20	17.00	1.20
	2.4	0.17	40	6.00	0.43
	2.07	0.15	39	5.31	0.38
	1.67	0.12	22	7.59	0.55
First	1.07	0.08	12	8.92	0.67
	1.2	0.09	17	7.06	0.53
	0.87	0.06	16	5.44	0.38
	1.07	0.08	12	8.92	0.67
	1	0.07	20	5.00	0.35
	0.87	0.06	16	5.44	0.38
	0.27	0.02	7	3.86	0.29
	0.33	0.03	8	4.13	0.38
	2.53	0.18	22	11.50	0.82
	2.33	0.16	15	15.53	1.07
	1.2	0.09	30	4.00	0.30
	1.13	0.08	21	5.38	0.38
Second	1.47	0.1	34	4.32	0.29
	1.4	0.1	34	4.12	0.29
	3.27	0.23	60	5.45	0.38
	1.67	0.12	22	7.59	0.55
	5.47	0.38	85	6.44	0.45
	4.8	0.34	56	8.57	0.61
Third	12	9	70	17.14	12.86
	10.07	0.7	62	16.24	1.13

CONCLUSION

This paper studied the effect of further reducing the need for mechanical ventilation and relying on the usage external windows and skylight openings to cool the building spaces by means of natural ventilation with compromising the occupants' thermal comfort. Two distinct scenarios for natural ventilation have been investifated, single-sided ventilation and cross ventilation, based on CIBSE design guide for "Natural Ventilation in Non-domestic Buildings". The results showed that the percent openable areas for single sided naturally ventilated spaces with all rooms doors kept closed at all times vary from 3.86 — 17.14 %, whereas it decreased to 0.29— 12.86 % with cross ventilation.

NOMENCLATURE

Qtotal= Total load (W/m2)

QP= People load (W/m2)

QL= Lighting load (W/m2)

QS= Total Solar load (W/m2)

QE= Equipment load (W/m2)

V= Flow rate (m3/s)

Subscripts

P = People

L = Lighting

S = Solar

E = Equipment

REFERENCES

G. R. Annan, N. Ghaddar, K. Abou Ghali. *Natural Ventilation in Beirut Residential Buildings for Extended Comfort Hours.* International Journal of Sustainable Energy, 2014.

CIBSE. *Natural ventilation in non-domestic buildings.* 2005.

CIBSE. *www.cibse.org/venttool.*

Study on Improving Ventilation Rate Through Aerosol Protective Clothing by using Electrospun Nanofibers

Nagham Ismail

Nesreen Ghaddar, PhD
Member ASHRAE

Kamel Ghali, PhD
Member ASHRAE

Ali Tehrani, PhD

ABSTRACT

Over recent years, researchers seek to prevent the thermal discomfort of aerosol protective clothing wearer by providing sufficient ventilation accompanied with high filtration rate. This is achieved by using electrospun nanofibers deposited on a nonwoven substrate. The reason is that electrospun nanoweb provides high surface area and very small weight while keeping an acceptable air penetration and ventilation rates.

The ventilation phenomenon through clothing plays an essential role in bringing the thermal comfort of the aerosol protective clothing wearer. Therefore, it is important to investigate the ventilation rate provided by protective clothing fabricated with electrospun nanofiber and study the effect of different parameters in improving it. Several studies investigated experimentally the air permeability of different nanofiber mats of high filtration. However, analytical approach that relates the processing parameters of the fabrication of electrospun nanofiber with the associated ventilation rates and filtration efficiency is missed.

In this study, an analytical model derived from integrating different published models is developed to predict the final characteristics of the nanoweb as function of the electrospinning parameters. The model allows relating the final properties (filtration efficiency, air permeability, ventilation rates) of the nanoweb created to different processing parameters that characterize the electrospinning operation. Such model is beneficial in saving time and experimental cost by excluding the need of repeating the electrospinning experiments to achieve some desired nanoweb properties. This is followed by a parametric study showing the influence of the electrospinning parameters on the filtration and the ventilation of the electrospun nanoweb.

Key-words: *Electrospinning; ventilation; aerosol filtration.*

INTRODUCTION

An aerosol is a colloid of fine particles or liquid droplets in a gas (could be air) like dust, air pollutants, smoke… An aerosol particle is tiny and ranges in size from a few nanometers (smaller than the width of smallest viruses) to several tens of micrometers (equal to the diameter of human hair). Aerosol groups include sulfates, organic carbon, black carbon, nitrates, mineral dust, and sea salt. Nowadays, aerosols production is increasing especially from fossil

Nagham Ismail is a graduate student in the Department of Mechanical Engineering, American University of Beirut, Beirut, Lebanon. **Nesreen Ghaddar** is a professor in the Department of Mechanical Engineering, American University of Beirut, Beirut, Lebanon and the corresponding author, email: farah@aub.edu.lb. **Ali Tehrani** is an assistant professor in the Department of Chemical Engineering, American University of Beirut, Beirut, Lebanon. **Kamel Ghali** is a professor and chair in the Department of Mechanical Engineering, American University of Beirut, Beirut, Lebanon.

fuel combustion (sulfate aerosols), biomass burning (organic carbon and black carbon), automobiles, incinerators, smelters, and power plants. Therefore, there is an increasing concern of occupational exposure of aerosols particles in workplace (Ling et al., 2012). This necessitates the use of protective clothing garments to prevent workers from the dermal exposure of harmful aerosols particles. Conventional protective clothing system needed for minimizing the exposure of workers to hazardous aerosol particles are characterized by their heavy weight and high cost. In fact, conventional protective clothing system does not only provide the worker with protection but unfortunately it also subjects the worker to heat stress because of its low air permeability which reduces clothing air ventilation (Nunneley et al., 1989). A compromise between high contaminant impermeability, good breathability and relatively high air permeability is well established by the nanotechnology. Indeed, nanotechnology includes the use of nano-sized materials for fabricating very thin fibers capable of forming a highly porous mesh characterized by a large surface-to-volume ratio to improve the clothing protection for better worker comfort (Ramakrishna et al., 2006). The fabrication of nanofibers is achieved by the electrospinning process which consists of producing nanofibers fabric by subjecting a polymer solution to an electric field (Caroll et al., 2006). Electrospinning is a promising and straightforward technique because it produces continuous fibers with diameters in the range of nanometers to a few micrometers. The nanofiber web posseses high surface area to volume ratio, high porosity, and relatively high permeability making them excellent candidates for coating conventional woven/nonwoven fabrics enhancing the fabric protection from liquid and gaseous contaminants while maintaining its breathability.

Many studies considered the use of electrospun nanofibers in the protective clothing. The performance of such protective clothing is measured by studying the efficiency in preventing the penetration of aerosols particles. Consequently, researchers took two ways: experimental (spurny et al. 1998) or mathematical that deals with fabric as filter prtecting the worker from aerosols penetration by providing sufficient barrier to particles (Kim et al. 2001). It is shown that the filtration efficiency is mainly related to the fiber diameter of the electrospun nanofiber which is in its turn related to the electrospinning parameters. Therefore, the electrospinning parameters influence the performance of the electrospun nanoweb by means of the filtration efficiency. In the other hand, the comfort of the wearer is related to the air permeability of the electrospun nanoweb. Prediction of the air permeability is studied in the literature and it is also related to the fiber diameter (Hosseini et al. 2010). Therefore, it is of interest to relate the processing parameters of the electrospinning to the final characteristics of the nanoweb such as filtration efficiency and air permeability (or ventilation rates). This can be done experimentally but it needs time and cost. However, developing a therorectical model that predicts the electrospun nanoweb characteristics from the electrospinning parameters can offer an in-depth insight into the physical understanding as a useful tool from managing contributing factors toward increasing production rate and decreasing experimental cost and time.

In this study, an attempt to derive a complete model that relates the processing parameters of the electrospinning to the final electrospun nanoweb characteristics is made. The characteristics are useful in finding the efficiency of the nanoweb in preventing the aerosol particles penetration and enhancing the air permeability. This dual effect is important because it provides a compromise between the aerosol filtration and the comfort of the protective clothing wearer by means of increasing the ventilation rate.

MATHEMATICAL FORMULATION

In order to relate the processing parameters of the electrospinning with the final characteristics of the electrospun nanoweb, different models of different published studies are used. The first model (electrospinning model) predicts the final fiber diameter. The second model (morphological model) is a combination of several equations to predict the porosity and the air permeability of the electrospun nanoweb at a fixed thickness. The third model (aerosol filtration model) predicts the aerosol filtration efficiency of the electrospun nanoweb used as function of the porosity and the fiber diameter predicted from the first and second model. The fourth model (the ventilation model) uses the air permeability to predict the mass flow rate of air entered through the electrospun nanoweb.

1. Electrospinning Model

Figure 1 presents the physical model of electrospinning. A solution of polymer-solvent is pumped to a metal syringe where a voltage (2-20 KV) is applied between the needle and the collector. This voltage creates an electric field that creates in its turn a tangential force on the charges distributed at the jet surface. The tangential force causes the jet to emanate from the tip of a needle. The motion of the fiber is straight for a relatively short distance and then becomes erratic due to bending instability.

While electrospinning is a simple process to perform, the prediction of the final fiber diameter involves a complex interplay between fluid dynamics, electrodynamics, and rheology. Given this complexity, it is difficult to derive a mathematical model for the electrospinning. Therefore, Frifrikh et al. (2003) present a simple expression that relates the final fiber diameter to the influencing parameters. The model prediction arises from a force balance between the surface tension and the electrostatic charge repulsion. The expression is then corrected by the concentration of the polymer in the solution:

$$d = C^{0.5}\left(\gamma\bar{\varepsilon}\frac{Q^2}{I^2}\frac{2}{\pi(2\ln\chi - 3)}\right)^{1/3} \tag{1}$$

Where d is the fiber diameter, C is the polymer concentration in the solution, γ is the surface tension, $\bar{\varepsilon}$ is the air permittivity, Q is the solution flow rate, I is the electrical current, χ is the characteristic length related to the wavelength over the radius of the stable jet region. χ is usually is the range of 10-1000. In this study, χ is taken as 10.

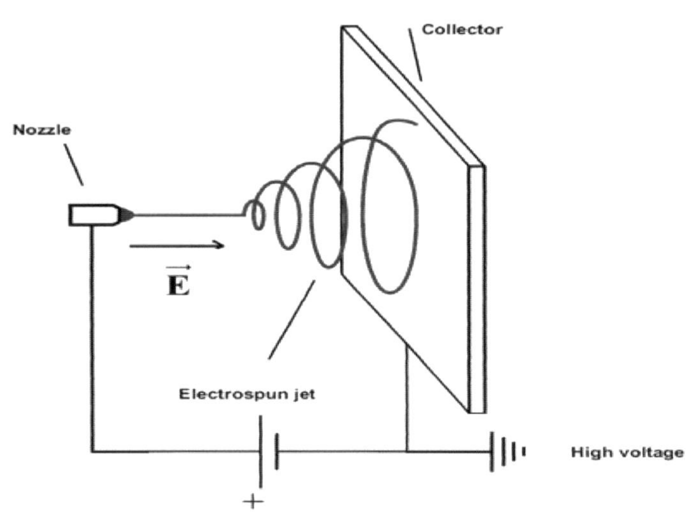

Figure 1: Schematic of the electrospinning process

2. Morphological model

According to Matulevicius et al. (2014), the solidity α is the volume of solids per unit volume of medium. The solidity is given by: (2)

$$\alpha = \frac{w}{\rho_f \bar{z}}$$

Where a is the solidity, w is the basis weight, ρ_f is the fiber density usually used as the polymer density, $\bar{z}$ is the thickness of the web supposed here to be 10^{-3} m.

If we assume that there are no polymer losses and that the solvent is totally evaporated, the basis weight w is related to the electrospinning time te, the concentration of the polymer C, the flow rate Q, the solution density ρ and is given by:

$$w = \frac{\rho C Q t_e}{A_c} \quad (3)$$

The air permeability of the nanoweb is based on the model of Hosseini et al. (2010). The intrinsic permeability (m^2) is given by:

$$k = \frac{3d_f^2}{40\alpha}(-\ln(\alpha) - 0.921) * c_r$$

Where c_r is a correction factor used in order to incorporate the slip effect and is given by:

$$c_r = \frac{(ku + 1.996k_{nf} * (-0.5\ln(\alpha) - 0.25 + 0.25\alpha^2))}{ku * (1 + 1996k_{nf})} \quad (4)$$

Where k_{nf} and ku are respectively the Knudsen and the hydrodynamic factor of Kuwabara flow and are given by:

$$k_{nf} = \frac{2\lambda}{d_f} \quad (5a)$$

$$ku = -0.5\ln(\alpha) - 0.75 + \alpha - 0.25\alpha^2 \quad (5b)$$

Where λ is the mean free path of the gas molecules.

Thus, the air permeability (m/s) is deduced from the intrinsic permeability by multiplying it by the air density ρ_{air} and the gravity g divided by the dynamic viscosity of air η_{air}.

$$AP = \frac{k * \rho_{air} * g}{\eta_{air}} \quad (6)$$

3. Aerosol filtration model

Lee et al. (1982) developed a theoretical study of aerosol filtration by fibrous filters. They found an improved expression for the diffusion and interception filtration efficiencies and validate it with experimental data. This expression, found by summing the diffusion filter efficiency and the interception filter efficiency, is based on a boundary layer approach using the Kuwabara flow field. Combining the diffusion and interception mechanisms efficiencies is based on the assumption that only one mechanism is pre-dominant. For example, in the case of nanoparticles, the predominant mechanism is diffusion because small particles are submitted to random displacement due to Brownian motion which enhances the collision probability with fibers. In the case of microparticles, interception is the predominant mechanism when the microparticle trajectory meets the fiber. The efficiency of the fiber in collecting the aerosol particles is:

$$\eta = 1 - (1.6(\frac{1-\alpha}{ku})^{1/3} Pe^{-2/3} + (\frac{1-\alpha}{ku})\frac{R^2}{1+R}) \quad (7)$$

Where Pe is the pecklet number and is given by the product of the flow velocity U and the fiber diameter d_f over the diffusion coefficient D, R is the interception parameter and it is the ratio of the particle diameter d_p over the fiber diameter d_f. Pe and R are given by:

$$Pe = U d_f / D \quad (8)$$

$$R = d_p / d_f \quad (9)$$

4. Ventilation model

The ventilation model is taken from Ismael et al. (2013) where the human clothed segment is modeled as two concentric cylinders spaced by an air layer as shown in figure 2. The Human segment studied in this section is the human trunk of rasdius R_s covered with a clothing cylinder of radius R_f. The ventilation model is explained in details in the literature. Therefore, the equations will not be repeated in this study. However, the ventilation model will be simplified by one simple equation to show the direct relation between ventilation rate of the human trunk VR, the air permeability AP, and the wind velocity U:

$$VR = f(AP, U) \quad (10)$$

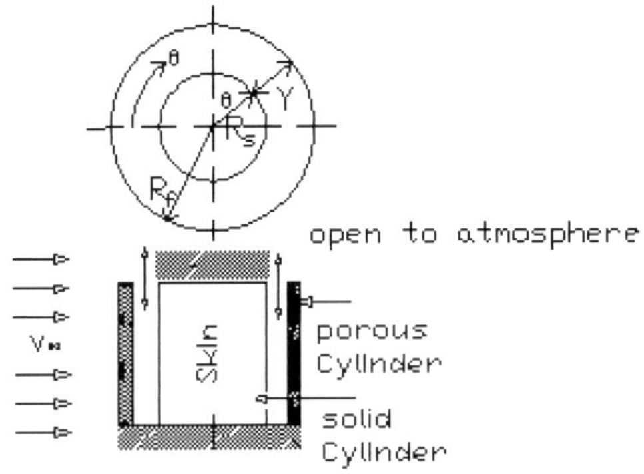

Figure 2: Schematic of the clothed trunk ventilation

RESULTS AND DISCUSSION

The polymer used in this study is the Polyacrylonitrile (Mw= 150,000 g/mol), the solvent is N, N-dimethylformamide (≥99.8%, A.C.S. spectrophotometric grade). Different processing parameters are used in the electrospinning process; therefore, different fiber diameter nanowebs are developed. The porosity, the web density, and the permeability are predicted for each set of processing parameters. From these predicted morphological properties, the filtration efficiency and the ventilation rates are predicted. Table 1 shows the morphological properties of the nanoweb for different electrospinning parameters. However, Table 2 shows the filtration efficiency of the nanoweb and the ventilation rate provided at $U=5cm/s$ and a particle diameter of $0.1\ \mu m$.

Table 1: Morphological properties as function of the processing parameters

	Fiber diameter (nm)	Web density (g/m^2)	Solidity (%)	Permeability (mm/s)
$Q=250m^3/h$, $I=40$ nA, $te=1$ hour	2.16	0.3854	0.03	0.02
$Q=500m^3/h$, $I=56$ nA, $te=1$ hour	272	0.774	0.06	0.0194
$Q=800m^3/h$, $I=71$ nA, $te=1$ hour	319	1.231	0.1041	0.0153
$Q=500m^3/h$,$I=56.2$ nA, $te=2$ hours	272	1.5414	0.1302	0.0087
$Q=800m^3/h$, $I=71$ nA, $te=2$ hours	319	2.466	0.2083	0.0069

Table 2: Filtration efficiency and ventilation rates as function of the processing parameters

	Filtration efficiency (%)	Ventilation rate (l/mn)
$Q=250m^3/h$, $I=40$ nA, $te=1$ hour	87.2	5

Q=500m^3/h, I=56 nA, te=1 hour	89.22	3.4
Q=800m^3/h, I=71 nA, te=1 hour	90.35	2.2
Q=500m^3/h,I=56.2 nA, te=2 hour	88.45	1.9
Q=800m^3/h, I=71 nA, te=2 hour	89.62	1.2

It is shown that the increase of the flow rate of electrospinning leads to an increase in fiber diameter which in its turn contributes to an increase in solidity (Table 1). This increase in solidity increases the filtration efficiency of the electrospun nonoweb (Table 2). In the other hand, an increase in solidity affects the air permeability in an inverse manner (Table 2). An increase in the electrospinning time leads to an increase in the web density which in its turn increases the solidity (Table 1). However, the filtration efficiency does not increase consequently (Table 2). Thus, there are some optimal processing parameters that maximize the filtration efficiency. In the other hand, the increase in web density decreases the air permeability. Thus, it is shown that the nanoweb used allows air to penetrate providing relatively high filtration efficiency which is not the case of the conventional protective clothing because it is usually impermeable.

LIMITATION AND APPLICABILITY

The limitation of the combined model developed in this study is related to the electrospinning process. For instance, the correlation used to predict the final fiber diameter of the electrospun nanoweb (equation 1) is applicable to small polymer concentration (<12%). Furthermore, the collector rotational speed must not exceed an rpm of 600 because this will lead to an additional decrease in the fiber diameter that is not taken into account in equation 1.

This study shows the importance of incorporating the electrospun nanofiber on the protective clothing field in order to provide an acceptable confort level for the wearer as well as good air quality by getting rid of hazardeous toxic particles. The performance of the electrospun nanofiber (filtration and ventilation) could be predicted without any need of performing experiments. The applicability of electrospun nanofiber is usually recommended in toxic environments where the worker is exposed to both heat stress and hazardeous environment.

CONCLUSION

A mathematical model was developed to predict the morphological properties of the electrospun nanoweb as a function of the electropsinning processing parameters. The model was deduced from different published models: electrospinning model, morphological model, aerosol filtration model, and ventilation model. The developed complete model is a predictive tool that helps in optimizing the filtration of aerosol particles and the ventilation through electrospun based protective clothing. This is due to large surface-to-volume ratio and high porosity of electrospun nanoweb which provides an optimization between filtration and ventilation.

ACKNOWLEDGMENT

The authors would like to acknowledge the financial support of the Lebanese National Council for Scientific Research.

REFERENCES

C.P. Carroll CP, YL. Joo, Electrospinning of viscoelastic Boger fluids: Modeling and experiments, Physics of Fluids 18 (2006) 053102.

H.T. Kim, C. H. Jung, S. N. Oh, K. W. Lee, Particle removal efficiency of gravitational wet scrubber considering diffusion, interception, and impaction, Environmental Engineering Science 18 (2001) 125-136.

J. Matulevicius, L. Kliucininkas, D. Martuzevicius, E. Krugly, M. Tichonovas, J. Baltrusaitis, Design and Characterization of Electrospun Polyamide Nanofiber Media for Air Filtration Applications, Journal of nanomaterials (2014).

K. R. Spurny, CM. Jan, Nicolai Albertowich Fuchs: the pioneer of aerosol science: biography. Delft University Press (1998).

K.W. Lee, B. Y. H. Liu, Theoretical study of aerosol filtration by fibrous filters, Aerosol Science and Technology 1.2 (1982) 147-161.

N. Ismael, N. Ghaddar, K. Ghali. Predicting segmental and overall ventilation of ensembles using an integrated bioheat and clothed cylinder ventilation models. Textile Research Journal (2014): 0040517514535868.

S.A. Hosseini, H. Tafreshi, Modeling permeability of 3-D nanofiber media in slip flow regime, Chemical Engineering Science 65 (2010) 2249-2254.

S. A. Nunneley, Heat stress in protective clothing: interactions among physical and physiological factors, Scandinavian journal of work, environment & health (1989) 52-57.

S. Fridrikh, J. Yu, M. Brenner, G. Rutledge, Controlling the fiber diameter during electrospinning, Phys Rev Lett 90 (2003) 144502

S. Ramakrishna, K. Fujihara, W.E. Teo, T. Yong, Z. Ma, R. Ramaseshan, Electrospun nanofibers: solving global issues, Materials today 9 (2006) 40-50.

T. Y. Ling, J. Wang, Y.P. David, Numerical modeling of nanoparticle penetration through personal protective garments, Separation and purification technology 98 (2012) 230-239.

Investigating the Performance of Ceiling Mounted Personalized Ventilation System Assisted by Chair Fans: Assessment of Thermal Comfort and Indoor Air Quality

Bachir El Fil
Student Member ASHRAE

Nesreen Ghaddar, PhD
Member ASHRAE

Kamel Ghali, PhD
Member ASHRAE

ABSTRACT

This study investigates the performance of a ceiling mounted personalized ventilation (PV) system when assisted with chair-mounted fans. Detailed computational fluid dynamics (CFD) simulations were performed to study the velocity, temperature, and CO 2 fields around the microclimate of the occupant. The CFD model was integrated with a validated bio-heat model to determine the corresponding segmental skin temperature and local and overall comfort and sensation. The CFD model was validated experimentally using a thermal manikin in a climatic chamber. Segmental skin temperature, velocity field, and CO2 field were validated experimentally. The predicted values and the measured values showed relatively very good agreement. This study revealed that he chair fans configuration to achieve best thermal comfort and best IAQ occurred at different fan heights. The fan height and flow rate for best thermal comfort were 40 cm above floor level and 10 L/s, respectivelyThe use of chair mounted fans permitted achieving energy savings up to 17% when compared with conventional mixing ventilation system.

INTRODUCTION

Indoor air quality and thermal comfort are two critical factors that researchers focused on because they have a direct influence on the occupants' health and performance. People living in urban regions spend more than 90% of their time in closed indoor spaces (Dockery and Spengler 1981). These indoor spaces must be conditioned to provide the occupants' thermal needs. The thermal environment and air quality in any indoor room is not always optimal for energy conservation goals. Ventilation systems condition the indoor space to provide thermal comfort and to remove contaminants that are normally generated by occupants, building material, or furniture. The trend in current building design is to come up with energy efficient systems for heating, ventilation and air conditioning (HVAC). Localization of thermal comfort and indoor air quality is only viable via complex systems. Confining thermal comfort and air quality tends to reduce energy consumption because it directly supplies conditioned air to the breathing zone of the occupant rather than having it supplied via diffusers in the conditioned space. For instance, personalized task ventilator systems are commonly used in office spaces since this system does not need any exposed ducting (Cermak et al. 2006a). However these technologies are expensive, their payback period is uncertain, and it is difficult to retrofit these systems to existing air distribution units. Bolashikov et al. (2009) showed that the use of desk mounted fans was successfully able to control the convective plume arising from the occupant and therefore delivering more fresh air horizontally. To solve the problem of the ceiling-mounted PV system, Makhoul et al. (2012a) proposed a ceiling mounted coaxial nozzle where the primary nozzle supplies fresh air surrounded by concentric secondary nozzle supplying return air at the same velocity. Makhoul et al. (2012) used desk mounted fans to enhance the effectiveness of PV system as was proposed by Yang (2009). The desk fans provided a controlling mechanism of the convective flow around the occupant thus hindering free convection and showed good performance. However, the success of the desk fans in suppressing the human thermal plumes is

Bachir El Fil was a Master's student at the American University of Beirut Dr..Nesreen Ghaddar and Dr. Kamel Abou Ghali are professors of mechanical engineering at the American University of Beirut.

limited. It is effective only when the occupant is close to the desk. Chair fans offer a better solution since they are attached to the chair, and therefore they will pull down the thermal plumes regardless of the person's position.

Previous researchers have tackled the importance of chair fans for an individually controlled comfort (Watanabe et al. 2009). Moreover, Habchi et al. (2015) investigated empirically the effect of chair fans on aiding displacement ventilation (DV) system which posed less challenge than the case of the downward PV jet. No work has been reported in the literature that addresses the optimized use of chair fans to enhance the performance of the ceiling mounted PV system.

Nonetheless, when it comes to implementing chair fans to assist the single jet PV system, two parameters play a major role in determining the local comfort, sensation, and ventilation effectiveness at the breathing zone of the occupant. These parameters are (i) the elevation of the fans i.e. height at which they are fixed on the chair and (ii) the flow rate at which the air is being sucked by the fan. Therefore, the objective of this work is to investigate the height and the total flow rate of the chair fans resulting in the good of thermal comfort and indoor air quality. This paper investigates by modeling and experimentation the enhancement in the performance of ceiling mounted single jet PV nozzle placed at the center of the supply diffuser when assisted by chair fans for controlling the natural plumes generated by human body.

SYSTEM DESCRIPTION

The localized HVAC system under consideration is depicted in Figure 1 and is composed of an air distribution system represented by a ceiling-mounted personalized ventilation system. Directly beneath the PV, an office chair with fans is located. The ceiling-mounted personalized ventilation system consists of a single nozzle that supplies fresh air to provide the occupant with breathable air of high quality, and four peripheral diffusers that supply recirculated air at an angle of 45°. These angled diffusers permit the formation of a canopy effect around the occupant, thus helping in confining the flow. The canopy allows maintaining a desired air temperature in the microclimate of the occupant lower than that of the macroclimate air temperature. In addition, the chair mounted fans are placed on each side supplying a uniform flow rate. The fans have a single degree of freedom, in which they are allowed to move up-and-down vertically. Only two parameters are considered for this study, the fan flow rate (Q_f) and the fan height (H). On the other hand, the carbon dioxide gas, which is used as a tracer gas in this study, is injected to the room through a tube to mimic a typical CO_2 flow rate generated in an office space.

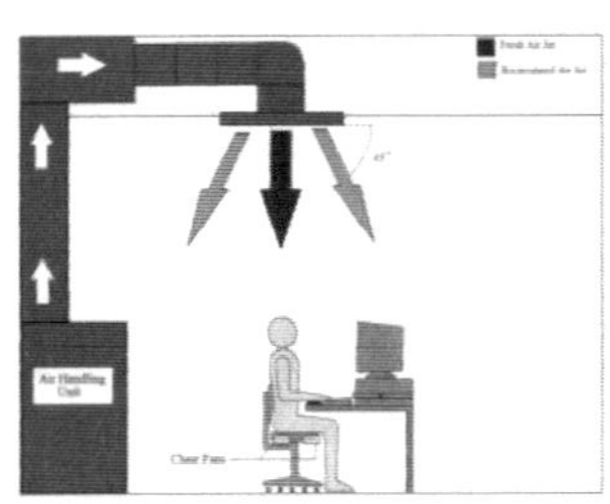

Figure 1: Schematic diagram of the proposed localized PV system

METHODOLOGY

The objective of the proposed fan-equipped chair combined with ceiling mounted personalized ventilation system is to control the convective flow resulting from the rising plumes of the occupant and to find an optimal height and flow rate for the chair fans to ensure good indoor air quality and thermal comfort at minimum energy consumption. The relatively complex interaction occurring between the rising thermal plumes and the ceiling mounted PV jet acting downwards rendered the modeling process more complicated, therefore the CFD simulation is used to capture the intricate interaction between the rising plumes, PV downward jet, and the chair fans. Therefore, to assess the performance of the proposed system in maintaining localized thermal comfort and air quality, 3-D CFD simulations are performed using the commercial software ANSYS Fluent©. The 3-D CFD model is integrated with a validated bio-heat model (Salloum et al. 2007; Ghaddar et al. 2011). The bio heat model is used mainly to assess localized thermal comfort and sensation that are attained by the occupant under specific PV parameters. Salloum et al. (2007) bio-heat model and Zhang et al. (2010a, 2010b, 2010c) comfort models are integrated in the CFD model to predict the segmental skin temperature for the CFD model and the segmental comfort and sensation. The findings of the CFD simulations are validated experimentally. The experiments are conducted on a thermal manikin seated under the PV diffuser in the center of the room.

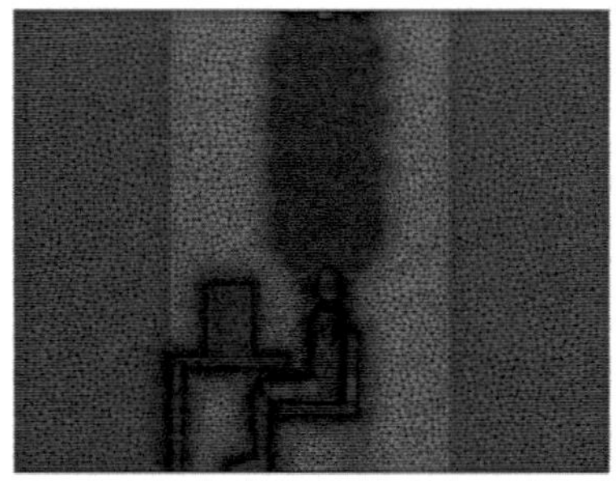

Figure 2: The mesh of the computational domain

CFD model

Coupling the PV system with chair fans induces complex flow interaction between the air-jets present in the room. The fans are sucking the rising thermal plume, while the nozzle is supplying a vertical jet of fresh air to the office space. CFD modeling has shown to be very effective in capturing the physics of complex flow problems; moreover CFD modeling has shown good prediction of the air flow pattern and species distribution in the space for different HVAC configurations (Makhoul et al. 2013; Habchi et al. 2015). Hence CFD is a suitable tool to investigate the performance of the ventilation system in an indoor environment, and it is used in this work to study the performance of coupling the PV system with chair fans. Moreover, since temperature gradients in the conditioned space are not high, radiation is neglected in this study for its insignificant effect.

Airflow modeling

A detailed 3D-CFD model is developed to accurately predict the air entrainment by the rising convective thermal plume and their interaction with suction jets due to chair fans. To attain robust CFD results, it is essential to model the physics of the flow i.e. the turbulence, buoyancy, and boundary layers near surfaces. Moreover, a fine grid is developed near the surfaces to accurately capture the shear-layer entrainment, and to capture all the thermal plumes present in the space. A tetrahedral unstructured mesh (as shown in Figure 2) is generated using an element size of 2 cm on the surfaces. In addition, spheres of influence above the occupant are created to capture the velocity boundary layers and the rising plume. The air in the room is modeled as a continuous fluid hence it is appropriate to simulate using the Eulerian approach. There are plenty of different turbulence models available in ANSYS Fluent; each of them solves the Navier Stock equations by modeling them in a specific way. Some turbulence modeling methods are quite advanced (such as the large eddy simulation (LES)); however, the computational cost of such methods is really expensive. One of the widely used turbulence models is the realizable k-ε turbulence which was also selected in addition to the enhanced wall treatment option. The realizable k-ε model correctly predicts the flow having round jets, circulations, and separation (Makhoul et al.). The ideal gas law is used to model the density variation when it comes to account for buoyancy. The CFD software discretizes the Navier-Stokes equations; the discretization scheme selected for this work is the second order upwind scheme to solve for the momentum, k, ε, and energy equations. The Pressure Staggering Options (PRESTO!) scheme is used for the pressure (Makhoul et al. 2013). For the velocity and pressure coupling the Semi-Implicit Method for Pressure

Linked Equations (SIMPLE) algorithm is found most appropriate for this work. For airflow modeling, the supply velocity that is normal to the outlet of the nozzle, the turbulence intensity and velocity are specified. The exhaust is modeled as a pressure outlet with zero gage pressure. For the chair fans, the "fan" boundary condition is imposed at the inlet of the chair fan at a specified pressure jump (Makhoul et al. 2013; Habchi et al. 2015). Thermally, heat fluxes are imposed on the walls and the ceiling to mimic a real office internal load. Different occupant activities impose different metabolic rates, thus having an effect on the rising thermal plume, for this current work; in an office space the occupant is assumed to be performing "light work". Carbon dioxide (CO_2) is used as tracer gas to assess for indoor air quality. The tracer gas is injected in the ceiling peripheral diffusers to simulate recirculated air; moreover a constant dose of CO_2 was injected to the room to mimic a typical CO_2 release in an office. Besides, the under relaxation factors are carefully manipulated to dampen the scaled residuals to speed up convergence.

The described mesh treatment resulted in a total number of elements equal to 2,830,080 elements for the entire computational domain. Numerical convergence is judged using mainly three criteria: (1) scaled residuals reported by Fluent should reach a minimum of the order 10^{-5}; however convergence is not reached until (2) the net heat flux of the computational domain is below 1% of the total heat gain and (3) the CO_2 concentration at the breathing zone stabilizes.

Air quality modeling

Air is modeled as a continuous fluid, and the Eulerian approach is adopted. To investigate the level of mixing between the supplied air and the stagnant air in the room (contaminated air), the ventilation effectiveness is assessed using a tracer gas, which in this study is carbon dioxide gas. The fresh air nozzle in the ceiling has a CO_2 concentration of a mean value of 400 ± 50 ppm, a typical value of outdoor CO_2 concentration. It is important to note that the amount of fresh air coming in to the room is equal to the amount of air leaving the room through the exhaust; this way no CO_2 accumulation remains in the room. A typical CO_2 emission of 0.31 L/min/person is used in an office space (Dougan et al. 2004). Ventilation effectiveness is defined as ε_v as follows:

$$\varepsilon_v = \frac{C_R - C_B}{C_R - C_F} \tag{1}$$

where C_R is the CO_2 concentration of the return or recirculated air, C_B is the CO_2 concentration at the breathing zone of the manikin, and C_F is the CO_2 concentration in the fresh air supplied to the workspace. Typically all concentrations are measured in ppm. The CO_2 concentration of the air supplied by the primary nozzle is in the range of 300-500 ppm according to the ANSI/ASHRAE Standard 62.1-2013 (Shell et al. 2001; ANSI/ASHRAE 62.1-2013). The equation above refers to the Air Quality Index (AQI) at the breathing zone of the manikin. The breathing zone is defined to be a sphere of radius 1 cm located at a distance of 2.5 cm from the nose of the manikin. The near head temperature and velocity are defined to be at a distance of 5 cm offset from the head of the occupant. The species transport model is enabled to help calculating the carbon dioxide concentration field, thus assessing the indoor air quality. Moreover, the turbulent Schmidt number is defined by being proportional to the ratio of viscous diffusion rate over the mass diffusion rate. The value adopted was 0.95.

Bio-heat Model

The CFD model is coupled with a validated bio-heat model to calculate the skin temperature of the thermal manikin at each segment of its body. The model is coupled with Zhang's thermal comfort models mainly to assess the local comfort based on the resulting segmental temperature (Zhang et al. 2010a, 2010b, 2010c). The bio-heat model divides the thermal manikin into eleven segments. The segmental parts are: head, chest, abdomen, back, buttocks, claves, thighs, upper arms, lower arms, hands, and feet. The physiological response is simulated at each segment to predict the thermal comfort based on Zhang's model of comfort (2010a, 2010b, 2010c). At first, the skin temperature is chosen to be that of a typical human being skin temperature in a "thermo-neutral" condition, then the CFD simulation runs until convergence is reached. From the CFD model, the air temperature at the vicinity of segmental body parts and the segmental heat transfer coefficient are obtained

Experimental Protocol

The experimental setup was constructed in a controlled climatic chamber ventilated by a ceiling mounted personalized ventilation system. Primarily, the experiment aimed to validate the CFD model in terms of indoor air quality and thermal comfort. The validation consists of comparing the measured and predicted segmental skin temperatures, velocity and temperature near head of the occupant, and ventilation effectiveness. The climatic chamber of dimensions 3.4 m × 3.4 m × 2.8 m is used in the experiments and is constructed of highly insulated ceiling and floor and has an internal layout of a typical office. The air distribution system consists of a single nozzle 9 cm in diameter that supplies fresh air, and four peripheral diffusers that supply recirculated air at an angle of 45°. This angled diffuser flow will aid in creating a canopy effect around the occupant, thus aiding in the localization of the flow.

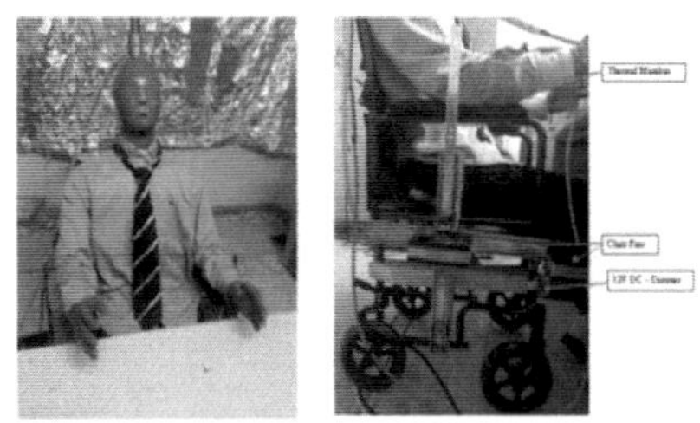

(a) (b)

Figure 3: Photos of (a) the thermal manikin used in the experiment and (b) the chair fan mechanism

The thermal manikin (shown in Figure 3) has its own software whereby it is coupled to maintain a constant core temperature while its skin temperature and heat flux generated vary with different ambient conditions. Prior to every experiment, the manikin must be initialized to a typical human body in neutral conditions. The thermal manikin is dressed with a typical office outfit which is a long sleeved shirt, and pants. The clothes worn by the manikin have an overall resistance of 0.6 clo, which is the typical clothing resistance in an office. A desktop computer generating 93 W (Makhoul et al. 2013) is placed on the desk in front of the manikin, as represented in Figure 1. The return duct was seeded with 0.0035 L/s (Makhoul et al. 2013) flow rate of carbon dioxide to represent return air. A constant CO_2 dose of 0.31 L/min/occupant is used in the experiment to mimic a typical CO_2 office load that is emitted from equipment, furniture, and occupants (Dougan et al. 2004). The thermal plume generated by the thermal manikin produces nearly 75 W, in addition to the lighting load that was almost 10 W/m^2. The air conditioning of the space is provided via two Honeywell 2.64 kW air handling units. One of the air handling units (AHU) supplies fresh air to the PV nozzle, while the other AHU supplies return air to the peripheral diffusers. The experimental setup allows variation in fans' height and flow rate. The fan mechanism allowed variation of the height ranging between 30 cm to 50 cm above the ground. On the other hand, the flow rate of the fan is controlled by a 12 V DC-dimmer. The total flow rate of the fans can vary between 5 L/s to 15 L/s.

On the other hand, air quality is measured by taking CO_2 samples and measuring their CO_2 concentration using FIGARO CDM4161- CO_2 sensors with an accuracy of ± 20 ppm. Six CO_2 sensors are installed and a set of 11 K-type thermocouples having an accuracy of ± 0.5 °C manufactured by Omega are distributed around the body parts of the thermal manikin. For model validation, four experiments are conducted. All experiments are conducted while maintaining the room temperature (temperature of the macroclimate) at 26 ± 0.5 °C. To ensure steady state conditions, each experiment is run at least three hours, and the data is recorded when the readings of CO_2 concentration, temperature fields, and velocity stabilize.

Parametric Study

The validated CFD model is applied to a typical office of dimensions 3.4 m × 3.4 m × 2.6 m. The parametric study serves to find the optimal combination of fan flow rate and height that results in the best ventilation effectiveness and thermal comfort. In the parametric study, the total load in the space is kept similar to a typical office load and is equal to 60 W/m^2. The fresh flow rate is set to the newly updated ASHRAE recommendation of 8.5 L/s (ANSI/ASHRAE 62.1-2013; ASHRAE 2009). The temperatures of the fresh air and the recirculated air supplied by the peripheral diffusers are set to be 16 °C while return room air is set at 26 °C. For each simulation, the room is divided into two zones: macroclimate and microclimate. The fan height was studied in between range of 30cm and up to 50cm above the floor. The main function of the chair fan is to pull down the occupant thermal plume and to aid the downward travel of the fresh air to enhance the breathing air quality while avoiding possible thermal draft. Therefore, it is important not to place it lower than 30 cm because the occupant thermal plume would not have been fully developed and it will be away from the fresh air jet. Although a fan height above 50 cm will improve the breathing air quality, but it will induce thermal draft due to higher air velocities would occur close to the head. Table 1 presents all the simulation cases that are conducted to find the optimal height and flow rate.

Table 1: Simulation Cases

Simulation Cases	Fan Height (cm)	Total Fan Flow rate (L/s)
Set 1	30	5, 10, 15
Set 2	40	5, 10, 15
Set 3	50	5, 10, 15

RESULTS AND DISCUSSION

CFD validation

Since the CFD model is coupled with the bio-heat model, it is necessary to validate the predicted results capture the human-response to the localized environment in a PV + CF system. To validate the CFD model, the segmental skin temperature, the near head temperature and velocity, and the ventilation effectiveness are measured experimentally, and they are compared to those predicted by the numerical model. It is also important to validate the flow field from the PV jet due to the use of the momentum method for modeling the diffuser's jet. Four experiments are conducted at different fan elevations at 30 cm, 40, and 50 cm in addition to having chair fans turned OFF. The predicted segmental skin temperature and ventilation effectiveness are compared to those measured experimentally. All experiments reported good agreement between the predicted data and the measured values.

Parametric Study

Air quality comparison

In this section, the effect of fan height and flow rate on ventilation effectiveness is studied. The low temperature of the supplied PV jet has a positive effect on the air quality of the inhaled air due to the buoyancy forces generated by the cold jet that is acting downwards. The ventilation effectiveness at three different heights having various fan flow rates ranging from 5 L/ to 15 L/s. It is clear that as the fan level increases while maintaining relatively low fan flow rates, ventilation effectiveness increases remarkably. However when fan flow rates exceed 10 L/s a counter effect on the ventilation effectiveness is observed due to mixing effect of recirculated air. This is illustrated by comparing between the effectiveness at a total fan flow rate of 10 L/s and 15 L/s at every fan height. It is important to note that the mixing effect is dominant at high fan levels due to the fact that at relatively high fan level, the thermal plumes generated by the occupant (at trunk level of the body) tend to be stronger (i.e. having higher velocities). This will generate a recirculated flow from the contaminated region to the microclimate region of the occupant driven by the plumes' higher upward velocity. Therefore, the fans' high flow rate aids in increasing the mixing effect. Nonetheless, it is important to note that despite this "mixing" effect, the ventilation effectiveness is 18.16% at fan level of 50 with a total fan flow rate of 15 L/s and this is almost double the effectiveness when chair fans are turned OFF (9.31%). The results are tabulated in Table 2.

Table 2: Comparison of the ventilation effectiveness PV+CF

Total Fan flow rate (L/s)	Fan Height = 30 cm	Fan Height = 40 cm	Fan Height = 50 cm
5	11.84	12.45	15.49
10	13.67	21.69	23.39
15	10.45	17.39	15.94

Thermal comfort comparison and Energy Analysis

Thermal comfort of the occupant is directly affected by the height and flow rate of the chair mounted fans. Figure 5 shows the perceived thermal comfort and ventilation effectiveness by the occupant when varying fan heights and flow rates. When comparing the effect of the chair mounted fans on the induced thermal comfort, a net enhancement is attained when the chair fans are turned ON. It is noticed that as the fan level increases while maintaining low fan flow rates, thermal comfort also increases. Conversely at total fan flow rate of 10 L/s, the perceived thermal comfort was 0.76, 1.15, and 0.96 at fan height of 30 cm, 40 cm, and 50 cm respectively. It is noticed that at high fan level (50 cm) and high fan flow rate (~ 10 L/s) the thermal comfort decreases mainly due to the thermal draft caused by the cold sensation at the head. Figure 6 shows the contours of the CFD plots of the simulation case at fan level of 40 cm.

Energy analysis

In order to calculate the energy savings associated with the option of assisting the ceiling mounted PV system with chair fans, an energy analysis is performed to estimate the cooling load for three selected cases: Fan level 50 cm at a total fan flow rate of 10 L/s (highest attained IAQ) and Fan level 40 cm, at a total fan flow rate of 10 L/s (highest perceived thermal comfort). However in order to achieve the same indoor air quality, the equivalent amount of fresh air is used in the calculations. For the energy analysis calculations, it is assumed that the outdoor ambient air has a dry bulb temperature of 32 °C and 75% relative humidity. Therefore, the fresh air needs to be dehumidified and sensibly heated to the desired conditions. The fan of the PV system has a power of 15 W, and the nominal power of the chair fans is 20 W. These nominal powers are then added to the electrical cost of the chiller. A ty

pical chiller COP of 3.5 was selected for energy calculations. The energy savings were up to 17%.

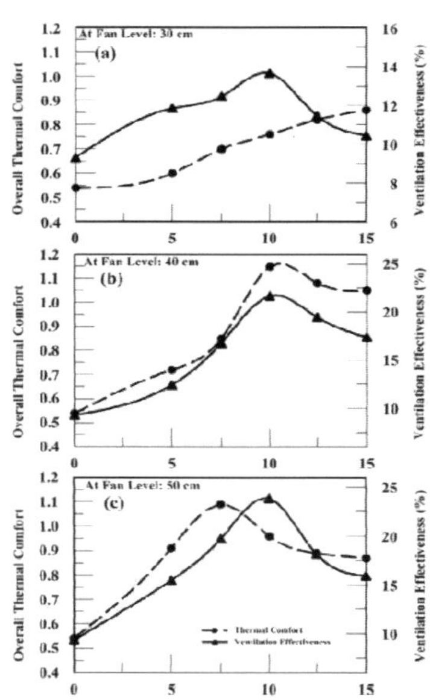

Figure 5: Overall thermal comfort and Ventilation effectiveness as function of the total fan flow rate at fan level

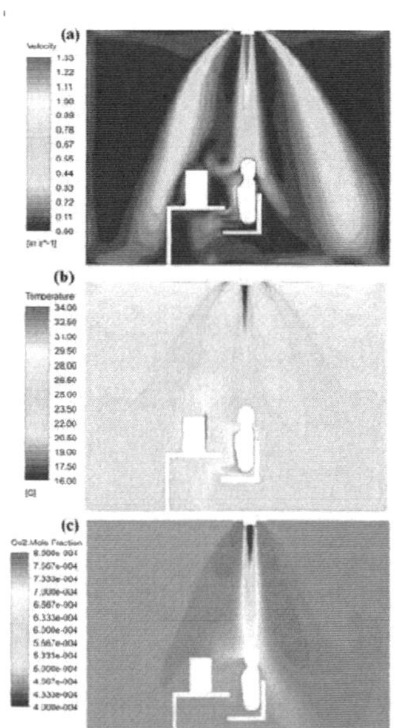

Figure 6: displays the contour plots of (a) velocity, (b) temperature, (c) CO2 concentration fields

Conclusion

Two main criteria are to be considered in every energy efficient-healthy building design, (i) providing good indoor air quality and (ii) maintaining good thermal comfort. In addition to these parameters, the proposed system tends to decrease the electrical cost. Ceiling mounted personalized systems are considered one of the promising solutions to solve ducting problems, chair fans have proved to be very effective in enhancing the performance of the ceiling mounted PV system. Installing chair fans to control the human convective thermal plume has shown a remarkable enhancement in terms of indoor air quality and thermal comfort. When operating the chair fans, the ventilation effectiveness (23.39%) increased by almost 2.5 times that when chair fans were turned OFF (9.31%). The fan height and fan flow rate have a dual effect on the thermal comfort and indoor air quality, where the best IAQ is achieved at different fan configuration than that of thermal comfort. A peak in energy savings is achieved at the best IAQ case reaching 17% when compared with mixing ventilation.

REFERENCES

ANSI/ASHRAE. 2013. Standard 62.1, *Ventilation for Acceptable Indoor Air Quality*. Atlanta: American Society of Heating, Air- Conditioning and Refrigeration Engineers.

ANSYS Software: ANSYS Inc. http://www.ansys.com/..

ASHRAE. 2009. Psychometrics. In 2009 *ASHRAE Handbook of Fundamentals*. Atlanta. American Society of Heating, Air- Conditioning and Refrigeration Engineers

Bolashikov, Z.D., A.K. Melikov, and M. Krenek, 2009. Improved performance of personalized ventilation by control of the convection flow around occupant body. *ASHRAE Transactions* 115(2), Paper ID: LO-09-038.

Cermak, R. and A.K. Melikov, 2006. Air quality and thermal comfort in an office with underfloor, mixing and displacement ventilation. *International Journal of Ventilation* 5(3): 323–32.

Dougan, D. and L. Damiano, 2004. CO_2-Based demand control ventilation. *ASHRAE Journal* 47:52-54.

Epoxy Thermal Manikin. Instruments for Textile & Biophysical Testing. Seattle, WA, [Online]. Available: http://www.thermetrics.com/sites/default/files/product_brochures/NEWTON%20Manikin_Spec%20Sheet.pdf. Accessed on July 10, 2015.

Ghaddar, N., K. Ghali, and S. Chehaitly, 2011. Assessing thermal comfort of active people in transitional spaces in presence of air movement. *Energy and Buildings* 43(10):2832-2842.

Habchi, C., K. Ghali, N. Ghaddar, and A. Shihadeh, 2015. Chair fan fan-enhanced displacement ventilation for high IAQ: Effects on particle inhalation and stratification height. *Building and Environment* 84: 68-79

Kulkarni, V., N. Sahoo, and S. Chavan, 2011.Simulation of honeycomb screen combinations for turbulence management in a subsonic wind tunnel. *J Wind Eng Indust Aerodyn* 99:37-45.

Makhoul, A., K. Ghali, N. Ghaddar, and W. Chakroun, 2013a .Investigation of particle transport in offices equipped with ceiling-mounted personalized ventilators. *Building and Environment* 63:97-107.

Makhoul, A., K. Ghali, and N. Ghaddar, 2013. Low-mixing coaxial nozzle for effective personalized ventilation. *Indoor and Built Environment* 24 (2): 225-243.

Makhoul, A., K. Ghali, and N. Ghaddar, 2012. The energy saving potential and the associated thermal comfort of displacement ventilation systems assisted by personalized ventilation. *Indoor Built Environ* 22:508-19.

Makhoul, A, K. Ghali, and N. Ghaddar, 2012. Desk fans for the control of the convection flow around occupants using ceilings mounted personalized ventilation. *Building and Environment* 59: 336-348.

Salloum, M., N. Ghaddar, and K. Ghali, 2007. A new transient bioheat model of the human body and its integration to clothing models. *International Journal of Thermal Sciences* 46: 371-84.

Watanabe, S., T. Shimomura, and H. Miyazaki, 2009. Thermal evaluation of a chair with fans as an individually controlled system. *Building and Environment* 44:1392-8.

Yang, B., C. Sekhar, and A.K. Melikov, 2010. Ceiling mounted personalized ventilation system in hot and humid climate - An energy analysis. *Energy and Buildings* 42:2304-2308.

Zhang, H., E. Arens, C. Huizenga, and T. Han, 2010a. Thermal sensation and comfort models for non-uniform and transient environments: Part I: Local sensation of individual body parts. *Building and Environment* 45(2):380-8.

From Moderate Mediterranean Climate to Extreme Gulf Conditions

Heba Nazer **Ayoub Abu Dayyeh, Ph.D**

ABSTRACT

This paper will attempt to present a brief introduction of the Fawaliyya Residence; being a sustainable building of zero-carbon emission and an important pilot project in the overall green economy of construction in Jordan. It will explain how the Fawaliyya Residence was built in a green and sustainable manner; discussing orientation of structure with respect to passive architectural design, thermal insulation techniques, innovative solutions applied to reduce air infiltration, introduction of solar thermal hollow circular tube technology, photo-voltaic panels for electricity production, grey water management, as well as rainwater harvesting techniques.

Furthermore, the authors will offer revised experiential reflections based on on-site monitoring for the residence to compare with the theoretical results in the original assumptions of these strategies. Thereafter, examining the posibility of applying the design criteria learnt from moderate climate conditions of the Eastern Mediterranean to extreme conditions, particularly that of the Gulf Region. This can hopefully be done through passive design, improving thermal conditions, orientation of elevations, flexibility of single façade units and energy management. Finally, a conclusion would attempt discussing the implications of our findings on similar real-life situations as a guideline approach to adopt when designing buildings in such climates. This will further reduce carbon emissions from the electro-mechanical heating and cooling loads, and result in improved energy efficiency, thermal comfort and savings in energy bills and in operation and maintenance.

Keywords: Cave Effect, Cold Joints, Decrement factor, Extruded Polystrene, Foam Concrete, Green Buildings, Passive Design, Thermal Comfort, Thermal Mass, Time lag ,Trombe Wall, Mean Radient Temperature.

INTRODUCTION

Due to limited local energy resources and earlier dependency on subsidized imports of oil from Iraq and gas from Egypt, the country of Jordan ended up with 97 % of energy imported in 2013 at market prices which constituted almost 21% of the country's GDP in 2012 (Nepco annual report, 2014). This issue, as well as the ever increasing demand of energy coupled with the remarkable growth in population from foreign workers and refugees entering the country during the past several years, have all impacted the economy. The household sector in Jordan represents the foremost segment of the energy consumption amongst others with 43% existing electricity consumption and total share of primary energy basket equals to 20% (Nepco annual report, 2014). More pressure on the construction sector has raised the need to provide 200.000 additional dwellings till the end of 2015 (Al-Momani, 2008). Hence, it is essential to insure that these future constructions are built to provide occupants with satisfying thermal comfort with least environmental impact through feasible solutions that reduce excessive energy consumption of fossil fuel in

Heba Nazer is a faculty member at the Middle East Universiy. Amman, Jordan, nazerheba@gmail.com. **Ayoub Abu Dayyeh** is head of AD Chartered Energy and Green Buildings Consultant, Ayoub101@hotmail.com, Amman, Jordan.

response to the late COP21 Paris agreement on climate change in December 2015.

Jordan's climate is known for its moderate temperatures that can drop in winter a few degrees below freezing, mainly in January, whereas in summer it might surge up to 42°C on rare occasions. Rainfall is on average 400 – 500 mm annually in that region.

This research will introduce the Kamaliyya Residence, in Amman, the capital of Jordan, as a representative building that assures reduction in heating and cooling loads in buildings through providing comfortable thermal environments. The Kamaliyya residence has won the Golden prize for the Built Environment in the Middle East 2010, followed by the "Green Hero Award" from the Green Apple organization in UK 2011, as well as the appreciation emblem of the Jordanian Engineering Association in 2012. It consists of two floors, 300 m^2 each with a 60 m^2 supplementary roof space for services. The ground floor which is presented in figure 1 comprises of an open dining, a guest room, two bedrooms, one kitchen, and a bathroom. The architectural design of the house integrated various passive strategies with green sustainable technologies such as photovoltaic solar cells, geothermal energy, water solar thermal panels, and rain water harvesting which resulted with a zero carbon emission house.

(a) (b)

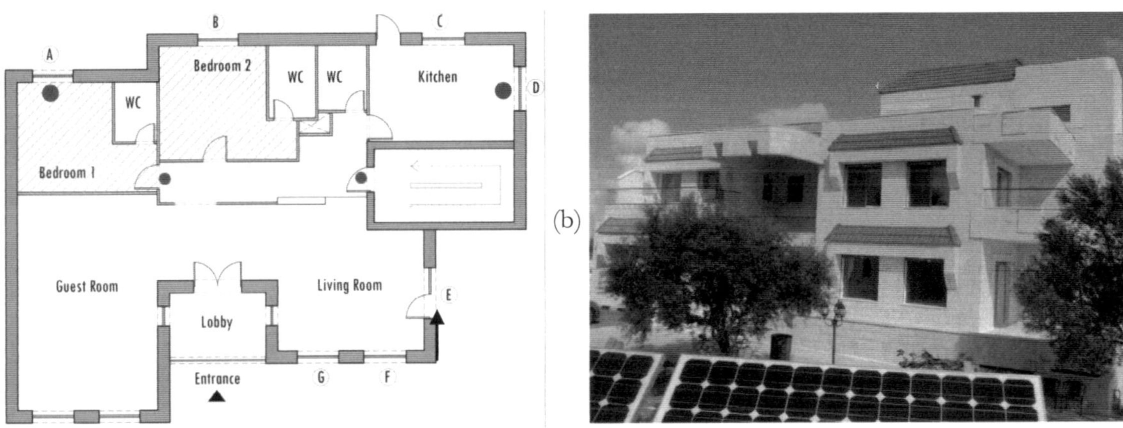

Figure 1 (a) Ground Floor Plan and (b) The Exterior of the House

PASSIVE STRATEGIES

The building was designed with great considerations to energy and water efficiency. It was built from local building materials, minimum GHGs production and at minimal environmental damage. The passive design strategies assumed were incorporated in the building and monitored afterwards to insure the compliance with environmental requirements and are divided as follow:

Orientation

Orientation is vital in passive design because it incorporates active systems, renewable energy systems and affects the amount of solar gain entering the building from exterior façades. Windows' surfaces should be carefully designed because they absorb, reflect, shade and radiate a significant amount of solar radiation (both light waves and heat waves-infrared), which can cause a rise in the temperature of the inside glazing surface to above (60°C) in summer conditions. The orientation of the structure in the Kamaliyya Residence was passively designed so as the longer elevation was extending in the direction East - West and generally facing south to absorb as much heat from the sun as possible. Openings on the west façade were eliminated to reduce air infiltration from exterior openings facing the predominantly strong winds from the west. Aditionally, all other windows and openings on the southern elevation were shaded enough to obstruct direct sunrays in peak hours as seen in figure 1(b).

To understand the orientation effect on glazing surface temperatures, all glazings were monitored on a warm winter day (February 14th, 2014) and outlined in Figure 2 and Table 1.

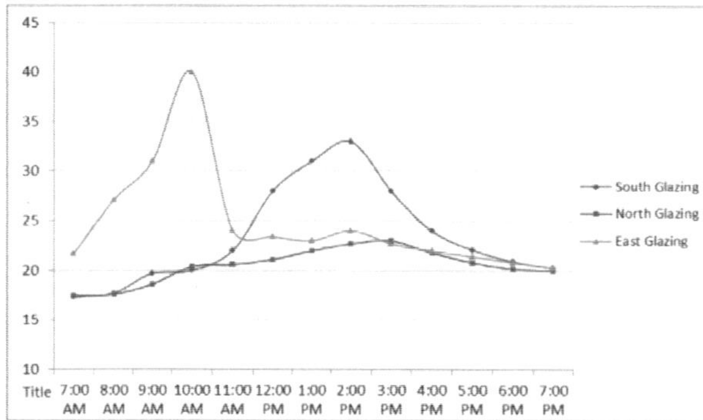

Figure 2 Heat Gain on Glazed Walls in February (Monitored Temperatures (°C))

Table 1. Monitored Temperatures on Glazed Walls

TIME	SOUTH	EAST	NORTH
7:00 am	17.3	21.7	17.5
8:00 am	17.7	27.1	17.6
9:00 am	19.7	31	18.6
10:00 am	20	40	20.4
11:00am	22	24	20.6
12:00pm	28	23.4	21.1
1:00pm	31	23	22
2:00pm	33	24	22.7
3:00pm	28	22.7	23
4:00pm	24	22	21.8
5:00pm	22.1	21.4	20.8
6:00pm	20.9	20.8	20.2
7:00pm	20.3	20.3	20

As presented in Figure 2 and Table 1 above, the northern elevation hardly has any differences in the glazing surface temperatures throughout the day because it lacks exposure to the direct sun path; therefore, shading is not essential there. However, the average increase in surface temperatures of the window`s glazing on the east elevation in winter during 4 hours (7 am-11 am), compared to 21.7 °C as the baseline reading at 7am, was: ((27.1-21.7) + (31- 21.7) + (40-21.7) + (24-21.7)) ÷ 4 = 8.8°C/hour.

On the southern orientation, the average increase in surface temperatures of the window`s glazing during 5 hours (11am – 4 pm), compared to 20 °C as baseline temperatures were 9 °C/hour in winter.

We conclude that both eastern and southern facades are important in energy gain in winter. The direct impact of the Sun on the eastern façade was surprisingly higher, although it persisted for a slightly shorter period of time compared to the south (40 °C in comparison to 33 °C), due to the incident angle of the Sun which is rather perpendicular in the east early in the morning. However, the incident angle of the Sun's rays on the southern façade would make an angle of 35 to 40° to the horizon in February (Amman has a latitude of 31° 57' north). This emphasizes the important role of designing the house in such a way to allow direct solar radiation through exterior openings in winter, as it heats up the glazed surfaces which in turn radiates heat inside the internal spaces. Furthermore, the Sun`s angle becomes much higher in summer (81 °C) on June 21st at noon time which warns designers in the Gulf Region and other hot areas from over-heating potentials of glass.

This phenomenon enlightens architects to thoroughly analyse the design of shading devices used to obstruct direct solar radiation entering the buildings in summer. In addition, the consideration of pyramid or trapizodal shapes in buildings can be a viable approach since solar radiation is partially reflected on inclined surfaces compared to vertical ones, which will consequently decrease the amount of absorbed heat gain into the building (AIA, 2012). This can reduce the load of air conditioning in the Gulf Region which is consuming 70% of the electricity production.

Building Envelope and Thermal Comfort

The most important element in hot climates is solar radiation, heat gain on the building envelope and the subsequent release of heat by radiation from the interior surfaces towards occupants (Koch, 2013). However, in cold climates, this is controlled by the adequate design of thermal mass in the building's envelope which involves materials that can be employed to store heat during the day and release it at night due to their high thermal inertia (trombe walls for example).

The main criterion adopted in the Kamaliyya house was the improvement of the building fabric as the most important factor that affects thermal performance. Although it was built using similar construction techniques as in the majority of dwellings in Jordan, the incorporation of thick layers of insulation have improved the thermal resistance of the building envelope. The roofs were covered with foam concrete of $600kg/m^3$ density and an average thickness of 15cm to cater for the cold joints of the R.C ribs. The insulation thickness in the walls was doubled with respect to Jordan`s Codes to 10cm thickness of thermal conductivity equal to 0.035 W/m.k, while the roof constituted of 25cm of extruded polystyrene ribs. This enabled the building to surpass the actual U-values required by the Jordanian Thermal Insulation Code with U-value of the roof equal to $0.2W/m^2K$ (code 0.55) and $0.13W/m^2K$ (code 0.57) for the walls (Abu Dayyeh, 2013).

In order to evaluate the effect of the insulation in preserving stable comfort temperatures in the interior of the house, the ambient temperatures and internal wall surfaces were monitored on a hot day in summer (figure 3a, table 2) and on a Winter day on February 9th 2014 (figure 3b, table 2).

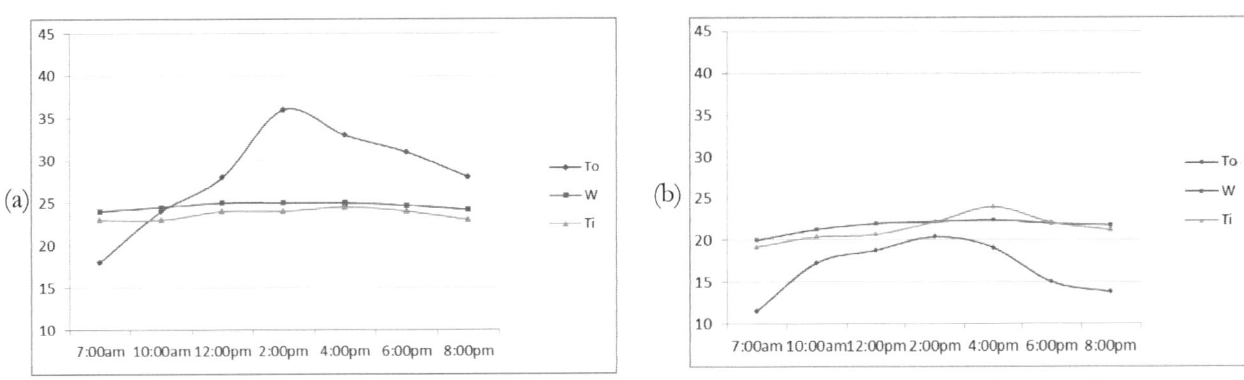

Figure 3 Indoor Surface Temperatures and Outdoor Temperatures in Summer (a) and Winter (b). (To: Outside temperature, Ti: Inside ambient temperature, W: Inside wall surface temperature)

Table 2. Indoor Surface Temperatures in Respect to Outdoor Temperatures

(Summer)Time	To	Wi	Ti	(Winter)Time	To	Wi	Ti
7:00am	18	24	23	7:00am	11.5	20	19.2
10:00am	24	24.5	23	10:00am	17.3	21.3	20.4
12:00pm	28	25	24	12:00pm	18.8	22	20.7
2:00pm	36	25	24	2:00pm	20.4	22.2	22.2
4:00pm	33	25	24.5	4:00pm	19.1	22.4	24
6:00pm	31	24.7	24	6:00pm	15	22	22.1
8:00pm	28	24.2	23	8:00pm	13.8	21.8	21.2

We infer from the readings that internal surface temperatures are stable in both seasons, and the difference between them and the ambient temperature is small which provides thermal comfort. In summer, the temperatures are around 11-12°C less than the exterior temperatures in peak hours. Similarly in winter, temperatures are 1.8-8.5°C higher inside than the exterior temperatures, and the difference between the Mean Radient Temperatures and internal ambient tempteratures does not exceed 1.5 °C in all seasons. This satisfies the requirements of a thermally efficient green building and improves the internl thermal comfort around 15% compared to the traditional building envelope that complies with Jordan's code (Nazer, Rodrigues, 2015).

In terms of thermal comfort, the internal and external temperatures were plotted on the ASHRAE Standard 55 model (figure 4a) which defines two comfort regions: 80% Acceptability, and 90% Acceptability. It was found that the Kamaliyya internal temperatures were within the comfort range in almost all times of the day. Furthermore temperatures were monitored on two consequent days in December when temperatures reached the minimum and presented in figure 4b. The results display that when the outside temperature oscillated between 5 – 11 °C (points B and D), on Friday, December 8, 2012, solar heat waves were sufficient to heat the internal space to a maximum of 23 °C (point C), while the drop in the internal ambient temperature overnight was limited to one degree only (point A). This is due to the "cave" effect that is expected to maintain constant temperatures in the interior environment in both seasons due to the thermal properties of the building envelope.

(a)

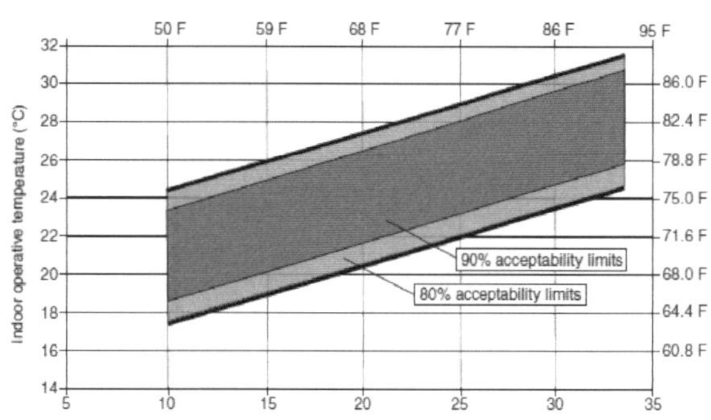

(b)

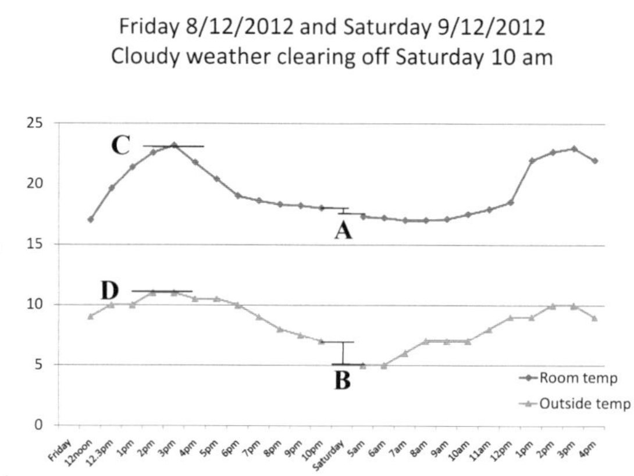

Figure 4 (a) ASHRAE Standard Thermal Comfort Model and (b) Indoor Ambient Temperatures in Respect to Outdoor Temperatures in Winter.

Air Tightness

All windows used in the Kamaliyya Residence are air tight at normal wind speeds; doors are also air tight with rubber reinforcement on the frames of the swing doors. No sliding doors exist on the premises to avoid high air infiltration. The only considerable leaks possible are from the main doors, which are limited to one on every floor.

The other sources of air leakage are from voluntary ventilation, such as opening doors and windows. An unexpected source of air leakages was discovered lately at an occasion when high speed winds were blowing in January. Almost all PVC electrical pipes connecting electric motors of window shutters to the main electric switches placed on the interior surfaces of external walls were actually blowing wind inside at a relatively exceptional rate. Since the whole system was "almost" air tight, it was possible to notice and hence be sealed accordingly.

Other air leakage possibilities were from cracks between window frames and between window frames and the cement plaster at the sides of the openings in the external envelope. It was noticed that these connecting lines between two different materials were cracking in the course of time due to heat and moisture movement as well as from continuous exposure to rain water, house cleaning chemicals and the Ultra Violet wave of the sunlight.

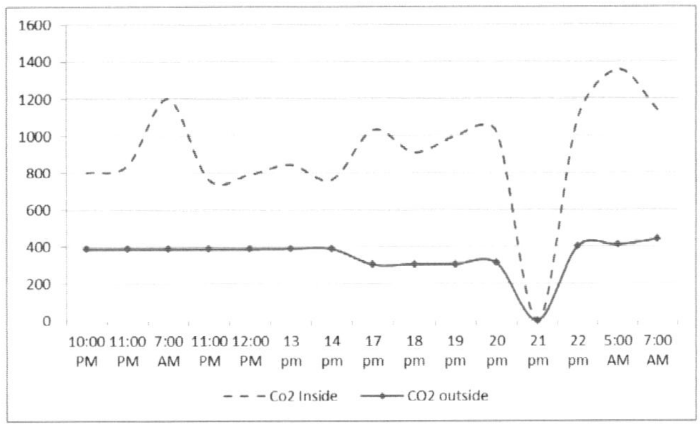

Figure 5 Monitored CO^2 Levels in the Bedroom on December 2014. (Note that the Readings at 9pm were Not Recorded Which Caused the Drop in the Figure Above).

Consequently, with the awareness of heat losses due to wind forced air infiltration and the continuous high air exchanges that incited an air-tightness design, rising levels of CO^2 were expected and hence monitored, particularly in bedrooms at night. However, CO^2 levels were monitored not to have exceeded 1358 ppm, for only two occupants sleeping overnight in the bedroom, as can be seen in figure 5 on Friday at 5am. An average Jordanian family is at least three times this number; hence further research on the impact of household individuals on concentrations of CO^2 and other GHGs is required.

During winter, increasing GHGs inside the house to an acceptable level is an asset due to the polarity effect between two or more elements that enhances the absorbance of the infrared heat wave from the electromagnetic light wave of the lights. However, in summer it seems to entail an adverse effect (Excessive warming) which was discussed earlier, therefore introducing fresh air into the building is important in summer to avoid over-heating through the potential polarity effect of trapped gases, particularly when building in hot climates such as the Gulf.

ACTIVE SYSTEMS AND RENEWABLES

All space heating needed for the Kamaliyya Residence is harvested from four sets of circular evacuated tubes, each set comprises of 4 units, 3 sets each consists of 12 tubes 1.2 m long and 5.7 cm diameter each, the fourth set consists of four units also but each unit consists of 16 tubes 1.2 meters long and 5.7 cm diameter for each of the tubes. Three of these sets are connected in series to boost water temperature on cloudy days. The best angle to harvest optimum sunrays intensity during winter was also investigated on the three similar sets of solar thermal collectors connected in series. Three angles were used inclined at 45, 60 and 75° to the horizontal. The results are shown in figure 6, which was the outcome of long observations and data collection. The case displayed during December 31st, 2010 is nominal for the total observations that were monitored; other readings were recorded in January and February with almost compatible results.

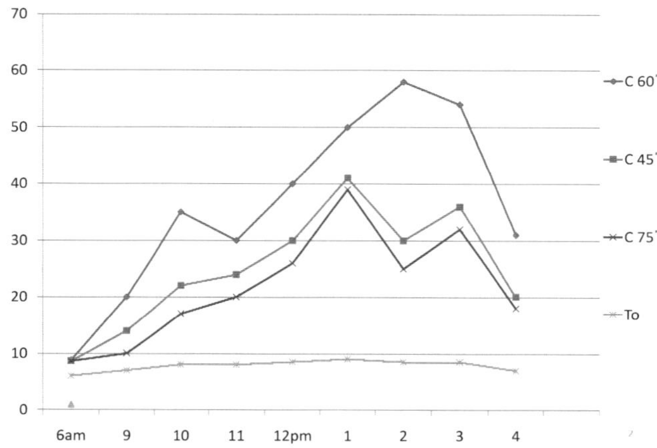

Figure 6 Vacuum Tubes Solar Panels at Different Angles to the Horizon

It is important to note that an average optimal angle of incidence of sunlight for the year round is recommended by the solar energy code of Jordan, at 45°C, which does not represent the best choice, because the angle in winter is much less in Amman, as it reaches 32°C on December 21st, however, this is the peak reached, at the instant of noon time, therefore if one aspires for the optimum angle it can be experientially confirmed at around 60° - 65° .This angle would not just yield enhanced efficiency from the solar system in winter but will definitely reduce over-heating of the solar systems in summer too, which is a common hazard in hot regions. The suggested angle will also make the units of the solar panels occupy less space on the roofs, although might aesthetically cause an issue for the architect in some cases.

This advantage solves a persisting problem in Jordan ever since a law was passed which came into effect on April 1st, 2014 enforcing the installment of solar thermal panels for all apartments which are less than 120 m^2 in net area. An average block of 300 m^2 roof area consists of 12 flats on average, making flat-plate solar heaters almost impossible to be installed because of the limited roof space, particularly because water storage tanks also exist on the roofs for water storage as municipality water is cut off for most days of the week in summer. Installing evacuated tube units is more practical because of the smaller size needed compared to flat plate solar heaters, however, if erected at a higher angle to the horizontal, evacuated tubes would make wonders in a limited space area and become a desirable solution.

CONCLUSION

Creating energy efficient buildings is an essential requirement in the energy crisis that Jordan is enduring in this era. Being in a climate with temperatures that vary between mediterranian yet hot climates, the considerations should be focused on both winter and summer equally. This paper has discussed a multiplicity of strategies that have been applied in the residential building of Kamaliyeh and monitored afterwards in both seasons, to assess the viability of the application. Glazing should be carefully designed according to orientation and area; east and south facades are more exposed to solar radiation, and can reach a temperature of 40oC even in winter. This raises the importance of the design formation in the Gulf region in a way to diffuse the solar radiation by managing the inclination of the building walls.

On the other hand, the building envelope with enhaced properties and u-values around 0.13 W/m^2 in Al-Kamaleyeh has verified its capability in preserving almost stable constant temperatures in the interior of the house in both seasons during day and night.

In winter, enhanced u-value reduces heat losses, which was evident in only one degree loss of internal ambient temperature over 8 hours during the whole night. In addition, in intense heat conditions, the higher U-value (with a considerable internal thermal mass), has proved valuable results in increasing the time lag to avoid the 2 pm peak heat wave, compared to modestly thermally insulated homes (U-value 0.6 to 1 W/m2.K) where the time lag can be 2-3 hours, transmitting the heat and raising the Mean Radiant Temperature on the internal surfaces around 5 pm.

Consequently, the "Decrement Factor" is decreased in summer as the ratio of the difference between outside surface temperatures and the difference between maximum and minimum internal temperatures is minimized which causes the internal surface temperatures to hardly rise by less that one degree. This 'cave effect' maintains internal constant temperatures emitting the needs for any mechanical heating and cooling methods that eventually reduces carbon emissions.

GHGs inside the house has proved to be a serious issue in summer at semi-arid climates, therefore, air quality must be monitored and fresh air intake considered seriously to dispose of critical gases, those of high polarity, in particular.

Thermal quality, glazing characteristics, angle of tilt or Sun tracking mechanisms, shading, ventilation and the introduction of renewable energies can be as important for the green economy as mitigating the damage done by climate change by reshaping the world`s environmental status through the reduction in GHSs.

Hitherto, we ought to perceive green economy today as a reflection of our environmental concerns, shifting from an anthropocentric behavior to an ecocentric stance that controls human action, not only for incurring the minimum amount of damage to the environment and resources, but also to facilitate the introduction of new methods capable of suggesting practices for a more resilient impact on the biosphere to avoid the long term damage we have already agreed upon as " the point of no return" which resulted in allowing an increase of 2 degrees Celsius by 2030, and looking forward to reduce it to 1.5, compared to pre-industrial levels as a point of reference. However, this approach to sustainable green buildings can only be approached in a collective manner and ought to be launched through nurturing young generations at an early age.

REFERENCES

Abu Dayyeh, A. 2013. Green Buildings, 1st Edition, United Arab Emirates: Zayed International Foundation for the Environment. P. 276.

Al Momani, H. M., & Ali, H. H. (2008). Sick building syndrome in apartment buildings in jordan. Jordan Journal of Civil Engineering, Volume 2, No. 4.

American Institute of Architects (2012). Society of Building Science Educators: "The Carbon Nuetral Design Project". http://www.tboake.com/carbon-aia/strategies1a.html. Accessed: February 9, 2016.

CO2 Monitors and other monitoring devices were presented as a gift to support the Kamaliyya project from: AMEGTECH-Alpha Middle East Green Technologies Ltd. Co.

Dhaher, M, 2010 "Kamaliyya Residence Construction Management. " Masters Thesis in Construction management", Reading University, U.K.

Google, "Thermal-comfort". http://www.isover-airtightness.com/Benefits/Thermal-comfort (Accessed April 27th, 2014).

Jaber, J. O. (2002). Prospects of energy savings in residential space heating.Energy and Buildings, 34(4), 311-319.

Koch-Nielsen, H. (2013). Stay cool: A design guide for the built environment in hot climates.

Mohsen, M. S., & Akash, B. A. (2001). Some prospects of energy savings in buildings. Energy conversion and management, 42(11), 1307-1315

Nazer, H, & Rodrigues, L., 2013. "Optimising Housing Design to Improve Energy Efficiency in Jordan", presented at SET Conference. Nottingham, UK, 2015.

Sengupta, J., Chapman, K. S., & Keshavarz, A. (2005). Window Performance for Human Thermal Comfort". ASHRAE Transactions, 111(1).

Application of Multi-Criteria Decision-Methods for the Selection of Renewable Energy Sources, Lebanese Case Study

Georges EL-JAMAL[1] **Mazen Ghandour, PhD**[2] **Hussein Ibrahim, PhD**[3]

ABSTRACT

Renewable energy and energy efficiency have been a significant issue for the policy of decision makers. In fact that resource of energy in earth is limited, renewable energy alternatives have started to replace the fossil energy alternatives. So, selection of renewable energy type for industrial or residential applications becomes a question should be solved. In general, evaluating energy systems is a complex analysis that can be defined as a multi-dimensional space of different indicators and objectives. Indeed, a variety of criteria can be applied to identify the suitability of technologies, whereas no ideal family of criteria has been defined in the literature. Hence, decision support for energy planning and management is required. The use of multicriteria decision method (MCDM) techniques provides a reliable methodology to rank alternative renewable energy resources. Preference Ranking Organization METHod for Enrichment Evaluation (PROMETHEE) and Analytic Hierarchy Process (AHP) are two widely methods used to take the appropriate decision to select the best solution among others can be applied. The above two methods are used in a Lebanese context case study in order to determine the priority of one type of renewable energy (wind, hydraulic, solar, geothermal, etc.) over the others and rank the different possible types by using many criteria like cost, environment, operational, site selection, efficiency, etc. The purpose of using of two methods is only for the confirmation of the results achieved. In the first phase AHP is utilized to determine the importance of the criteria and in the second phase PROMOTHEE is used to confirm the choice of the renewable energy technology.

INTRODUCTION

Energy is a major problem that disturbs the world. Fossil energy has a limited resource in earth. So, replacing this kind of energy with another cheapest, harmless and eternal becomes a strategic question. During the last years, renewable energy resources (RES) exploitation has gained a vast interest; many countries have committed themselves to include them in their energy supply systems. RES are capable of replacing fossil fuels at rival prices. Their contribution, to the current global energy supply is still small despite considerable technological advances in this field. Many operational parameters should be studied carefully, that could be through the inclusion of several technological, economic, environmental, risk, and social criteria in the planning of application of RES.

By using the MCDM, the decision-maker is required to choose among of quantifiable or non-quantifiable criteria. H. Polatidisdias et al. in 2006 developed a methodological framework to provide insights regarding the suitability of multi-criteria techniques in the context of renewable energy planning. B. Oztaysi et al. used in 2014 in their study fuzzy multi criteria decision- making methodologies, axiomatic design (AD) and analytic hierarchy process (AHP) were utilized for ranking both renewable and conventional energy alternatives. O. Demirtas et al. used AHP methodology, where results indicate that wind energy is the most appropriate renewable energy option in their study and proposed future researches could be conducted based on different multi-criteria decision-making techniques such as AHP, fuzzy ELECTRE or fuzzy TOPSIS for comparative purposes. D. Georgiou et al. evaluated in 2013 five

[1]**G. EL-JAMAL**, PhD student at Saint-Joseph and Lebanese University, Beirut, Lebanon, (e-mailgfjamal@yahoo.com)
[2]**M. GHANDOUR**, Professor at Lebanese University, Beirut, Lebanon, (e-mail: ghandour@ul.edu.lb)
[3]**H.IBRAHIM**, Techno Centre éolien, 70 rue bolduc, Gaspé, QC, Canada,(e-mail: hibrahim@eolien.qc.ca).

alternative energy generation topologies of reverse osmosis desalination process. The proposed topologies are assessed in terms of economic, environmental, technological and societal indices and compared using multi-criteria analysis, namely the AHP and PROMETHEE. S.D. Pohekar et al. were observed that AHP, is the most popular technique followed by outranking techniques PROMETHEE and ELECTRE.

MULTI-CRITERIA DECISION MAKING MCDM OVERVIEW

Multi-criteria decision making (MCDM) is a section of operation research models and known field of decision-making. These models can treat quantitative and qualitative criteria and analyze conflict in it where several classification and categorization exist. The main objective of MCDM is to select an alternative that has the highest score according to the set of the evaluated criteria which means the most suitable alternative among the others. A descriptive summary of the most commonly used multi-criteria decision-making methods is presented below:

1- Analytic hierarchy process (AHP): In AHP, the problem is constructed as a hierarchy breaking down the decision top to bottom.
2- Preference ranking organization method for enrichment evaluation (PROMETHEE): uses the outranking principle to rank the alternatives and performs a pair-wise comparison of alternatives in order to rank them with respect to a number of criteria.
3- The elimination and choice translating reality (ELECTRE)[8]: This method is capable of handling discrete criteria of both quantitative and qualitative
4- Multi-attribute utility theory (MAUT)[9]: The theory takes into consideration the decision maker's preferences in the form of the utility function which is defined over a set of attributes

MULTI-CRITERIA DECISION ANALYSIS IN RENEWABLE ENERGY

The application of MCDM in RES is divided into four categories:

1- Planning and policy refers to the assessment of a feasible energy plan and/or the diffusion of different RES option
2- Evaluation of different alternative energies or energy technologies. Choosing between alternatives could be for assessing the "best" energy to be utilized in electrical or thermal energy, or any other systems
3- Technology and project selection and allocation refers to site and technology selection, and decision support in RES harnessing projects
4- Environmental is concerned with the literature discussing alternative technologies from an environmental perspective and climate issues.

In general, evaluating energy systems is a complex analysis that can be defined as a multi-dimensional space of different indicators and objectives. The use of (MCDM) techniques provides a reliable methodology to rank alternative RES. Even with the large number of available MCDM methods, there is no better or worse method but only a technique that fits better in a certain situation. It is noticed that AHP is the most used methodology of all MCDM methods. This can be credited to its simple structure

Analytic Hierarchy Process (AHP)

The analytical hierarchy process (AHP) (Saaty, 1982, 1988, 1995) is well known and widely used and the application of AHP provides the following:

1- Choice of one alternative from a given set.
2- Ranking of the alternatives involved in order
3- Allocation resources among a set of alternatives
4- Comparing the processes in one's own organization with those of other best-of-breed organizations

Principles of AHP:

Hierarchy construction: A decision question, centered on the extent of contributions to an overall goal, is constructed and broken down into its components (criteria, alternatives, etc.), using a hierarchy (Fig. 1).

Priority setting: By comparing two by two the contribution of each element, the decision-maker uses a paired comparison mechanism as shown in Table 1 and using a scale of 1 to 9 as indicated in Table2, its standard element is $\mathbf{P_c(a_i;a_l)}$ i.e., the intensity of the preference (in terms of contribution to a specific criterion (**C**) of the row element ($\mathbf{a_i}$) over the column element ($\mathbf{a_l}$). In case the pairwise comparisons are completely consistent, the matrix "A" has rank 1 and Maxi = n, in that case, weights can be obtained by normalizing any of the rows or columns of A. The procedure described above is repeated for all subsystems in the hierarchy, in order to synthesize the various priority vectors, these vectors are weighted with the global priorities of the parent criteria and synthesized.

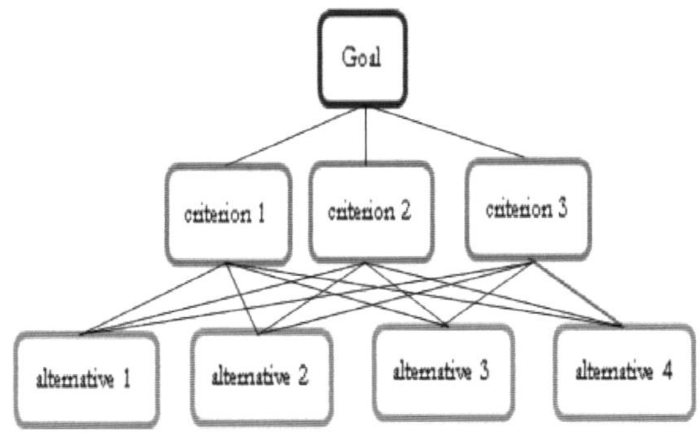

Fig. 1 AHP Hierarchy Construction

Logical consistency: In each pairwise comparison matrix (see Table. 1), a number of comparisons are redundant. In fact, in case of complete consistency, relation (1) holds:

$$P\,(a_i;\,a_j) \;=\; P_c(a_i;\,a_k) \times P_c(a_k\,;\,a_j) \quad whatever\ i;\ j;\ k \qquad (1)$$

When the pairwise comparison matrices are completely consistent, the priority (or weight) vector is given by the right eigenvector (**W**) corresponding with the highest eigenvalue ($\boldsymbol{\lambda_{max}}$), in that case, the latter is equal to the number of elements compared (**n**). In case the inconsistency of the pairwise comparison matrices is limited, slightly ($\boldsymbol{\lambda_{max}}$) deviates from "n".

The process is summarized as follows:

1- Decompose the decision-making problem into a hierarchy (table 1)
2- Make pair wise comparisons and establish priorities among the elements in the hierarchy (table 2)
3- Normalizing the resulting matrix
4- averaging the values in each row to get the corresponding rating
5- Choose the one with the highest score.
6- Synthesize judgments (to obtain the set of overall or weights for achieving your goal)
7- Evaluate and check the consistency of judgments, CR should be less than 0.1

$$CR = \frac{CI}{CI^*} \qquad (2)$$

$$CI = \frac{\lambda max - n}{n-1} \qquad (3)$$

n= nb of alternatives, λ_{max}=average of consistency measure, CI* =Random consistency indices (table 3)

By drawing the variations of CI* in function of alternative nb(AN) the Fig. 2 can be reached and a mathematical equation is elaborated $CI^* = 0.0012 \times (AN)^3 - 0.043 \times (AN)^2 + 0.5222 \times (AN) - 0.6304$ (4)

for (AN) ={1,2,3,4,5,6,7,8,9,10,11,12,13,14,15}With a R-squared =0.9766

Table 1. Hierarchy Table

	a_1	a_i	...	a_j	a_n
a_1	1				
...					
a_i		1		$P_c(a_i,a_j)$	
...					
a_j		$P_c(a_j,a_i)$		1	
...					
a_n					1

Table 2. Relative Weights

Scale of relative weights for pairwise comparison	
1	same importance
2	slightly more important
3	weekly more important
4	weekly to moderately more important
5	moderately more important
6	moderately to strongly more important
7	strongly more important
8	greatly more important
9	absolutely more important

The following table contains the results of the application of the previous formula

Table 3.Consistency Indices According To AHP Method and By Calculation

Nb of Alter	1	2	3	4	5	6	7	8	9	10	11	12	13	14	15
CI*	0	0	0.58	0.9	1.12	1.24	1.32	1.41	1.45	1.49	1.51	1.48	1.56	1.57	1.59
CI*cal	0.13	0.24	0.57	0.83	1.05	1.22	1.35	1.44	1.48	1.5	1.51	1.51	1.51	1.54	1.62
Error percentage			-2%	-8%	-7%	-2%	2%	2%	2%	1%	0%	2%	-3%	-2%	2%

Where the deviation between the two values of CI* is calculated and doesn't exceed 8%.
The importance of the equation (4) is the elaboration of a mathematical relationship between the alternative number and the random consistency.

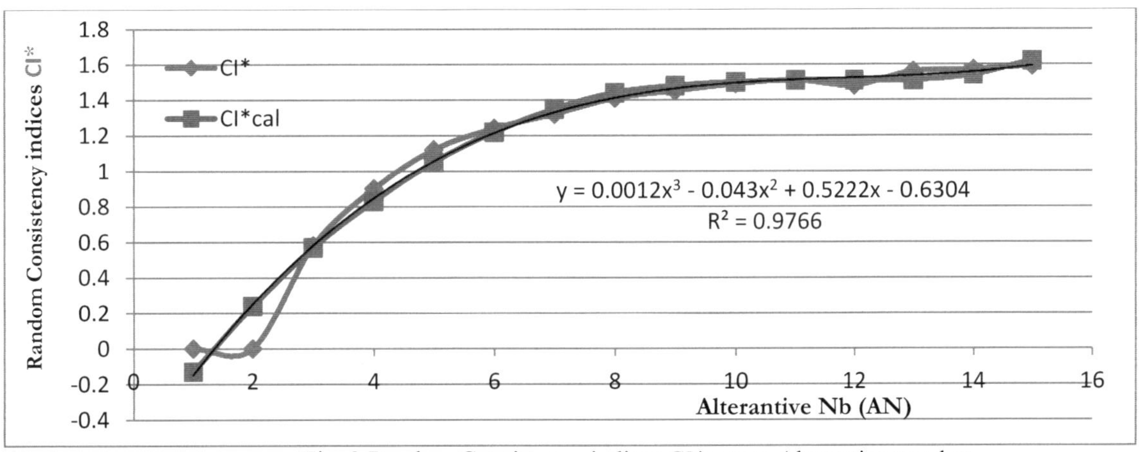

Fig. 2 Random Consistency indices CI*versus Alternative number

Preference ranking organization method for enrichment evaluation (PROMETHEE).

PROMETHEE tools is used to analyses the evaluation problem it's divided into tree sub tools:

- The PROMETHEE I partial ranking;
- The PROMETHEE II complete ranking;
- The GAIA plane.

PROMETHEE I partial ranking provides a ranking of alternatives. In some cases, this ranking may be incomplete.
PROMETHEE II insures a full ranking of the alternatives from the preferable to the unfavorable one.
GAIA The geometrical analysis for interactive aid plane visualize graphically the position of the alternatives in terms of contributions to the various criteria
PROMETHEE begin with the evaluation table. Where, the alternatives are evaluated with different criteria. The implementation of PROMETHEE requires two other types of data:

- Data on the weights of the criteria considered.
- Data on preference function, to compare the contribution of the alternatives of each separate criterion.

The weights can be determined according to various methods. The decision-maker should be able to weigh the criteria appropriately, the preference function (Pj) translates the difference between the evaluations (i.e., scores) obtained by two alternatives in terms of a particular criterion, into a preference degree ranging from 0 to1.

$$Pj(a;\ b) = Gj[fj(a)\_fj(b)],\quad 0 \le Pj(a;\ b) \le 1 \ (5)$$

PROMETHEE permits the computation of the following quantities for each stakeholder "r" (r = 1; . . . ; R) and alternatives "a" and "b":

$$\pi_r(a;\ b) = \sum_{j=1}^{k} Pj(a,b)Wr,j \qquad (6)$$

$$\emptyset r^{+}(a) = \sum_{x \in A} \pi r(x, a) \qquad (7)$$
$$\emptyset r^{-}(a) = \sum_{x \in A} \pi r(a, x) \qquad (8)$$
$$\emptyset r(a) = \emptyset r^{+}(a) - \emptyset r^{-}(a) \qquad (9)$$

For each alternative "a", belonging to the set A of alternatives, $\boldsymbol{\pi}$(a; b) is an overall preference index of "**a**" over "**b**", taking into account all the criteria, $\mathbf{\emptyset r^{+}(a)}$ and $\mathbf{\emptyset r^{-}(a)}$. These measure respectively the strength and the weakness of "a" vis-à-vis the other alternatives. $\boldsymbol{\emptyset r(a)}$Represents a value function, whereby a higher value reflects a higher attractiveness of alternative "a". We call $\mathbf{\emptyset r(a)}$ the net flow of alternative "**a**" for stakeholder "k".

DEVELOPMENT OF NEW TOOL (GFJ) [13]

In order to implement a renewable energy resources, a new tool names GFJ was developed as shown in Fig.3 a) (Algorithm of GFJ) where as first stage the actual energy profile is determined, in the second stage, the new reduced energy profile is developed after the intervention of the energy efficiency methods, by using of two multicriteria decision methods (AHP, Promothee) the suitable type of renewable energy is selected and implemented, where a new energy profile is displayed as a third phase.

The appropriate type of RES is selected in the MCDM section shown in Fig.3 b) as follows:

1- Selection of criteria (environment, cost, operating, efficiency…etc).
2- Selection of the types of RES (Biomass, Photovoltaic,
3- Geothermal, wind, Hydropower…etc) provided in the region of the project
4- Weights each RES type according to each criteria
5- Find the best RES for each methods (AHP, PROMOTHEE)
6- Test the conformity of the RES for the two methods
7- If no conformity, recheck the weights of the alternatives and criteria for each method.
8- Selection of RES

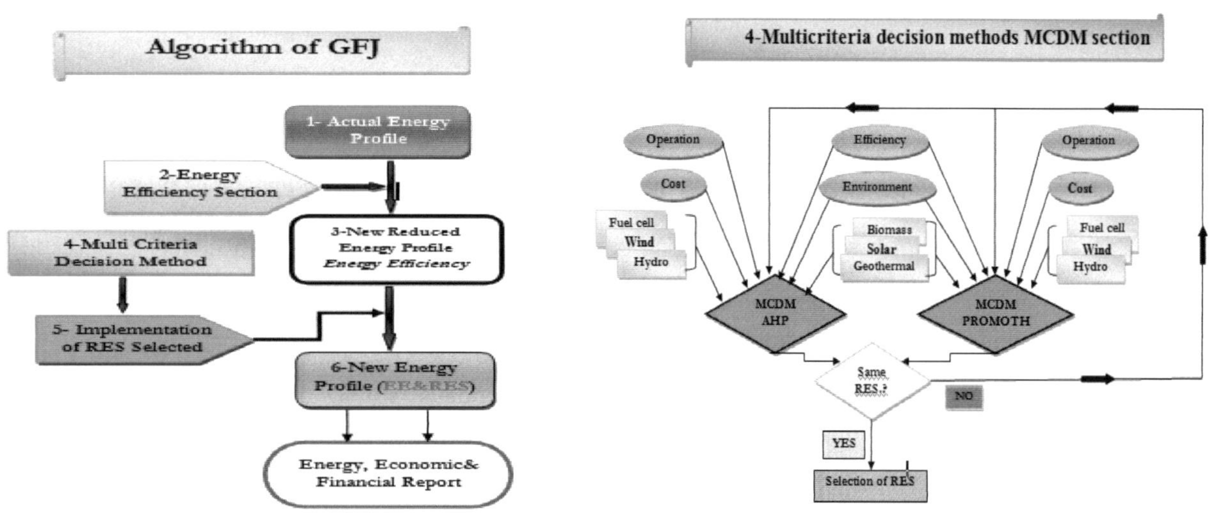

a) b)

Fig.3(a) Algorithm of GFJ part of Figure 3 and (b) Multicriteria decision methods section part of Figure 3.

CASE STUDY

The case study includes selection of RES can be implemented in the three facilities:

1- The two methods applied are PROMOTHEE and AHP in order to determine the suitable type of RES
2- All facilities are in Lebanon - Middle East, bordering east part of the Mediterranean sea, location 33 50 N, 35 50 E, area of 10452 km², climate mild to cool, wet winters with hot, dry summers; Lebanon mountains experience heavy winter snows. Population 5,882,562 (July 2014.). Electricity consumption 14.4 billion kWh (2011). 88.1%

of total installed capacity (2011) coming from fossil fuels, the others come from hydroelectricity. No petroleum resources exist; Lebanon has a precipitation of 750 mm and an insolation capacity of 4.8 kwh/day.m^2 and has some wind potential in north and south regions.

3- The first facility is a school composed of 3 buildings, 3 stories each 1400 m^2 area with external facilities, located in north of Lebanon near a river equipped of office, school and kitchen equipment. Four types of RES, which are PV, PV with storage, wind and hydroelectricity, were choose to tradeoff between them.
4- The second facility is a residence complex located in Marjaayoun south of Lebanon composed of three buildings each has four stories with total area of 6000 m^2, the location has a high potential of wind energy , wind velocity of 6-7 m/s, is predicted and near the basin of Litany river Also the four previous RES were selected.
5- The third facility is a hospital located in Beirut, composed of six medical care and offices buildings part of them working 24/24 H 7/7 D only PV with storage, PV off grid and wind were the RES selected despite Beirut doesn't has a good potential of wind power.
6- Some results of ranking of the different alternatives for the three cases are cited below (Table 5 to Table 10).

RESULTS

The results presented in table 5 to table10, are parts of achieved outcomes from the applications of the two methods of MCDM, AHP and PROMETHEE in order to rank the different type of RES can be applied in the facilities according to the criteria, environment, cost, operating, and efficiency. For example table 5 and 6 represents the results obtained of the school facility according to environments criteria by AHP and PROMOTHEE respectively, where PV panel without storage has the highest rank for both methods. In table 11 the comparison of all the results reached by the application of the two above methods.

Table 5.to 10 Indicates The Ranking Results For Different Facilities And RES

Table 5. AHP-Cost -Hospital

	PV	PV/S	WIND	CM
PV	1.00	2.00	3.00	3.03
PV/S	0.50	1.00	1.00	3.01
WIND	0.33	1.00	1.00	3.01
			CI=	**0.0090**
			CI*=	**0.5800**
			CR=	**0.0158**

Table 6. AHP - Environment -School

	PV	PV/S	HE	WIND	CM
PV	1.00	2.00	2.00	3.00	4.02
PV/S	0.50	1.00	1.00	2.00	4.01
HE	0.50	1.00	1.00	2.00	4.01
WIND	0.33	0.50	0.50	1.00	4.01
				CI=	**0.0040**
				CI*=	**0.9000**
				CR=	**0.0045**

Table 8. PROMOTHEE -Cost-Hospital

rank	action	Ø	Ø+	Ø-
1	PV	0.20	0.50	0.30
2	PV/S	0.00	0.40	0.40
3	WIND	-0.20	0.40	0.60

Table 9 PROMOTHEE-Environment-School

rank	action	Ø	Ø+	Ø-
1	PV	0.53	0.73	0.2
2	PV/S	0	0.4	0.4
3	WIND	-0.2	0.33	0.33
4	HD	0.33	0.2	0.53

Table 7. AHP-Efficiency- Residence

	PV	PV/S	HE	WIND	CM
PV	1.00	0.50	0.20	0.25	4.03
PV/S	2.00	1.00	0.33	0.50	4.04
HE	5.00	3.00	1.00	3.00	4.18
WIND	4.00	2.00	0.33	1.00	4.08
				CI=	**0.540**
				CI*=	**0.900**
				CR=	**0.030**

Table 10 PROMOTHEE -Efficiency- Residence

rank	action	Ø	Ø+	Ø-
1	HD	0.44	0.64	0,20
2	WIND	0,13	0,51	0,38
3	PV	0.10	0.40	0,41
4	PV/S	0.56	0.05	0.61

Table 11-Comparison Table of Ranking between AHP and PROMOTHEE Methods

FACILITIES/RES		AHP				PROMOTHEE				MISMATCH					
		Environment	Cost	Operating	Efficiency	Environment	Cost	Operating	Efficiency	FIRST RANK		FIRST & SECOND RANK		ALL RANKS	
										nb	%	nb	%	nb	%
Residence	**PV**	3*	3	1	4**	1	3	1	3	0		0		5	
	PV/S	2	4	2	3	2	4	3	4						
	HE	1	1	2	1	3	1	2	1						
	WIND	1	2	2	2	4	2	3	2						
Hospital	**PV**	1	1	1	2	1	1	1	2	2	6.8%	2	9.1%	3	31.8%
	PV/S	2	2	2	1	1	2	1	1						
	WIND	2	2	3	3	3	3	3	3						
School	**PV**	1	2	1	1	1	1	1	1	1		2		6	
	PV/S	2	2	2	1	2	2	2	1						
	HE	2	2	3	3	4	3	4	3						
	WIND	2	1	3	1	3	1	3	2						

* Numbers indicated are the rank of each type of RES after the calculation on each method applied

** The mismatched results obtained between the two methods are presented in red color

ANALYSIS OF THE RESULTS AND CONCLUSIONS

The implementation of RES in a project is a strategic decision and the selection of the suitable type of RES still a difficult problem should be solved. In general, evaluating energy systems is a complex analysis that can be defined as a multi-dimensional space of different indicators and objectives. The algorithm of GFJ is an efficient gait to develop an energy profile depends on EE. and RES. The introduction of MCDM application to select the best suitable type of RES is an innovative idea can be considered as a strategic step in GFJ algorithm instead of lot of techno-economic study to decide the suitable RES. Four criteria are used in the case study, others can be added, but the ranking may be modified. Weights of the criteria used in the case study may be changed between decision makers according to own priority of each. In order to examine the accuracy behavior of the two methods, AHP and PROMOTHEE, low potentials of RES. are used in the case study of hospital project like wind. In some facilities and criteria, two types of RES. have the same rank. The percentage of matching the first rank between the two methods is 93.2%. which considered a high coincidence, The mismatch between AHP and PROMOTHEE are 6.8%, 9.1% and 31.8% for the first rank, second rank, and all ranks respectively which represent an acceptable results for the MCDM results taken into consideration the decision makers weights . In School project only the need of electricity is during day for this reason PV panel off grid are the most efficient. No evidence of preference of one of the MCDM used in this study versus the other one.

NOMENCLATURE

CM = Consistency Measure
PV = Photovoltaic panel
PV/S= Photovoltaic with storage
WIND= Wind power
HE=Hydroelectric
EE = Energy efficiency
RES. = Renewable energy resources
MCDM= Multi-criteria decision methods
A HP= Analytic Hierarchy Process
PROMETHEE= Preference Ranking Organization METHod for Enrichment Evaluation

REFERENCES

Basar Oztaysi et al 2014 "Multi-Criteria Decision Aid for Sustainable Energy Prioritization Using Fuzzy, Axiomatic Design", 2014 IGI Global

Dimitris Georgiou et al December 2013 "Multi-Criteria Decision Making on the Energy Supply Configuration of Autonomous Desalination Units" AUA Working Paper Series No. 2013-6

Forman, Ernest H.; Saul I. Gass (July 2001). "The analytical hierarchy process—an exposition". Operations Research49

Georges.El-Jamal 2014 "Development of renewable energy mapping using multi-criteria decision methods and applicationof energy efficiency validation of the Lebanese context" Master 2 Research report –USJ-UL Lebanon

Heracles Polatidisdias ,A. Aralambopoulos et al. 2006 "Selecting an Appropriate Multi-Criteria Decision Analysis Technique for Renewable Energy Planning"

http://hodgett.co.uk/topsis-in-excel/

John Buchanan, Phil Sheppard, Ranking Projects Using the ELECTRE Method

J.P. Brans and P. Vincke (1985). "A preference ranking organization method: The PROMETHEE method for MCDM". Management Science

L. Saaty 1980 "The Analytic Hierarchy Process." New York: McGraw-Hill,.

Nijkamp et al.,1990; Eckenrode, 1965 "PROMETHEE and AHP: The design of operational synergies in multicriteria" analysis. Strengthening PROMETHEE with ideas of AHP

Ozgur Demirtas," Evaluating the Best Renewable Energy Technology Sustainable Energy Planning", International Journal of Energy Economics and Policy Vol. 3, Special Issue, 2013,

Rimal Abu Taha and Tugrul Daim Multi-Criteria "Applications in Renewable Energy Analysis, a Literature Review"

S.D. Pohekar, M. Ramachandran December 2003, "Application of multi-criteria decision making to sustainable energy planning—A review"

The National wind Atlas of Lebanon,25 January 2011, CEDRO

The world fact book, 2014

Thomas L. Saaty. 2008. "Relative Measurement and Its Generalization in Decision Making Why Pairwise Comparisons are Central in Mathematics for the Measurement of Intangible Factors The Analytic Hierarchy/Network Process"

Energy Efficient Building Design Optimization Using an Underground Thermal Labyrinth

Ghina R. Annan, ME
ASHRAE Member

Balsam A. Nehme, ME

ABSTRACT

Thermal Labyrinth Ventilation System (TLVS) is an energy efficient ventilation system that offers the advantage of reducing the building energy consumption. It is based on precooling or preheating the outdoor air through heat exchange. This paper studies the energy savings achieved in the construction sector by means of the underground thermal labyrinth. A case study in Algeria is investigated with virtual illustration using Integrated Environmental Solutions. The results of the analysis show that by using an underground thermal labyrinth the outdoor air lost a maximum of 7.8 °C during peak summer hours. The estimated annual energy savings associated with temperature savings on the fresh air load is nearly 12 %.The results of the simulations and the recommendations are also indicated.

INTRODUCTION

Ventilation is essential in the building sector in order to maintain a healthy indoor environment. Commonly, providing ventilation in extreme conditions of summer and winter necessitates the presence of a substantial amount of energy. Accordingly, thermal labyrinth ventilation system (TLVS) is introduced as an energy efficient ventilation system in an attempt of minimizing building energy consumption. A thermal labyrinth is a form of a ventilation system that takes in outdoor air through an underground concrete-shaped structure that is a part of the building construction. During summer and winter seasons, a TLVS can pre-cool and pre-heat the outdoor air, respectively, through heat exchange with the ground.

Thermal labyrinth ventilation system have been adopted in many projects in Germany, and Australia. In the Municipal Theater in Heilbornn, Germany, a labyrinth of total area of 140 m² was installed in the underground floor. Optimization of such system assisted in cooling outdoor air by 1-8°C in summer season and heating it by 2-4°C in winter season (Daniels, K. 1997). Another prototype is the thermal labyrinth occupying an area of 1600 m², with a distance of 0.6 m between concrete vertical walls, installed in the Federation square in Melbourne, Australia (McGowan, S, 2003). This system, which provides environmental climate control for the square glazed atrium, is capable of delivering air to the atrium at up to 12°C below the outdoor air temperature in the peak summer conditions.

An energy efficient building provides the comfort levels and required services with minimum energy use in an economical and environmentally sensitive manner. The objective of this study is to evaluate the energy savings due to a

Ghina R. Annan is a design engineer in the energy modeling, condensation and acoustics unit in the Department of Mechanical Engineering at Dar Al Handasah (Shair and partners), Beirut, Lebanon. Balsam A. Nehme is the head of the energy modeling, condensation and acoustics unit in the Department of Mechanical Engineering at Dar Al Handasah (Shair and partners), Beirut, Lebanon.

thermal labyrinth ventilation system (TLVS). A case study in Algeria is investigated, for which an underground labyrinth-shaped concrete structure as intake channels is installed in a convention center. The design strategy and geometry of the case study are illustrated along with the temperature profile and model setup. The simulation of the design is achieved using a virtual illustration of the model investigated by means of IES- VE. The results of the simulations with the recommendations are presented.

DESIGN STRATEGY

The convention center studied in Algeria has a total area of 4,721 m2. The underground labyrinth is installed below the mechanical room at basement level located at 7.3 m depth below ground. A volume flow of 37,000 l/s of outdoor air is taken in through an opening at ground level; then passes through a concrete duct known as an earth tube and finally reaches the constructed labyrinth. The system makes use of the available seismic support base for building concrete walls by providing channels for air conduit. By heat transfer with respect to cooler/ warmer soil temperature at a depth ranging between -5.8 m and -7.3 m below ground, the outside air is expected to lose heat during summer season and gain additional heat during winter season thus saving on building's energy consumption. Hence, thermal labyrinth provides thermal mass which remains at constant temperature with a value cooler than the ambient air in summer and warmer than the ambient air in winter. The concept of this heat exchange is shown in Figure 1. As air passes through the thermal labyrinth, it is cooled or heated through contact with the thermal mass. The building assembly and thermal labyrinth 3D and 2D model are also shown in Figure 2 and 3 respectively.

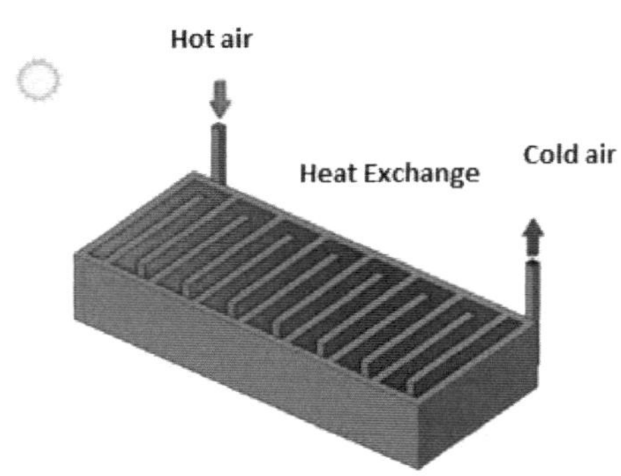

Figure 1: Thermal labyrinth ventilation system concept

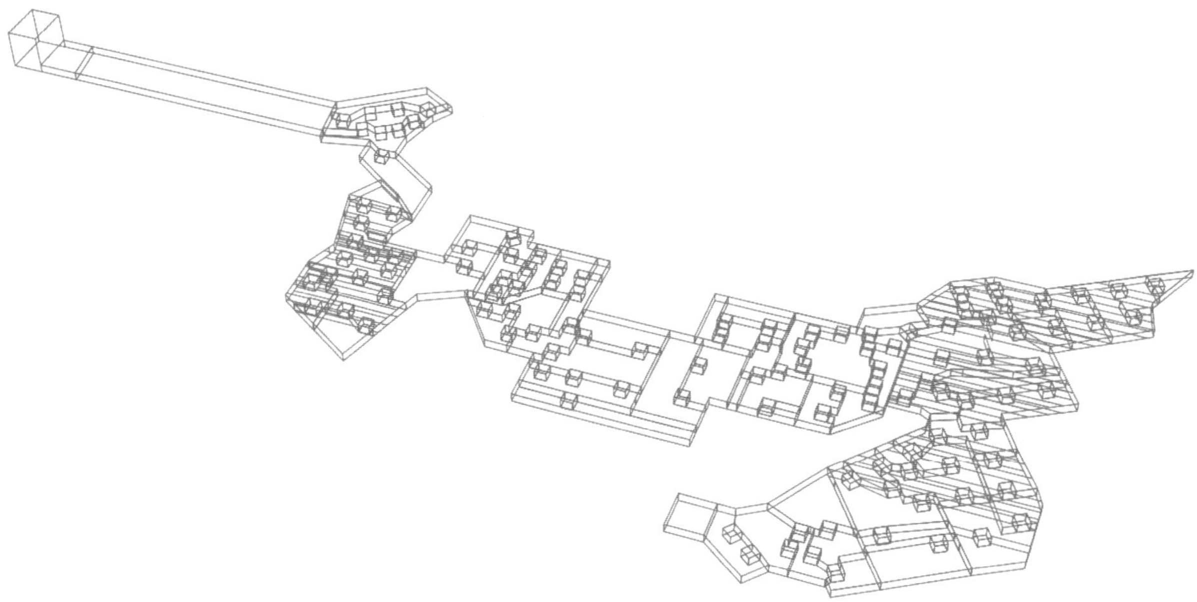

Figure 2: Thermal labyrinth ventilation system 3D

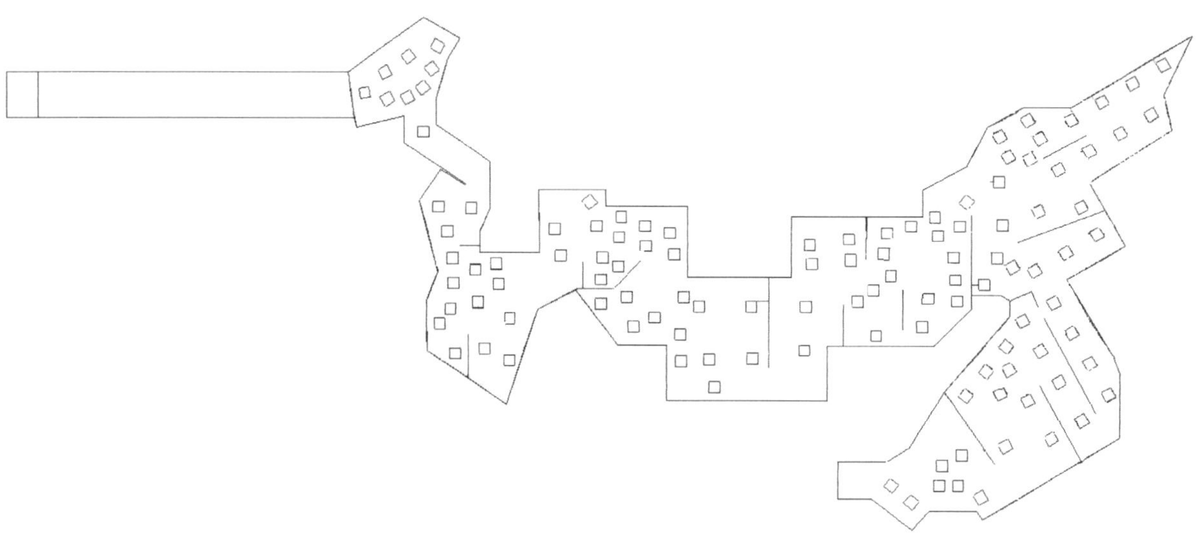

Figure 3: Thermal labyrinth ventilation system 2D design

TEMPERATURE PROFILE

The soil temperature is related to the ground surface temperature surrounding the labyrinth structure. The annual profile of the absolute value of ground temperature has been acquired from Meteonorm version 7 for Al Mouradiya, Algeria. Based on the calculated ground surface temperature, the temperature of the soil surrounding the labyrinth and heat transfer rate can be predicted. Assuming a homogeneous soil of constant thermal diffusivity, the soil temperature at any depth z and time t can be then estimated using the following equation (Kwang Ho Lee, Richard K):

$$T_{z,t} = T_m - A_s \exp\left[-z\left(\frac{\pi}{365\alpha_s}\right)^{\frac{1}{2}}\right]\cos\left[\frac{2\pi}{365}\left(t - t_0 - \frac{z}{2}\left(\frac{365}{\pi\alpha_s}\right)^{\frac{1}{2}}\right)\right] \quad (1)$$

MODEL SETUP

The walls of the labyrinth are set adjacent to a temperature profile equivalent to a depth of 5.8 m, the ceiling of the labyrinth is set adjacent to the mechanical room, and the floor of the labyrinth is set adjacent to a temperature profile equivalent to a depth of 7.3 m. Figure 4 shows the variation of soil temperature at ground level as well as depths of -5.8 m and -7.3 m below ground, correspondingly.

In compliance with ASHRAE standard 90.1-2010, Table A4.2, the below grade C-factor for walls underground without framing and insulation is set at C-6.48, , which is equivalent to U-3.25 W/m2.K for 200 mm concrete wall of density 1840 kg/m3 and solid grouted. The heat transfer coefficient for thermal labyrinth, it is set at 3.25 W/m2, assuming the inside air is moving with corresponding surface resistance of 0.044 m2.K/W, as per Table 1 of Chapter 26 of ASHRAE fundamentals standard 2013.

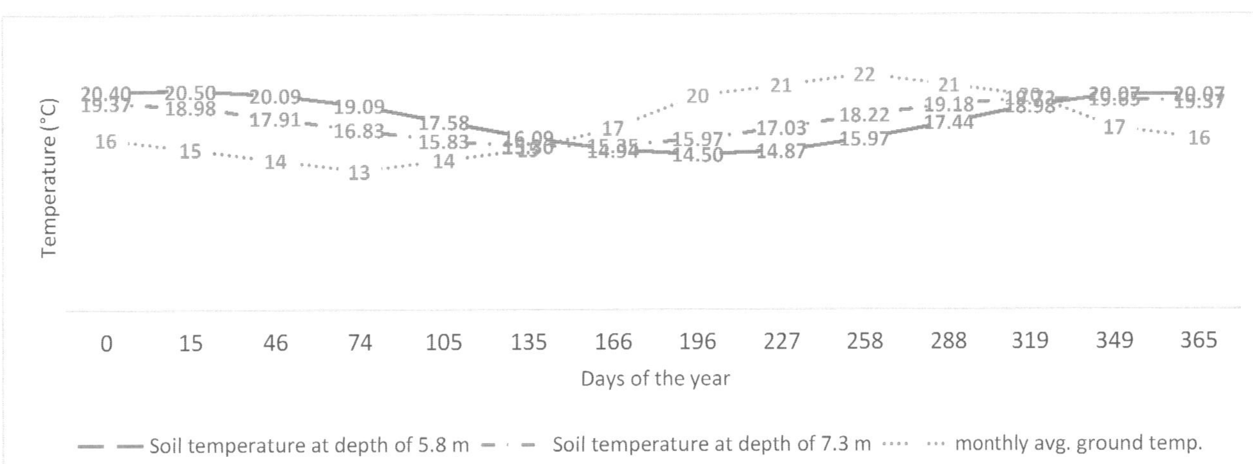

Figure 4: Variation of soil temperature as a function of days of the year

RESULTS AND RECOMMENDATIONS

Simulating the design using IESVE shows that the outdoor air lost a maximum of 7.8 °C during peak summer hours (external peak dry bulb temperature of 38.5 °C on August 3rd at 12 p.m.).

The base case assumes a flow of 37,000 l/s at a peak of 35.8 °C, treated all year round for cooling and dehumidification down to 12 °C. This requires an annual cooling energy of 1692 MWhr.

Upon inserting the proposed thermal labyrinth into the proposed ISE-VE software model and due to groud heat exchange the peak outdoor air temperature drops by 7.8 °C. Consequently, the annual energy required for cooling and dehumidifying the outdoor air at a lower temperature is reduced to 1489 MWhr. Accordingly, the saving in annual cooling energy due to the heat exchange across the thermal labyrinth are computed at 12%.

It can be concluded that since the thermal labyrinth is to be constructed right underneath the air conditioned basement of the palace, it is highly restricted by the seismic columns, with the requirement to keep at least three of the faces of each column spaced from the newly constructed concrete walls. As such, the distance between vertical concrete walls is of 4.7 m on average. This wide distance will not allow for the labyrinth system to capture the full extent of significant thermal characteristics of a similar concrete labyrinth setup with 0.6 m distance separating the concrete walls, as reported in the Melbourne thermal labyrinth case study. As such, the reduction in temperature during the peak summer season is restricted at 7.8 °C, despite the large available floor area within the seismic basement.

The concrete walls within the thermal labyrinth requires a contrcution cost of around 96,750 USD, assuming that the cost of reinforced concrete is currently estimated at a rate of 250 USD per m^3 in Algeria. The price of the flexible membrane to be added above the vertical concrete walls is estimated at a rate of 25 USD per m, thus totaling 26,350 USD for the entire proposed labyrinth system. On the other hand, energy savings due to heat transfer between the air travelling within the labyrinth and the adjacent soil are estimated at around 16,324 USD per year for an electricity rate of 0.077U kWh. The overall payback period of the labyrinth system is of 6 years.

CONCLUSION

This paper presented a study on the design strategy of a thermal labyrinth and its total energy savings. A case study in Algeria has been investigated with virtual illustration using IES-VE. The results of the analysis showed that by using an underground thermal labyrinth the outdoor air lost a maximum of 7.8 °C during peak summer hours. The estimated annual energy savings associated with temperature savings on the fresh air load is nearly 12 %. Hence, the use of Thermal labyrinth is effective in increasing building total energy savings.

NOMENCLATURE

Tz,t = ground temperature at time t and depth z (°C)
Tm= average soil surface temperature, set at 22°C and 13 °C for summer season
As = amplitude of the soil surface temperature variation (°C)
αs= soil thermal diffusivity (m^2/s; $m^2/days$)

Subscripts

z = depth
s = soil
t = time elapsed from beginning of calendar year (days)
t0 = phase constant of the soil surface (sec; days)

REFERENCES

Daniels, K. *The Technology of Ecological Building: Basic Principles and Measures, Example and Ideas,* Birkhauser Verlag, Basle Switzerland, 1997.

McGowan, S. *The Labyrinth cooling federation square,* the official Journal of Airah, 2003.

IES VE. 2015. Integrated Environmental Solutions
Kwang Ho Lee, Richard K. Implementation of an earth tube system into energyplus program, University of Illinois at Urbana-Champaign,
ASHRAE. 2010. *ASHRAE Handbook-Fundamentals*. Atlanta: American Society of heating Refrigeration and Air Conditioning Engineers, Inc.

New Concept of Power Generation from TEG Using the Exhaust Airflow of All-Air HVAC Systems and the Sun Irradiation

Mahmoud Khaled, PhD. **Khaled Chahine, PhD.** **Mohamad Ramadan, PhD.**

ABSTRACT

This work concerns a new concept of power generation from a thermo-electric generator TEG using the exhaust airflow of all-air Heating, Ventilating, and Air Conditioning HVAC systems and the sun radiation. As a first step, a parametric analysis of power generation from TEGs for different real configurations is done. The aim of the parametric analysis is to provide recommendations with respect to the applications of TEGs. To proceed, the one-dimensional steady-state solution of the heat diffusion equation is considered with various boundary conditions representing the real configurations. It is shown that the most promising configuration corresponds to the TEG module subjected to a hot fluid at one face and cold fluid at the other one. Then based on the parametric analysis, the new concept is drawn and analyzed using an appropriate thermal modeling. It is shown that for a solar radiation of 2000 W/m² and a space cooling load of 20 kW, a 40x40 cm² flat plate is capable of generating 120 W of electrical power.

INTRODUCTION

A method of applying energy management is through the recovery of waste heat, which is freely found in various applications starting from industries and reaching home appliances [1-5]. Excellent approaches in the direct conversion of waste heat energy into electrical power lie within thermoelectric power generation. A TEG is a solid state device that provides direct energy conversion from thermal energy into electrical energy due to a temperature gradient based on the "Seebeck effect" [6, 7]. On the other hand, Heating, Ventilating and Air Conditioning HVAC systems have changed from being luxury to essential need for people [8]. They involve many energy components that need to be more and more managed [9]. In this context, the present work suggests a new design that permits to couple the two energy domains described above: power generation from TEG and HVAC.

The proposed design uses a cubic oil tank having its bottom flat plate composed of several TEG modules in series and its upper surface transparent and exposed to the sun radiation. The bottom plate of the tank is cooled at its lower surface by the exhaust air of the HVAC system. Each of the TEG modules will then have its upper surface heated by oil (oil itself is heated by the transmitted solar radiation) and its lower surface cooled by air. This will induce a temperature gradient and then a potential difference (electrical power). To proceed, an appropriate thermal modeling of the design is developed and the performance is investigated in function of the solar radiation available and the cooling load of the conditioned space.

The remaining of the paper is organized as follows: section 2 presents the parametric analysis for different real thermal configurations a TEG must encounter, section 3 shows the thermal modeling of the system as well as the corresponding results, and finally section 4 draws the conclusions of the work.

Authors are faculty members at the Mechanical Department of the School of Engineering of the Lebanese International University

PARAMETRIC ANALYSIS AND RECOMMENDATIONS

This section concerns a parametric analysis of the TEG thermal behavior subjected to several conditions. The aim is to find recommendations as to obtaining significant temperature gradient through the TEG module. Although a temperature gradient is the trigger for a TEG to generate power, practically this gradient is not exactly nor necessarily obtained by fixing the temperatures at the two TEG surfaces. Moreover, fixing the temperatures at the two surfaces of the TEG requires external means requiring power. The most common boundary conditions that are encountered in practice will be exposed and treated below. For each case, the temperature gradient between the hot and cold surfaces of the TEG will be evaluated at steady state and parametrically analyzed. Figure 1 shows schematics of the different configurations that will be tested.

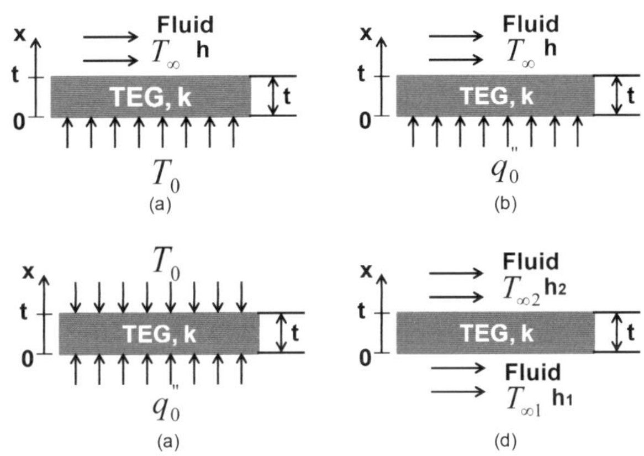

Figure 1 Schematics of (a) configuration 1, (b) configuration 2, (c) configuration 3, and (d) configuration 4.

In configuration one (Figure 1-a), the TEG of thickness t and thermal conductivity k is subjected to a constant temperature T_0 at its hot side and simultaneously cooled by air having a convective coefficient h and temperature T_∞. In configuration two (Figure 1-b), the TEG is subjected to a constant heat flux q_0'' at its hot side and simultaneously cooled by air having a convective coefficient h and temperature T_∞. In configuration three (Figure 1-c), the TEG is heated at its hot side with a constant heat flux q_0'' and simultaneously subjected to a constant temperature T_0 at its cold side. In configuration 4 (Figure 1-d), the TEG is heated at one side with air having a temperature $T_{\infty 1}$ and convective heat transfer coefficient h_1 while being simultaneously cooled with air having a temperature $T_{\infty 2}$ and convective heat transfer coefficient h_2 at the other side.

For the different conditions, the TEG is considered as a plane wall and the conduction through it as one-dimensional and steady. Then, the heat diffusion equation is reduced to:

$$\frac{d^2T}{dx^2} = 0 \qquad (1)$$

The temperature distribution and the temperature gradient in all configurations can both be obtained by solving the differential equation of the heat diffusion equation above, but will vary according to the boundary conditions. In all of the above configurations, integrating the reduced form of the heat diffusion equation with respect to x yields the below linear variation dependent on two constants A and B function of the boundary conditions:

$$T(x) = Ax + B \qquad (2)$$

Below, the boundary conditions, temperature distributions and temperature differences corresponding to the four configurations are exposed.

Configuration 1

$$T(0,\theta)=T_0 \tag{3-a}$$

$$-k\left.\frac{dT}{dx}\right|_{x=L}=h\left[T(t,\theta)-T_\infty\right] \tag{3-b}$$

$$T(x)=\frac{h(T_\infty-T)_0}{K+ht}x+T_0 \tag{3-c}$$

$$\Delta T=\frac{ht(T_0-T_\infty)}{k+ht} \tag{3-d}$$

Configuration 2

$$-k\left.\frac{dT}{dx}\right|_{x=0}=q_0^{"} \tag{4-a}$$

$$-k\left.\frac{dT}{dx}\right|_{x=L}=h\left[T(t,\theta)-T_\infty\right] \tag{4-b}$$

$$T(x)=-\frac{q_0^{"}}{k}x+T_\infty+\frac{ht+k}{hk}q_0^{"} \tag{4-c}$$

$$\Delta T=\frac{q_0^{"}t}{k} \tag{4-d}$$

Configuration 3

$$-k\left.\frac{dT}{dx}\right|_{x=0}=q_0^{"} \tag{5-a}$$

$$T(t,\theta)=T_0 \tag{5-b}$$

$$T(x)=-\frac{q_0^{"}}{k}x+T_0+\frac{q_0^{"}}{k}t \tag{5-c}$$

$$\Delta T=\frac{q_0^{"}t}{k} \tag{5-d}$$

Configuration 4

$$-k\left.\frac{dT}{dx}\right|_{x=0}=h_1\left(T_{\infty 1}-T(0)\right) \tag{6-a}$$

$$-k\left.\frac{dT}{dx}\right|_{x=L}=h_2\left(T(t)-T_{\infty 2}\right) \tag{6-b}$$

$$T(x)=\frac{h_1h_2\left(T_{\infty 2}-T_{\infty 1}\right)}{k(h_1+h_2)+h_1h_2t}x+T_{\infty 1}+\frac{kh_2\left(T_{\infty 2}-T_{\infty 1}\right)}{k(h_1+h_2)+h_1h_2t} \tag{6-c}$$

$$\Delta T=\frac{h_1h_2\left(T_{\infty 1}-T_{\infty 2}\right)}{k(h_1+h_2)+h_1h_2t}t \tag{6-d}$$

A parametric analysis based on the equations derived above is conducted to draw the main recommendations towards increasing the temperature difference across the TEG plate. These recommendations are summarized in Table 1.

Table 1. Summary of recommendations.

CONFIGURATION	DETAILS	RECOMMENDATIONS
1	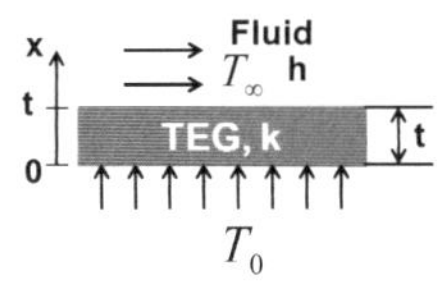	- The thickness of the TEG, the temperature prescribed at the hot surface of the TEG, and the convective coefficient should be high; - The fluid temperature at the cold side and the thermal conductivity of the TEG should be low.
2	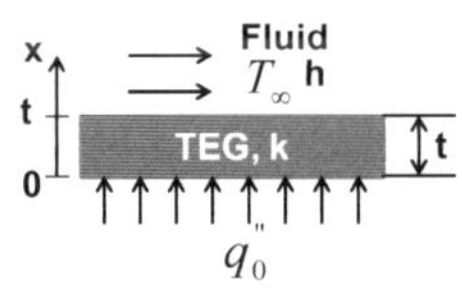	- The thickness of the TEG and the heat flux prescribed at the hot surface of the TEG should be high; - The thermal conductivity of the TEG should be low.
3	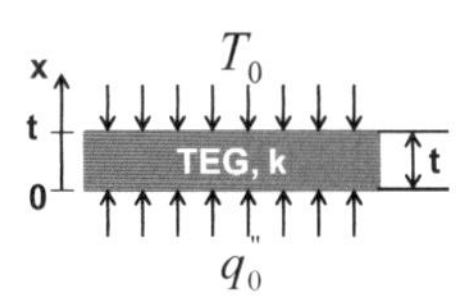	- The thickness of the TEG and the heat flux prescribed at the hot surface of the TEG should be high; - The thermal conductivity of the TEG should be low.
4	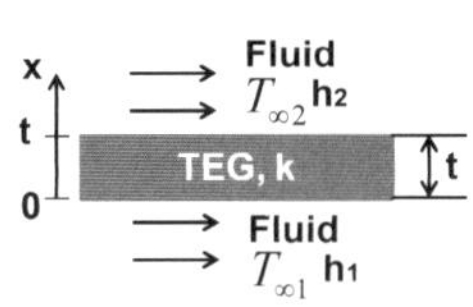	- The thickness of the TEG, the temperature of the hot fluid, the convective coefficients of the hot and cold sides should be high; - The thermal conductivity of the TEG and the temperature of the cold fluid should be low.

As conclusion, the most promising configuration is configuration 4 provided that the designer selects the appropriate couple of convective heat transfer coefficients since it has the most promising magnitude orders and corresponds at the same time to real configurations that can be encountered in engineering practice. All the configurations can be more promising if the thermal conductivity of the TEG can be lowered and its thickness increased which is related to the enhancement of the design of the TEG itself. Configurations 2 and 3 can be also promising if the applications involve high and extreme heat fluxes. Configuration 1 can also be more promising for applications where high temperatures are present. The following section will be devoted to the implementation of configuration 4 in a real case and its corresponding thermal behaviors.

NEW CONCEPT AND CALCULATIONS

The study presented in the previous section has shown that configuration 4 presents the highest gradient of temperature for the same conditions. As an application, a system that uses the solar radiation and the exhaust air flow of All-Air HVAC systems is suggested and investigated. The suggested system (Figure 2) consists in using a cubic oil tank having its bottom flat plate composed of several TEG modules in series and its upper surface transparent and exposed to the sun radiation. The bottom plate of the tank is cooled at its lower surface by the exhaust air of the

HVAC system. Each of the TEG modules will then have its upper surface heated by oil itself heated by the transmitted solar radiation and its lower surface cooled by air, which will induce a temperature gradient and then a voltage (electrical power).

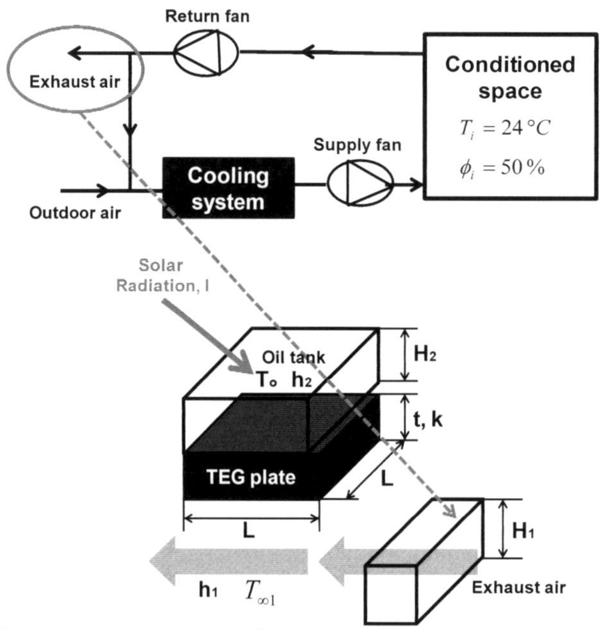

Figure 2 Schematic of the suggested concept of power generation.

The mass flow rate of air exhausted from the system is written as follows:

$$\dot{m}_e = E\dot{m}_s \tag{7}$$

where E is the fraction of exhausted air from supply air that depends on the type of conditioned space (apartment, industry, hospital, etc.) and $\dot{m}_s$ is the mass flow rate of air supplied to the conditioned space calculated from the following relation:

$$\dot{m}_s = \frac{\dot{Q}_{cooling}}{i_r - i_s} \tag{8}$$

where $\dot{Q}_{cooling}$ is the cooling load of the conditioned space, i_r the specific enthalpy of air in the room, and i_s the enthalpy of supplied air. According to ASHRAE standards [9], the indoor temperature and relative humidity are respectively 24 ^{0}C and 50 % which corresponds to an enthalpy of 48 kJ/kg. Then, with a typical value of the sensible heat factor of 0.8 and for a supply temperature of 13 ^{0}C, the enthalpy of supply air is 32 kJ/kg. Considering now the exhausted air flow below the TEG plate, the infinite velocity is calculated as follows:

$$u_{\infty 1} = \frac{4\dot{m}_e}{\rho \pi D_h^2} \tag{9}$$

where ρ is the air density and D_h the hydraulic diameter calculated as follows:

$$D_h = \frac{4S}{P} = \frac{2(H_1 L)}{H_1 + L} \tag{10}$$

Then, the Reynolds number at the edge of the bottom plate is calculated from the following relation:

$$\mathrm{Re}_L = \frac{\rho u_{\infty 1} L}{\mu} = \frac{4\dot{m}_e L}{\pi D_h^2 \mu} \tag{11}$$

where μ is the dynamic viscosity of air. Now to calculate the heat transfer coefficient h_1 of air at the lower surface of the TEG plate, the flow regime should be checked. If the Reynolds number Re_L is lower than 5x10^5, the flow is laminar all over the plate and then the convective transfer coefficient can be calculated from the following correlation [10]:

$$Nu_1 = \frac{h_1 L}{k} = 0.664\,\mathrm{Re}_L^{0.5}\,\mathrm{Pr}^{1/3} \tag{12}$$

where Nu_1 is the Nusselt number, k the thermal conductivity, and Pr the Prandtl number. If the Reynolds number Re_L is greater than 5x10^5, the flow is mixed (laminar then turbulent) over the plate and then the convective transfer coefficient can be calculated from the following correlation [10]:

$$Nu_1 = \frac{h_1 L}{k} = \left(0.037\,\mathrm{Re}_L^{4/5} - 871\right)\mathrm{Pr}^{1/3} \tag{13}$$

Consider now the energy balance on the oil side:

$$\dot{E}_{in} - \dot{E}_{out} + \dot{E}_g = \dot{E}_{st} \tag{14}$$

where $\dot{E}_{in}$ corresponds to the rate of energy in to the oil and equal to the transmitted solar radiation:

$$\dot{E}_{in} = \tau I L^2 \tag{15}$$

τ is the transmission coefficient of the transparent upper surface of the oil tank, I the solar radiation, and L the side of the oil tank. $\dot{E}_{out}$ is the rate of energy out from the system through the four walls of the tank with the ambient air and by convection through the base of the tank composed of the TEG modules. This component is written as follows:

$$\dot{E}_{out} = 4H_2 L U_{ins}\left(T_o - T_a\right) + L^2 U_b\left(T_o - T_{\infty 1}\right) \tag{16}$$

where H_2 is the height of the oil tank, U_{ins} is the overall heat transfer coefficient of the insulating material placed on the walls of the oil tank, T_o is the temperature of oil in the tank assumed uniform, T_a is the ambient temperature, $T_{\infty 1}$ is the air temperature below the bottom plate of the tank, and U_b is the overall heat transfer coefficient through the bottom plate of the tank and between the oil and air calculated from the following relation:

$$U_b = \frac{1}{\frac{1}{h_2} + \frac{t}{k} + \frac{1}{h_1}} \tag{17}$$

Then at steady state and without heat generation:

$$\dot{E}_{in} = \dot{E}_{out} \tag{18}$$

and so:

$$T_o = \frac{IL + 4H_2U_{ins}T_a + LU_bT_{\infty 1}}{4H_2U_{ins} + U_bL} \tag{19}$$

Then substituting the different parameters obtained above (h_1, $T_{\infty 1}$, h_2, T_o) in equation (6-d), the temperature difference across each TEG module is finally:

$$\Delta T = \frac{h_1h_2(T_o - T_{\infty 1})}{k(h_1 + h_2) + h_1h_2t}t \tag{20}$$

Figure 3 shows the variation of the temperature difference across each TEG module as well as the total power obtained by the suggested generator in function of the space cooling load for different solar radiation of 400, 800, 1200, 1600, and 2000 W/m^2. For this set of calculations, the room temperature is fixed at 24 ^{0}C, the fraction of exhausted from supply air E is set equal to 0.4, the height of the exhaust duct H_1 is taken equal to 0.1 m, the height of the oil tank H_2 is 0.1 m, the length and width of the bottom plate composed of TEG modules are equal to $L = 0.4\ m$, the heat transfer coefficient of oil h_2 is equal to 200 W/m^2.K [10], the overall heat transfer coefficient of insultation U_{ins} is equal 8.4 W/m^2.K [10], the ambient temperature T_a is set equal to 30 ^{0}C, and the thickness and conductivity of each TEG module are respectively 0.12 m and 0.3 W/m.K [11-12].

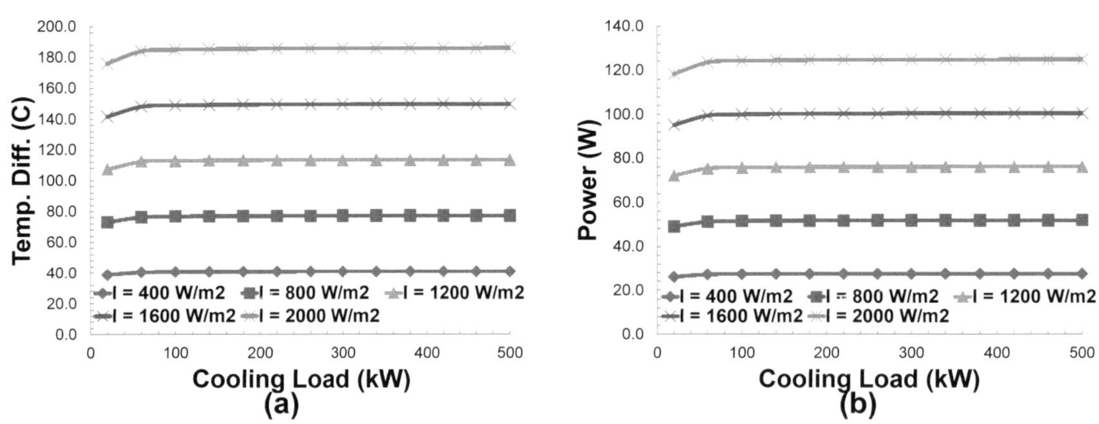

Figure 3 Variation of (a) temperature difference across each TEG module and (b) power generated by the assembly of TEG modules in function of the cooling load.

As shown in Figure 3, the temperature difference across each TEG module (Figure 3-a) and the power generated by the assembly (Figure 3-b) increase slightly with the cooling load and significantly (almost linearly) with the solar raidiation. As illustration for a solar radiation of 400 W/m2 when the cooling load varies from 20 to 500 kW, the temperature difference across each TEG module varies from 38.6 to 40.9 ^{0}C and the power generated by the assembly varies from 26 to 27.5 W. For a solar radiation of 2000 W/m2 when the cooling load varies from 20 to 500 kW, the temperature difference across each TEG module varies from 175.9 to 186 ^{0}C and the power generated by the assembly varies from 118.2 to 125 W.

Indeed when the cooling load increases, the mass flow rate of supply and then of exhausted air increases. This will increase the convective heat transfer coefficient at the lower face of the bottom plate and then will increase the temperature difference across each TEG module as shown in the parametric analsysis section above. However, increasing the convective heat transfer coefficient at the lower surface of the bottom plate will increase the heat losses of the oil in the tank and then decrease its temperature. This is why the effect of the cooling load on the temperature difference across each TEG module and the power of the assembly is dampened.

CONCLUSION

In this manuscript, a new concept of power generation from TEGs using the exhaust airflow of all-air HVAC systems and the sun radiation is presented. An appropriate thermal modeling of the system is exposed as well. It was shown that for a solar radiation of 2000 W/m^2 and a space cooling load of 20 kW, a 40x40 cm^2 flat plate is capable to generate 120 W of electrical power.

REFERENCES

Ramadan M, Khaled. M, El Hage H. 2015. Using Speed Bump for Power Generation –Experimental Study. Energy Procedia 75, 867–872.

Khaled. M, Ramadan. M, El Hage H. 2015. Parametric Analysis of Heat Recovery from Exhaust Gases of Generators. Energy Procedia 75, 3295–3300.

Ramadan, M., Gad El Rab, M. and Khaled, M. 2015. Parametric analysis of air–water heat recovery concept applied to HVAC systems: Effect of mass flow rates. Case Studies in Thermal Engineering, 6, 61-68.

Khaled. M, Ramadan. M, Chahine. K, Assi. A. 2015. Prototype implementation and experimental analysis of water heating using recovered waste heat of chimneys. Case Studies in Thermal Engineering, 5,127-133.

Johnson I. and Choate W. T. 2008. Waste Heat Recovery: Technology and Opportunities in U.S. Industry. BCS, Incorporated, March.

Bell L. E. 2008. Cooling, Heating, Generating Power, and Recovering Waste Heat with Thermoelectric Systems. Science 321, 1457.

Vining C. B. 2009. An Inconvenient Truth about Thermoelectrics. Nature Materials 8, 83.

Nagengast, B. 1999. Early Twentieth Century Air-Conditioning Engineering. ASHRAE Journal, March.

McQuiston, F.C., Parker, J.D., and Spitler, J.D. 2005. Heating, Ventilating and Air Conditioning: Analysis and Design. 6th Ed. John Wiley & Sons.

Incorpera, F.P. and DeWitt, D.P. 2007. Fundamentals of heat and mass transfer. Sixth Edition. John Wiley & Sons.

Spices P., Pollak M., and Mateu L. 2015. Handbook of Energy Harvesting Power Supplies and Applications edited by Peter. CRC Press Taylor and Francis Group.

Synder G.J. and Toberer E.S. 2008. Complex thermoelectric materials, Nature Materials 7, 108-116.

Validation of Key Performance Parameters for Low Carbon Buildings in the Hot Humid Climate of the Gulf

Yahya Al Shamsi

Student Member ASHRAE

ABSTRACT

All Gulf Cooperation Council (GCC) countries are listed in the top 20 countries for highest annual energy consumption and CO_2 emissions per capita. In the past few decades, numerous residential projects constructed in the region have led to higher energy demand for the residential sector. In addition, the growing rate of urbanization and changes in lifestyles caused a significant increase in domestic energy consumption and in Oman, during the period from 2012 to 2018, peak electricity demand is expected to grow at 8% per year compared to the global average rate of 2.5%. This rapid increase in energy consumption has highlighted the need for energy solutions to overcome its excessive consumption. Since the residential sector consumes 48% of total electricity, the main source of energy in Oman, reducing domestic energy use has the potential to reduce the country's total energy consumption. This research aims to evaluate experimentally the key Low Carbon Building (LCB) strategies and criteria best suited to the economic, climatic and social constraints for the hot humid climate of Oman that can be implemented in their future residential sector. For the purpose of this research, the energy performance of reference exemplar state of the art low carbon buildings available in the country, will be compared to selected examples of conventional buildings located within 15 km from the reference buildings. Six key performance parameters of energy consumption in the LCB; lighting, domestic electronic appliances, air conditioning (HVAC), hot water, refrigerators and wash machine, were monitored and recorded, whereas the energy consumption for the same parameters in the conventional buildings were calculated by energy audit. Building energy consumption patterns and performance were monitored and calculated. Based on the data collected, the domestic building energy consumption and energy performance index (EPI) for residential buildings were calculated. Then the modeled energy saving and CO_2 emission reductions that may be achieved by implementing LCB strategies on conventional building were evaluated using IES virtual environment software.

INTRODUCTION

Oman is one of the six Arab States that form the Gulf Cooperation Council (GCC) in the Arabian Peninsula. These countries are well known of their production and large reservoirs of crude oil and gas. The proven reserves of oil and gas in these countries are 40% and 23% respectively of the total global proven reserves (Reiche, 2010). As a result of rising oil prices in the 1970s, GCC countries used the revenue gained from the oil industry to build infrastructure and new modern cities. Thus, the oil wealth is reflected in considerable changes in life style patterns and standards of living. In the past 40 years, many new cities and residential developments were created to accommodate the rapidly increasing population. Similarly, the construction industry changed to meet the modern demands of local societies. New construction materials to the region were introduced and building design and features changed to suit the changes in life styles. The growing rate of urbanization, and changes in life patterns have resulted in huge domestic energy use characterized by the high rate of increase in energy consumption per capita. In 2012, the annual electricity report presented by Oman Power and Water Procurements Company stated that the average annual increases in electricity were 7.1% and 8.4% for the two main electricity utilities. The company expected the growth of electricity demand to be between 8% and 10% for the period from 2012 to 2018 (SAOC, 2012). In 2012 the electricity energy consumed by the residential sector was 48% of the total electricity produced, whereas an energy report presented by Earth Trends Organization in 1999 showed that the residential energy consumption was only 9%. Hence, demand for electricity will keep growing in the absence of domestic building energy saving strategies.

CO_2 emissions produced by buildings account for nearly 40 percent of recorded global emissions. Hence, the energy performance of buildings received increasing concern from researchers, government, and non-governmental organizations in the past decades. Many researchers believe that improving the energy performance of buildings can contribute to solving global climate change if we intend to build and rebuild more energy efficient buildings. The use of low energy technology at the operational stage, with the awareness of the communities on energy consumption, alongside the implementation of practical regulations and standards, will substantially reduce overall energy consumption in the building sector.

Nowadays the construction industry in the Gulf Countries is considered to be one of the largest expanding sectors, with an annual worth of more than $5 billion, and it includes more than 2.7 million companies of various sizes (Construct Arabia, 2012). The construction industry is growing rapidly, and these countries are facing continued demand for housing, especially in major cities. This is due to rapid increases in population and the expanding oil industry sector that attracts many people to migrate to these countries (A. Al-Mulla, 2013). The continued demand for housing units has resulted in a shift in the industry to new methods of construction to sustain its ability in constructing the required buildings. Therefore, building design, materials and methods of construction are influenced by the construction practices of modern industrial countries. In some cases, low rise buildings have been replaced by high rise buildings to overcome the rapid increase in required housing units. According to data from the Skyscraper Centre, 11 of the top 30 tallest towers completed in the world are located in GCC countries. Consequently, construction of new dwellings has increased at a high rate and this rate may continue for the next few years to meet the rate of increasing population growth. According to Oman census data, residential building construction permits increased from 18 230 to 33 264 between 2010 and 2012, making an increase of 55% in two years.

Since fossil fuel is the main energy source in GCC countries, and the consumption of fossil fuel is one of the major drivers a deteriorating for climate, it has raised the awareness to reduce local energy consumption. The Intergovernmental Panel on Climate Change (IPCC) stated that the energy consumption in the developing countries is responsible for 40 % of greenhouses gas generation (IPCC, 2007). In fact, these values can be considered more in Gulf Countries due to a more rapid consumption of energy for their accelerated development. Hence, emissions of CO_2 associated with energy will remain at its current high values or may even increase as the use of fossil fuel continues to rise. According to World Development Indicators from the World Bank, all GCC countries are listed in the top 20 highest energy consumption per capita per annum. Oman was ranked in 10th position in this list, whereas residential building consumes 48% of the total electricity produced in the country. By considering the fact that residential buildings were designed without considering their energy consumption, most of this energy is consumed for space cooling and meeting thermal comfort requirements. Therefore, this paper presents the problems related to the absence of sustainable architectural design and shows the need to consider main elements of low carbon building to reduce energy consumption of buildings based on the evidence of multiple case study simulation and analysis.

THE NEED TO PROMOTE LOW CARBON SOLUTIONS IN THE GULF STATES

In gulf countries, where the development and construction activities rapidly follow trend, the construction industry and their associated energy use became major contributors to many environmental statistics such as the higher CO_2 emission per capita. Therefore, designers and contractors are required to consider the importance of sustainable development and how it can be incorporated into the construction industry sector. The UN Agenda 21 (CIB & UNEP-

IETC, 2002) stated that the majority of developing world is in a process of fast construction. Therefore, a greater understanding of sustainable development is needed in these countries in general and in the gulf countries in particular because of the issues associated with the environmental impact of rapid building construction in the GCC region. The increasing population with increased housing demand, their increased energy needs, with the finite conventional energy supplies makes the issue of energy consumption in buildings a challenging issue for gulf countries. Energy use in buildings has became more critical in the absence of government incentives such as subsidies and grants to promote energy efficiency measures or low cost strategies such as the adoption of low carbon buildings design. It seems to be a better option for these countries to reduce energy use and associated energy requirements without affecting the internal comfort conditions requirements of occupants through the use of building energy guidelines that promote and encourage the adoption of low carbon building in residential sector. Unfortunately, the associated knowledge-base in the gulf region, and in particularly for Oman is poor. In addition, the concept of low carbon building is still generic, with evaluation of the key performance parameters that define low carbon building in this particular climate. Therefore, the need exists to understand these parameters and evaluate their effectiveness when implemented as part of low carbon building strategies for this region. This paper will focus on validating the effectiveness of key low carbon strategies for residential buildings in order to promote their adoption in the Sultanate of Oman. It is generally believed that a such target involves many parties including decisions makers, researchers, construction industry participates and general public to generate solution exclusively suitable for Oman.

Low Carbon Building Characteristics

Low carbon buildings can be defined as any type of building that from design and operation uses less energy than average traditional building of a similar size. In practice this includes the use of design, architecture, energy efficient strategies, low-energy appliances, and, landscaping to reduce its energy demand. Further, low carbon buildings often use passive solar design to meet its required energy efficiency level. Therefore, the use of solar energy, either in passive or active form or are very common in LCBs.

Buildings achieve high energy performance through three main strategies: energy demand reduction, use of energy efficient house appliances and the use of renewable energy. The energy demand reduction can be achieved to a very significant extent through design, orientation, construction materials, use of proper shading and the application of insulation. Whereas energy efficiency in building achieved through the use of rated and smart low energy home appliances that use less energy than conventional appliances. Finally, the use of available renewable energy will reduce energy consumption from conventional sources. The common main elements of best practice low carbon building includes construction methods and materials, operations and management of energy inside building. This paper will evaluate the effects of building materials, orientation and glazing type as an elements of good practice low carbon residential building on the total energy consumptions and CO_2 emissions. The energy consumption of the main home tasks included in this evaluation are lighting, domestic electronics, air conditioning (HVAC), hot water, refrigeration and washing. These type of energy loads, commonly known as measured load, normally consumes the largest part of the total home energy consumption in the gulf countries.

METHODOLOGY

The methodology used in this paper consisted of comparing a sample of state of the art low carbon residential building exemplars with two conventional dwellings of the same size. These houses were subjected to detailed analysis and simulation in order to determine the effects of the main performance parameters that characterize low carbon building and estimate the specific annual energy consumption for typical Omani dwellings (kWh/m^2). The selection of conventional building considered the number of occupants in each property and the building sizes were typical Omani style house typologies. The selected reference LCB is located in the campus of Higher College of Technology (HCT) in Muscat. The reference LCB is located in the heart of Muscat and represents the state of the art for low-carbon design strategies and energy performance. The building is a two-story 240 m^2 residential building constructed as a research building located in the campus of HCT in Muscat, Al-Khuwair. The design of this building is similar to typical 3-bed rooms Omani houses with a sitting room, living room, three toilets with shower and kitchen. A number of low carbon techniques were employed by the design team in this building to ensure low energy consumption. The building was comprised of individual zones reflect the different thermal characteristics of the building area in terms of both the physical properties and the properties of the air. These zones correspond largely to the cooling loads. The building is constructed from reinforced Nudura Concrete covered by a canopy carrying PV panels that also provides shading for the roof.

The sample conventional buildings were selected from 50 residential buildings reviewed in this research. Each building consisted of three bedrooms, three toilets, sitting room and family room. All houses were occupied by families close to the average size of an Omani family, typically 7 people per house. For each house, the actual architectural layout and construction materials were studied to inform the building description required for the thermal modelling phase. In addition, the monthly electricity bills were obtained from local utility company per annum. The age of both buildings was less than 10 years old taking into account that more than 40% of residential buildings in Muscat were constructed after 2004. A summary description of case study buildings is shown in Table 1.

LBC energy consumption measured by a data-acquisition system recorded energy use for 6 main home tasks: lighting, domestic electronics, air conditioning (HVAC), hot water, wash machine and refrigerators; whereas energy consumption in conventional building calculated by energy audit. Since energy used in the conventional buildings was mainly electricity, and according to International Energy Agency report (2010), generating 1 kWh of electricity in Oman produces 794 grams of CO_2 emission, allowing CO_2 emissions to be calculated accordingly, whereas the EPI for conventional on buildings calculated by dividing energy consumed by building gross area.

Table 1. Discretion of Case Study Buildings

Heading Number One	Building A	Building B	LCB
Area (m^2)	212	240	220
Occupancy (No.)	6	7	3
Long Axis Orientation from North (°)	75	60	90
Floor Plan Shape	Rectangular	Square	Rectangle
Construction Materials	Concrete	Concrete	Nudura Concrete
% Glazing	10	17	22
External wall U Value (W/m^2K)	2.1	2.1	0.233
Internal wall U Value (W/m^2K)	2.1	2.1	0.68
Roof U Value (W/m^2K)	2.2	2.2	0.339
Floor U Value (W/m^2K)	2.5	2.5	0.568
Glazing U Value (W/m^2K)	2.7	2.3	1.9

IES-VE dynamic thermal analysis software was used in this study to simulate each conventional building individually and compared with recorded energy consumption, whereas the monitored LCB were modelled to predict the value of each LC design feature on the energy consumption. Then a dynamic simulation was carried to find out the impact of LCB features on energy consumption in conventional buildings. These features include building fabric, type and sizes of glazing and orientation.

RESULTS AND DISCUSSION

Breakdown of the energy consumption per tasks shows that 73% of the total annual energy is used for space cooling. This can be seen from the high variation in energy bills between summer and winter. Based on energy audits conducted on 50 typical Omani houses, the average energy use for air-conditioning is 73% of total energy consumption. This result is seems to be similar in most GCC countries where researchers from Kuwait and Oman found similar values (F. Farraj, 2010), (H. Al-Hinai,.1993). The energy consumption in domestic building in Oman expected to grow more due to urbanization, subsidized tariffs and the increase use of energy intensive home appliances in the absence of regulated labelling system and energy use regulation.

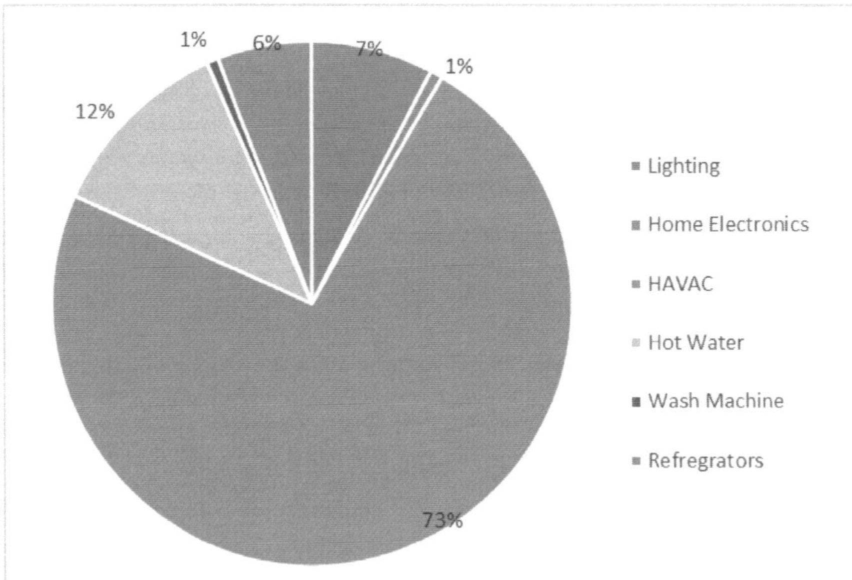

Figure 1 Typical Omani Household Energy Consumption Breakdown

Energy audits on conventional buildings were considered acceptable as energy consumption of these tasks are in the range of 96% to 87% of the total energy bill. The energy consumptions measurement of LCB shows less use of energy for HVAC system, lighting and hot water compared to conventional buildings (Figure 2). Reference LCB designed and operated as net energy consumption therefore, its net CO_2 emission was assumed to be neutral, whereas its EPI from modelling was 96 kWh/m^2. Applying LCB features to conventional buildings modeled using IES virtual environment software showed a reduction in energy use and its associated emissions of up to 39% (Figure. 3).

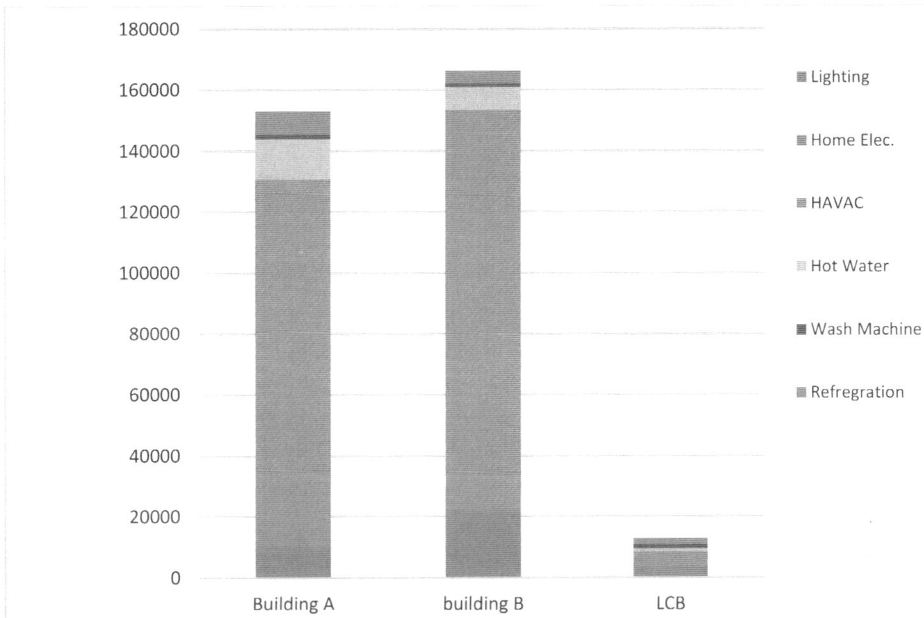

Figure 2 Energy Measurements and Calculations for Reference buildings in kWh

Building Energy Performance and Modelling

Figure 1, clearly shows that the main energy consuming tasks in a typical Omani house are HAVAC and lighting. HAVAC system energy consumption is influenced by the building design, materials, orientation and occupants use; whereas lighting can be reduced by making use of available natural lighting. Therefore, the parametric modeling of sample conventional buildings focused on building materials, cooling loads, lighting and orientation of the building to investigate the possible reduction of energy use associated with these parameters. The annual energy consumption in for simulated conventional dwellings with LCB features were found to be lower than modelled values for conventional buildings. These reductions varied by parameter, for each building. The reductions in energy consumption were between 27% and 39%, whereas the range of energy reduction due to orientation in both houses was between 2% and 10%. This variation in reduction of energy use between the two buildings can be referred to the layout of the buildings and the difference in total built area. The type of glazing was found to be responsible for a reduction of 3% and 4% in house A and B respectively. This showed limited effects compared to the effects of the fabric due minimal differences in windows U-values between conventional houses and LCB compared to the differences in U values for building materials.

Table 2. Calculated and Measured Energy Consumption and CO_2 Emissions of Case Study Buildings

References	Building A	Building B	LCB Exemplar
Lighting (kWh)	3.8	2.8	2.9
Home Electronics (kWh)	1	1.1	0.4
HAVAC (kWh)	67.2	103.4	18.7
Hot Water (kWh)	13.2	15.6	0.8
Washing Machine (kWh)	0.2	1.4	0.9
Annual CO_2 Emissions (kg)	24311.5	30682.5	N/A
CO_2 Emissions / Capita	4051.9	4383.2	N/A
CO_2 Emissions / m^2	114.7	127.8	N/A
Energy Performance Index kWh/m^2	144.4	161	96

For the reference LCB the detailed inputs data for the building envelope were used to model the hourly energy consumption for each room? for the simulation period. For this building, at its current orientation, the main energy consumption was in the guest sitting room and master bedroom on 25th of July. This resulted from the solar gains experienced by these two rooms at their current orientation.

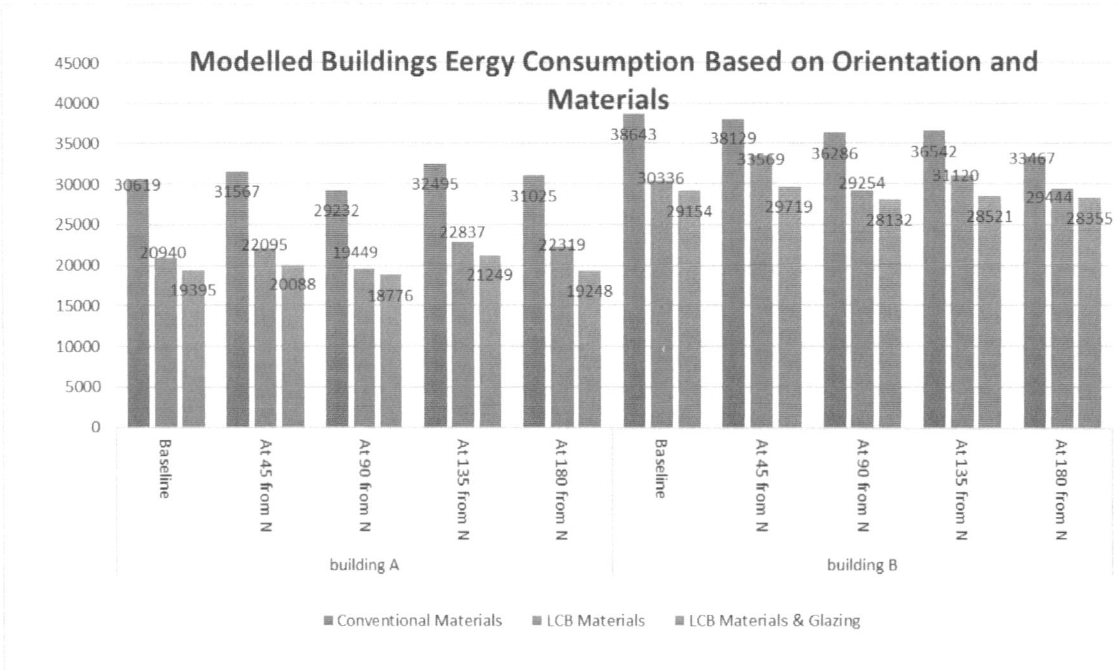

Figure 3 Results of Annual Energy Consumption for Modelled Conventional building

CONCLUSION

This paper evaluated and examined the main performance parameters of LCB in hot humid climates, using Sultanate of Oman as a case study region. It evaluated the energy consumption patterns based on simulations of multiple case study analyses, with respect to actual utility electricity bills and direct energy measurements. Dynamic thermal modeling software was (IES-VE) employed to simulate and compare the energy consumption of three reference houses. The simulation results were supported by data collected from official utility bills, actual layout plans, field measurements, energy audits and interviews with occupants of each dwelling. The results show the potential to reduce energy use and associated CO_2 emissions of domestic building by up to 39% through the introduction of LBC design strategies. This, will help to reduce the rapid demand for energy in domestic building which will reduce overall country energy consumption. Therefore, this study concludes recommendations for adopting LCB feathers in the construction of new dwellings.

ACKNOWLEDGMENTS

I would like to acknowledge my sincere gratitude to The Research Council (TRC) in Oman for providing required measurements and data for reference LCB. Further, to Ministry of Labor and Vocational Training represented in HCT for funding this research.

REFERENCES

A. Al-Mulla, G.P.M., D. Al-Nakib, A. ElSherbini, F. Alghimlas, H. Al-Taqi, Y. Al-Hadban, 2013. *Enhancement of building operations: A successful approach towards national electrical demand management*. Energy Conversion and Management, 76.

A.J. Marszala, P.H., J.S. Bourrelle, E. Musall, K. Voss, I. Sartori, A. Napolitano, 2011. *Zero Energy Building – A review of definitions and calculation methodologies. Energy and Buildings*, 43, p. 971.

ASHRAE, 2009. *ASHRAE handbook: Fundamentals*. N.E, Atlanta: American Society of heating, refrigerating and air-conditioning Engineers, Inc.

CIB., UNEP-IETC., 2002. *Agenda 21 for Sustainable Construction in Developing Countries-A Discussion document*. Pretoria: CSIR building and construction technology.

ConstructArabia, 2012. *Updated Building Regulations in the GCC*. ventures.

Doukas, H., K. D. Patlitzianas, A. G. Kagiannas and J. Psarras (2006). *"Renewable energy sources and rationale use of energy development in the countries of GCC: Myth or reality?"* Renewable Energy 31(6): 755-770.

ESRU., 1998. ESP—*A Building and Plant Simulation System-User guide*. Glasgow: Energy Simulation Research Unit, University of Strathclyde.

Farraj F. Al-ajmi, D.L.L., 2010. *Indoor thermal conditions and thermal comfort in air-conditioned domestic buildings in the dry-desert climate of Kuwait. Building and Environment* 45 p. 6.

H. Al-Hinai, W. J. B. S. D. P. (1993). *"Vernacular Architecture of Oman: Features That Enhance Thermal Comfort Achieved Within Buildings."* Applied Energy 4: 233 258.

Iwaro, J., Mwasha, A., 2010. *A Review of Building Energy Regulation and Policy for Energy Conservation in Developing Countries*. Energy policy, 38, pp.7744-7755.

Kottek, M., Grieser, J., Beck, C., Rudolf, B., Rubel, F., 2006. *World map of the Koeppen-Geiger Climate Classification Update.* Meteorologische Zeitschrift, 15, pp.259-263.

Menezes, A., Cripps, A., Douchlaghem, D., Buswell, R., 2012. *Predicted vs. Actual Energy Performance of Non-Domestic Buildings: Using Post-Occupancy Evaluation Data to Reduce the Performance Gap*. Applied energy, 97, pp.335-364

Nicol, F., 2004. *Adaptive thermal comfort standards in the hot-humid tropics*. Energy and Buildings, 36, pp.628-637.

Niu, S., Ding, Y., Niu, Y., Li, Y., Luo, G., 2011. Economic growth, energy conservation and emission reduction: A comperative analysis based on panel data for 8 Asian-Pacific countries. Energy Policy, 39, pp. 2121-2131.

Petersen, S., Svendsen, S., 2010. Method and simulation program informed decisions in the early stages of building design. Energy and Buildings, 42, pp.1113-1119.

Ragette, F., 2003. Traditional domestic Arabic architecture of the Arab region. Sharja: American University of Sharja.

Reiche, D. (2010). *"Energy Policies of Gulf Cooperation Council (GCC) countries—possibilities and limitations of ecological modernization in rentier states."* Energy Policy 38(5): 2395-2403.

SAOC, Oman Power and Water Procurements Company. (2012)."*7YEAR STATEMENT (2012 – 2018).*" (6).

Shao, L., G. Q. Chen, Z. M. Chen, S. Guo, M. Y. Han, B. Zhang, T. Hayat, A. Alsaedi and B. Ahmad (2014). *"Systems accounting for energy consumption and carbon emission by building."* Communications in Nonlinear Science and Numerical Simulation 19(6): 1859-1873.

Weytjens, L., Verbeeck, G., 2010. Towards 'Architecdt-Friendly' Energy Evaluation Tools. *Proceedings of the 2010 Spring Simulation Multi-conference*, 2010 San Diego, CA, USA. Article No. 179.

World Bank., 2012. World Bank list of economies [online] available from: http://data.worldbank.org/about/country-classifications/country-and-lending-groups [accessed 11 March 2014].

The Numerical Design of a Micro Heat Exchanger Thermoelectrically Cooled and Operating in the Continuum Regime

Iyad Fayssal **Ahmad Ziadeh** **Bader Makki**

Hadi Farah **Mohammad Alameh** **Fadl Moukalled**

ABSTRACT

Small scale heat exchangers constituted of low hydraulic diameter micro-channels are regarded as future promising devices to be adopted extensively in cooling applications due to their high convection coefficients and lower pumping power demands. However, the mathematical modeling of such devices remains challenging due mostly to the nature of the flow regime in the micro-channel. The aim of this research is to build a computational-based strategy that will aid manufacturers in designing small scale heat exchangers guaranteeing their operation in the continuum regime. A 3D numerical model of a thermoelectric cooled micro heat exchanger was constructed consisting of a thermoelectric plate (23mm x 23mm) mounted on the top of 92 staggered rectangular PDMS micro-channels (150µm x150µm) separated by 100µm thickness PDMS fins. In order to induce the thermoelectric cooling effect, a lumped mathematical model is implemented as a boundary condition to the energy equation relating the heat flux at the cold surface of the thermoelectric to the temperature difference of its cold and hot sides. This boundary condition is implicitly specified with the boundary temperatures being outcomes of the solution. A generic C code is developed and coupled with fluent solver for serial and parallel computations to iteratively resolve the cold heat flux at the boundary. A conjugate heat transfer problem is formulated with the finite volume method adopted to solve the set of governing mass, momentum, and energy equations in the fluid zone (air) and energy in the solid zones (PDMS plate and fins). The method starts by imposing a fixed pressure at the outlet sections of the domain. A uniform velocity profile with arbitrary temperature is supplied at the inlet section of the heat exchanger. The hydrodynamic and thermal fields and axial distribution of Knudsen number (Kn) are then predicted in the interior domain. Different profiles in Knudsen number variation are observed depending on the simulated operating conditions. According to the spatial distribution of Knudsen number, the model parameters are updated for cases when the flow regime falls outside the continuum. As a future target, a design map relating the inlet air temperature, outlet pressure, and Reynolds number will be generated as an engineering tool via an integrated low computational cost methodology.

INTRODUCTION

Thermoelectrics, solid state Peltier devices made up of semi-conductor material that convert temperature gradient to electric current or vice versa, have been widely applied at the macro scale level. Various numerical and experimental based investigations were reported in the literature. Briefly mentioning some, Milani et al. (2011) investigated using thermoelectric the possibility of cooling humid air to produce fresh water. A similar study was conducted by Yildrim et al. (2014) in which thermoelectric coolers were used for humidification and dehumidification of air. Jradi et al. (2012) conducted an experimental and theoretical investigation of an integrated thermoelectric-photovoltaic system to study air dehumidification and fresh water production. Li et al. (2011) utilized thermoelectric coolers to improve power dissipation of light-emitting diodes (LEDs). On the other hand, Liu et al. (2014) implemented thermoelectric coolers in ventilation systems instead of hydronic pipes while Khattab and El Shenawy (2006) used solar thermoelectric generators to power thermoelectric coolers. Thermoelectrics have been also implemented extensively in Automobile Industry. For instance, Choi et al. (2007) used thermoelectric coolers to regulate car seat temperatures, while Orr et al. (2014) exploited the heat dissipated from car exhaust to generate power using thermoelectric generators.

While thermoelectrics have been integrated extensively at the macro scale level, limited engineering applications have adopted thermoelectrics at the micro level, in specific their integration in microchannels. Microchannels have been extensively used in several applications due to their high convection coefficients and low power pumping demands. Francescon et al. (2013) evaluated the performance of silicon micro-channel cooling devices for their use in the cooling of high energy physics detectors. Their cooling performance has been found to be officially selected as a thermal management device of the Gigatracker detector of the "NA62" experiment at CERN (Conseil Européen pour la Recherche Nucléaire). Microchannels are usually characterized by hydraulic diameters of less than 1 mm. Understanding the flow regimes in microchannels is quite interesting and is dominantly determined by Knudsen number (Kn). The Knudsen number is a dimensionless number that denotes the ratio of mean free path to a characteristic physical length scale and helps determine the flow regime in the channel. For small Knudsen numbers ($Kn < 10^{-3}$), the flow is said to be in the continuum regime and thus it is modeled following the traditional Navier-Stokes equation. For higher Knudsen numbers ($Kn > 10^{-3}$) the flow is no longer in the continuum regime and the conventional Navier-Stokes starts breaking down. Reporting on some general studies on modeling flow in micro sized channels, Papautsky et al. (1999) developed an alternative way to model fluid flow in micro channels using the micro polar fluid theory and validated their results with experimental data. A slip model on a macro scale was devised by Laurent et al. (2011) to study the slippage of a gas along mobile rigid walls in the sphere-plane confined geometry. Issa and Lakkis (2014) developed reduced-order models using analog circuits to study the unsteady low-Mach-number of ideal gas flows in two-dimensional rectangular microchannels subjected to first-order slip boundary conditions. In their study, they have investigated the accuracy of the proposed models for steady state and unsteady flow for different values of Mach (Ma) and Knudsen (Kn) by reporting the dynamic response of volume flow rate for different pressure functions. Harley et al. (1989) experimentally studied the effect of reducing channel characteristic length on flow reaction in a micro channel. Similarly, Pfahler et al. (1990) conducted a series of experiments on small channels (of orders of microns) with decreasing cross-sectional areas to determine the limit of the scale length upon which continuum breaks down.

While a general model for slip is not yet available, many studies have been conducted on obtaining particular slip models. Fan and Xue (1998) demonstrated the validity of the DSMC method in computing the flow regime in areas where the Knudsen number is relatively high ($Kn > 0.1$). This was later compared to the work of Arkilic et al. (1997) who modified the Navier Stokes equations to compensate for the slipping that occurs along the boundaries. A common slip model for gases was also studied by Bahrami et al. (2009), in which a first order Maxwell slip boundary condition was applied in a micro channel to investigate the effects of various channel shapes and different gases.

Guo et al. (2007) derived an extended form of the Navier-Stokes equations to account for hydrodynamic changes in the Knudsen layer to study the effect of the collisions between the gas molecules and bounded walls. Jie et al. (2000) investigated the flow in a micro channel by incorporating a second-order slip model and solved it using the LU-TVD implicit algorithm. Their model yielded very good results when compared to experimental data and other reported numerical results.

Since this work involves heat transfer in microchannels, few studies concerned with heat transfer processes in micochannels are discussed. Garcia-Hernando et al. (2009) conducted an experimental investigation to study the hydrodynamic and heat transfer performance in two micro-heat exchangers, characterized by microchannels of 100 × 100 and 200 × 200 μm square cross-sections, using deionized water as the working fluid. Another experimental work by Dang & Teng (2011) was incorporated to study the influences of geometric configurations on the performance in microchannel heat exchangers. A computational and experimental study in a rectangular-shaped microchannel counter-flow heat exchanger was conducted by Dang et al. (2010) to analyze the fluid flow and heat transfer characteristics under different physical parameters.

The current investigation aims at building a computational-based strategy that will aid manufacturers in designing small scale heat exchangers guaranteeing their operation in the continuum regime, not reported in earlier studies. Details of the physical model, mathematical formulation, solution methodology and predicted results are discussed though the rest of this manuscript.

PHYSICAL MODEL

The methodology starts by constructing the physical model of the micro heat exchanger consisting of a thermoelectric plate (23mm x 23mm) mounted on the top of 92 staggered rectangular PDMS micro-channels (150μm x150μm) separated by 100μm thickness PDMS fin. A plate sheet made of PDMS of thickness 1 cm is then introduced between the thermoelectric cold side and the microchannels (Figure 1). Air passes through the microchannels and is cooled due to the cooling effect induced from the cold side of the thermoelectric.

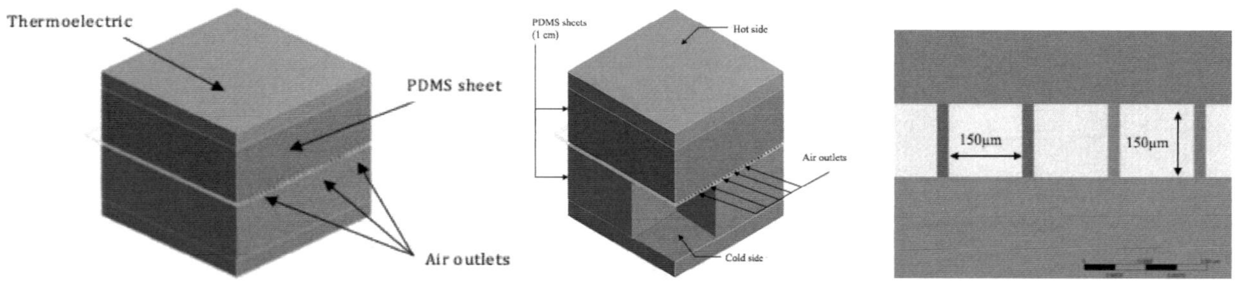

(a) (b)

Figure 1: (a) 3D Physical Model of the Simulated Domain; (b) Front View of the Geometry.

MATHEMATICAL MODEL

The steady state flow in the microchannels of the heat exchanger is modeled as a one phase incompressible flow. To resolve the hydrodynamic and thermal fields of air inside the microchannels, the set of continuity, momentum, and energy equations are applied. In addition, the heat transfer in the solid zone necessitates solving the energy equation in that region. Details of the integrated governing equations for the mathematical model are described briefly in what follows:

Governing Equations in Fluid Zone

Continuity

The continuity equation resembling mass balance over the domain is given by

$$\nabla.(\rho\, \boldsymbol{u}) = 0 \qquad (1)$$

Momentum

The momentum equation is given by

$$\nabla.(\rho \boldsymbol{u}\boldsymbol{u}) = -\nabla\ p + \nabla.(\bar{\bar{\tau}}) \qquad (2)$$

The deviatoric viscous stress tensor for a Newtonian fluid (air) is expressed as

$$\bar{\bar{\tau}} = \mu\{\nabla \boldsymbol{u} + (\nabla u)^T\} \qquad (3)$$

μ is the air dynamic viscosity, [Pa.s]

Energy

The energy equation is given by

$$\nabla.[\boldsymbol{u}(\rho E\ + p)] = \nabla.\left(K_f \nabla\, \mathrm{T}\right) + S_h \qquad (4)$$

K_f is the thermal conductivity coefficient of fluid air, [W/m K]

Governing Equations in Solid Zones

The temperature distribution in the PDMS fins and sheets are resolved by the following diffusion equation

$$\nabla.(K_s \nabla\, \mathrm{T}) = 0 \qquad (5)$$

K_s is the thermal conductivity coefficient of the solid material (PDMS), [W/m K]

BOUNDARY CONDITIONS

In this section, only the type of boundary conditions will be presented without referring to their actual values. The reason is that the objective of this research is to integrate advanced numerical modeling to design a micro heat exchanger to operate in the continuum regime. Hence, the boundary condition values are part of the operating conditions upon which their limit should be "numerically" designed to guarantee the operation of the heat exchanger in the continuum regime. The inlet section of the heat exchanger is modeled with a prescribed uniform normal velocity and fixed temperature. On the other hand, the outlet sections of the microchannels are modeled with a specified static gauge pressure. For the cold side of the thermoelectric, it was modeled with a heat flux related to the cold and hot sides of the thermoelectric. Due to the conservativeness property of the Finite Volume Method, the boundary condition at the fluid-solid interface is implicitly satisfied.

SOLUTION ALGORITHM

A multi-block structured grid with conformal mesh at the fluid-solid interfaces was generated to model the fluid and solid zones of the physical domain. In order to induce the thermoelectric cooling effect (i.e., the heat flux at the cold side, Qc), a lumped mathematical model is implemented as a Neumann type boundary condition to the energy equation to solve for the temperature distribution in the solid zone. The model relates the heat flux (Qc) at the cold surface of the heat exchanger to the temperature difference of its cold and hot sides. Since the temperature field is unknown, this boundary condition is iteratively updated with the boundary temperatures being outcomes of the solution. For that purpose, a generic C code is developed and coupled with the FLUENT solver for serial and parallel computations to iteratively resolve the cold heat flux at the boundary. A conjugate heat transfer problem is formulated with the finite volume method adopted to solve the set of governing mass, momentum, and energy equations in the fluid zone (air) and energy in the solid zones (PDMS plate and fins). The pressure and velocity coupling was resolved by the segregated SIMPLE algorithm under steady state conditions.

RESULTS AND DISCUSSIONS

The assessment of the flow regime (i.e. whether it is in the continuum regime or not) is accomplished by predicting the axial distribution of the Knudsen number (Kn) in the heat exchanger micro-channels. However, this is challenging due to the dependence of Kn on flow properties (Kn is proportional to T/P, where T is the fluid temperature and P is its static pressure) which are not previously known and are part of the solution prediction. Further, these properties both have a decreasing trend as depicted in Figures 2a and 2b, which makes the design of such micro heat exchangers and selection of its operating conditions cumbersome. This is clearly reflected on the variation of Kn where different profiles of Knudsen number variation are observed depending on the prescribed operating conditions under which the heat exchanger is simulated. For instance, for an inlet air temperature fixed at 353.15 K and outlet gauge pressure of 0 Pa, it is shown that for high Reynolds numbers ($Re > 90$), the pressure has a greater effect on the Knudsen number thus forcing it to increase all over the channel (Figure 3). However for $Re < 60$, the temperature has a greater effect on the Knudsen number, forcing it to decrease from the inlet to the exit of the micro-channel (Figure 3). For $60 < Re < 90$, Knudsen distribution has a parabolic distribution (Figure 3). According to the distribution of Knudsen number, the model parameters are updated for cases when the flow regime falls outside the continuum (cases where results of Kn falls above the continuum limit shown in red bold line in figure 3). The ultimate target of this research is to provide a numerical strategy that could be integrated in the design of mico-channel heat exchangers to guarantee their operation in the continuum regime. Later, design maps will be generated that helps identifying regions of allowable range of parameters that could be varied with certain limits to achieve continuum regime in the micro heat exchanger. This work is ongoing and results will be reported in a future paper.

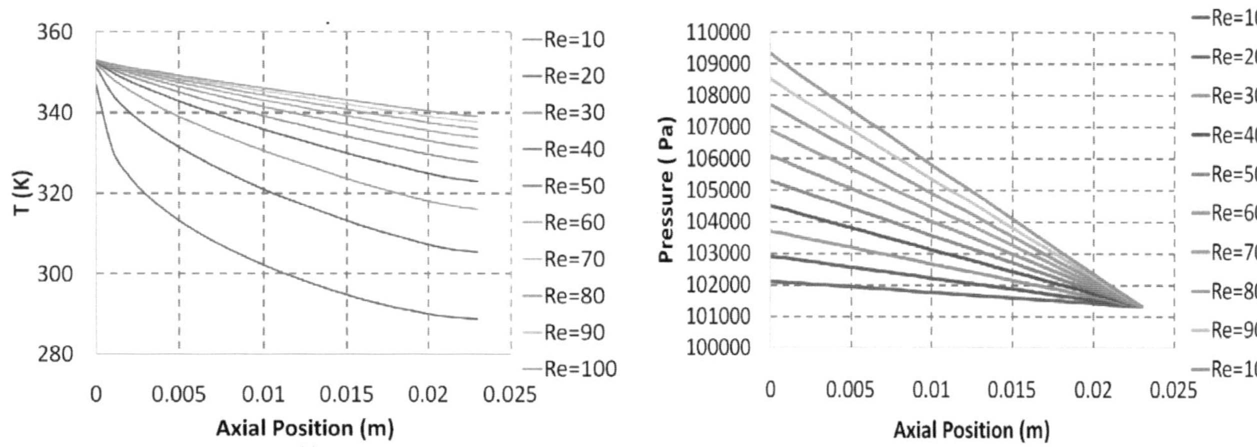

(a) (b)

Figure 2: (a) Axial Temperature Distribution for different Reynold's number; (b) Axial Pressure Distribution for different Reynold's number.

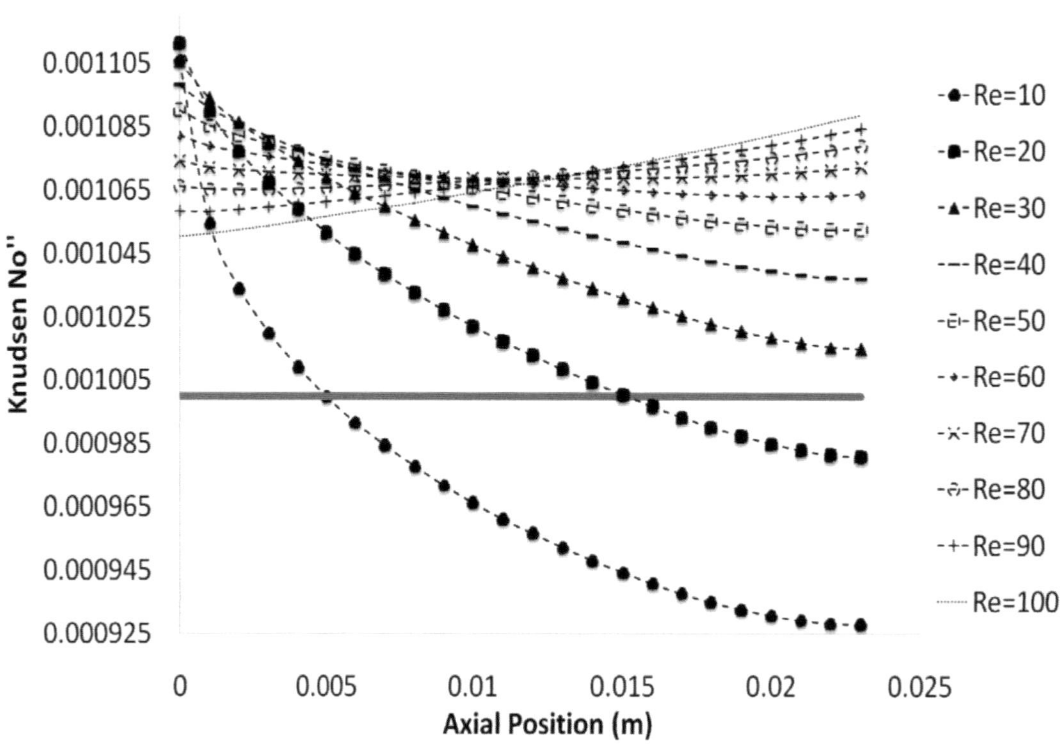

Figure 3: (a) Axial Distribution of Knudsen Number (Kn) for different Reynold's numbers.

CONCLUSIONS

This research aimed at building a computational strategy that will aid manufacturers in designing small scale heat exchangers guaranteeing their operation in the continuum regime. A 3D numerical model of a thermoelectric cooled micro heat exchanger was constructed consisting of a thermoelectric plate (23mm x 23mm) mounted on the top of 92 staggered rectangular PDMS micro-channels (150µm x150µm) separated by 100µm thickness PDMS fins. A plate made of PDMS of thickness 1 cm was also introduced between the thermoelectric cold side and the channels. The thermoelectric cooling effect was induced via a lumped mathematical model implemented as a boundary condition to the energy equation to solve for the temperature distribution in the solid zones. The model related the heat flux (Qc) at the cold surface of the heat exchanger to the temperature difference of its cold and hot sides. Due to the implicit nature of this boundary condition, it was iteratively updated with the boundary temperatures being outcomes of the solution. A conjugate heat transfer problem was formulated and solved via a finite volume based numerical technique. The method starts by imposing a fixed pressure at the outlet sections of the domain. A uniform velocity profile with arbitrary fixed temperature is supplied at the inlet section of the heat exchanger. The hydrodynamic and thermal fields are then predicted in the interior domain. The assessment of the flow regime (i.e. whether it is within the continuum or not) is accomplished by predicting the axial distribution of the Knudsen number (Kn) in the heat exchanger micro-channels. Different profiles of Knudsen number variation were observed depending on the simulated operating conditions. For instance, for an inlet temperature fixed at 353.15 K, it was shown that for high Reynolds numbers (Re > 90), the pressure had a greater effect on the Knudsen number thus forcing it to increase all over the channel. However for Re < 60, the temperature had a greater effect on the Knudsen number, making it to decrease from the inlet to the exit of the micro-channel. For 60 < Re < 90, Knudsen number had a parabolic distribution. Based on the distribution of Knudsen number, the model parameters were updated for cases when the flow regime falls outside the continuum regime. While extensive numerical simulations were conducted in this study to design the limit of the heat exchanger operating parameters to guarantee its operation in the continuum regime, work is still ongoing to design the operating parameters through single simulation via an integrated low computational cost methodology. This will aid reducing computational cost of solution prediction and allow translation of advanced computational fluid dynamics techniques and their application in the design process. Further, design maps will be generated in future studies which will help identifying regions of allowable range of parameters that could be varied within certain limits to achieve continuum regime in the micro heat exchanger.

ACKNOWLEDGMENTS

The financial support provided by the University Research Board (URB) of the American University of Beirut is gratefully acknowledged.

NOMENCLATURE

ρ	Density (kg/m^3)	T	Temperature (K)	Subscripts	
$\boldsymbol{u}$	Velocity (m/s)	K	Thermal conductivity $(m/W\ K)$	f	Fluid
p	Pressure (Pa)	μ	Viscosity $(Pa\ s)$	s	Solid
$\bar{\bar{\tau}}_{ij}$	Stress tensor (N/m^2)	S_h	Source term in energy equation (J/m^3)		
E	Total energy (J/m^3)	Re	Reynolds' number		

REFERENCES

A. Francescon, A. Mapelli, G. Nuessle, P. Petagna, A. Pezous, P. Renaud, G. Romagnoli, 2013. Application of micro-channel cooling to the local thermal management of detectors electronics for particle physics. *Microelectronics Journal* 44(7):612,618.

Arkilic, E. B., Schmidt, M. A. and Breuer, K. S. 1997. Gaseous Slip Flow in Long Microchannels. *IEEEJ. Microelectromech. Syst.* 6(2):167,178.

B. Orr, B. Singh, L. Tan and A. Akbarzadeh, 2014. Electricity generation from an exhaust heat recovery system utilising thermoelectric cells and heat pipes. *Appl. Therm. Eng.* 73(1):588,597.

C. Yıldırım, S. K. Soylu, İ. Atmaca and İ. Solmuş, 2014. Experimental investigation of a portable desalination unit configured by a thermoelectric cooler. *Energy Conversion and Management* 85:140,145.

Dang, T. and Teng, J., 2011. The effects of configurations on the performance of microchannel counter-flow heat exchangers–An experimental study. *Appl. Therm. Eng.* 31(17-18):3946,3955.

Dang, T., Teng, J., & Chu, J., 2010. A study on the simulation and experiment of a microchannel counter-flow heat exchanger. *Appl. Therm. Eng.* 30(14-15):2163,2172.

D. Milani, A. Abbas, A. Vassallo, M. Chiesa and D. A. Bakri, 2011. Evaluation of using thermoelectric coolers in a dehumidification system to generate freshwater from ambient air. *Chemical Engineering Science* 66(12):2491,2501.

García-Hernando, N., Acosta-Iborra, A., Ruiz-Rivas, U., and M. Izquierdo, 2009. Experimental investigation of fluid flow and heat transfer in a single-phase liquid flow micro-heat exchanger. *International Journal of Heat and Mass Transfer* 52(23-24): 5433,5446.

H. Choi, S. Yun and K. Whang, 2007. Development of a temperature-controlled car-seat system utilizing thermoelectric device. *Appl. Therm. Eng.* 27(17-18):2841,2849.

Ian Papautsky, John Brazzle, Timothy Ameel, A.Bruno Frazier, 1999. Laminar fluid behavior in microchannels using micropolar fluid theory. *Sensors and Actuators A: Physical* 73(1-2):101-108.

J. Laurent, A. Drezet, H. Sellier, J. Chevrier, and S. Huant, 2011. Large Variation in the Boundary-Condition Slippage for a Rarefied Gas Flowing between Two Surfaces. *Phys. Rev. Lett.* 107(16):164501-1,164501-5.

J. Li, B. Ma, R. Wang and L. Han, 2011. Study on a cooling system based on thermoelectric cooler for thermal management of high-power LEDs. *Microelectronics Reliability* 51(12):2220,2215.

J. Harley, H.H. Bau, 1989. Fluid flow in micron and sub-micron size channels. *Proceedings of IEEE, MEMS* pp. 25,28.

J. Pfahler, J. Harley, H. Bau, J. Zemel, 1990. Liquid transport in micron and submicron channels. *Sensors and Actuators A: Physical* 22(1–3):431,434.

Jie, D., Diao, X., Cheong, K. B. and Yong, L. K., 2000. Navier-Stokes Simulations of Gas Flow in Micro Devices. *J. Micromech. Microeng* 10(3):372,379.

Leila Issa and Issam Lakkis, 2014. Reduced-Order Modeling of Low Mach Number Unsteady Microchannel Flows. *J. Fluids Eng.* 136(5):051201-1, 051201-9.

M. Bahrami, A. Tamayol and P. Taheri, 2009. Slip-Flow Pressure Drop in Microchannels of General Cross Section. *J. Fluids Eng.* 131(3):031201-1, 031201-8.

M. Jradi, N. Ghaddar, K. Ghali, 2012. Experimental and theoretical study of an integrated thermoelectric–photovoltaic system for air dehumidification and fresh water production. *Int. J. Energy Res.* 36:963,974.

N. M. Khattab and E. T. El Shenawy, 2006. Optimal operation of thermoelectric cooler driven by solar thermoelectric generator. *Energy Conversion and Management* 47(4):407,426.

Q. Fan and H. Xue, 1998. Compressible effects in microchannel flows [MEMS]. *In Electronics Packaging Technology Conference, Proceedings of 2nd* pp. 224 ,228.

Z. Liu, L. Zhang and G. Gong, 2014. Experimental evaluation of a solar thermoelectric cooled ceiling combined with displacement ventilation system. *Energy Conversion and Management* 87:559,565.

Z. L. Guo, B. C. Shi and C. G. Zheng., 2007. An extended Navier-Stokes formulation for gas flows in the Knudsen layer near a wall. *Europhysics Letters (EPL)* 80(2): 24001-1,24001-6.

THE PERFORMANCE OF SOLAR-WIND ENERGY TOWERS USING A LAGRANGIAN-EULERIAN MULTIPHASE MODEL

Lina Hijazi[1] **Iyad Fayssal**[2] **Marwan Darwish**[3], **PhD** **Fadl Moukalled**[4], **PhD**

ABSTRACT

State of the art computational fluid dynamics (CFD) techniques have become indispensable for understanding physical phenomena, and for modeling and optimizing the performance of engineering devices involving fluid flow and heat and mass transfer mechanisms. Solar wind energy towers have been recently regarded as a bold new approach to the United States and other nations for their ability to provide clean and sustainable energy, overcoming the burdens of alternative energy sources. An energy tower is a tall tower, at least 400 m high, located in a hot and dry region whereby cool water sprayed at top will evaporate and create a downdraft. The produced high velocity airflow will drive turbines at the bottom of the tower and generate electricity. A 2D numerical model of a prospected tower is constructed to study the impact of different operating parameters including spray droplet diameter, spray flow rate, and ambient conditions on the performance of the tower. A Lagrangian-Eulerian multiphase technique is adopted for simulating droplet transport and evaporation. The model is first validated with published measured data extracted from a small scale PDEC experimental test facility at the Conphoebus Institute in Sicily. Results show that the model is able to accurately predict the evaporation phenomena and that an exit velocity of $4.5m/s$ *can be achieved over a 20m height tower.*

INTRODUCTION

In response to the 1970's energy crisis, and the need to overcome dependence on fossil fuels, alternative renewable energy conversion devices have become the object of interest worldwide. The building sector accounts for 40% of total primary energy consumption, while residential and commercial buildings consume 60% of the worlds' electricity (UNEP 2009). One mechanism to reduce reliance on fossil fuels in buildings is an Energy Tower. One commonly known form, suitable for hot and dry regions, is the PDEC Tower (Passive Downdraft Evaporative Cooling Tower). It is composed of a hollow tower, a few meters high, with a spray system at the top, whereby hot dry air enters the tower by means of a wind catcher and cools by evaporation. The cooler and denser air descends and exits the tower to cool and ventilate spaces.

An alternative function of the Water Spraying Energy Tower was first advocated by (Carlson 1975), which is today referred to as the Solar Wind Energy Tower. He suggested initiating a downdraft in a hollow duct by spraying water at

[1] Graduate Student, Department of Mechanical Engineering, American University of Beirut, e-mail: lmh17@mail.aub.edu, corresponding author.
[2] PhD Candidate, Department of Mechanical Engineering, American University of Beirut
[3] Professor, Department of Mechanical Engineering, American University of Beirut, Bechtel Building, Room 324
[4] Associate Dean and Professor, Department of Mechanical Engineering, American University of Beirut, Bechtel Building, Room 312

high altitude thereby creating high intensity artificial wind that passes through turbines located at the outlet. Wind power is then converted to hydraulic power, which is then converted to electric power by means of hydraulic to electric converters. (Zaslavsky et al. 2011) later contributed to developing the idea of the Energy Tower by tackling shortcomings in Carlsons Patent. In order to assess the performance and realistic potential of any device, full-scale experimental measurements are required whereby the full complexity of the problem is by default integrated. Till date, no working prototype of the Solar Wind Energy tower has been constructed due to the massive investment required; therefore full-scale measurements are not available. Alternatively, Computational Fluid Dynamics (CFD) has become a popular approach to understanding physical phenomena and has become an indispensable tool for the design and optimization of energy conversion devices involving fluid flow and heat and mass transfer mechanisms.

Optimization is best performed in a simulated environment with a validated model. For simulation of droplet transport and evaporation, two approaches are generally used in the CFD community. In the first approach, Lagrangian-Eulerian method (LE), droplets are modeled using discrete evaporating particles (Lagrangian), whereas the flow field is modeled using multi-species equations in an Eulerian flow field, that is expressed in terms of a set of Navier Stokes and Species equations. In the second approach, Eulerian-Eulerian (EE), a two phase flow field is used that accounts for volume fraction occupied by each of the phases in the domain.

(Subramaniam 2013) highlights the superiority of the LE method over the EE Method in capturing and accurately simulating the evaporation process. (Montazeri 2015) studied the evaporative cooling phenomena by water spraying systems using FLUENT to enhance thermal comfort in buildings and validated results with Wind-Tunnel measurements presented by (Sureshkumar, 2008) . The study showed that the LE method was able to accurately predict the evaporation phenomena. In fact, the majority of CFD studies in literature involving fluid flow and evaporation adopted the LE approach. For instance, (Kang and Strand 2013) simulated a 2D PDEC tower under turbulent flow using a LE method in FLUENT to study the impact of ambient conditions and droplet characteristics on exit conditions. (Cook et al 2000) similarly used CFD to model airflows in PDEC towers, while using CFX instead of FLUENT. The injection rate was calibrated by means of a sensitivity analysis to suit target conditions at the outlet. (Kalantar 2009) evaluated the cooling performance of a wind tower through a general purpose mathematical model in c++ and FLUENT. The model was 3D and predicted the conditions at the exit of the tower as well as evolution of the flow field along the centerline for a hot and dry condition.

When spraying water at high altitude with the objective of creating high intensity artificial wind to drive turbines, it becomes essential to describe the main driving buoyancy force. The assumption that the flow is incompressible is no longer valid since the pressure, temperature and density are no longer constant. Till date, CFD studies have been conducted on a solar chimney power plant, which operates in reverse of the Energy Tower (Zhou et al. 2009; Guo et al. 2015) yet no studies have ever been conducted on a large scale Energy Tower. (Omer et al. 2008) devised an optimization algorithm, given ambient conditions and tower height, which can identify the optimum discharge rate that will maximize net power output. (Abhinava et al. 2013) studied the intensity of downdrafts using mathematical formulations on a 500m energy tower. It was shown that the magnitude of the final output velocity is dependent on the negative available potential energy, precipitation loading, and maximum horizontal momentum. (Hassid et al. 2012) studied the effects of coalescence in energy towers. Results show that coalescence is substantial in the tower and its effects cannot be neglected. (Bauer and Gasser 2012) presented a 1D mathematical model to describe the physics of the energy tower. The model was run on a series of test cases by varying temperature, relative humidity and tower diameter for a 1200 m height tower. Due to the lack of an existing prototype, the results presented are only theoretical.

The objective of this research is to present a CFD model of an Energy Tower using ANSYS FLUENT 14.5. A full buoyancy model is employed using a user-defined function which is expressed in terms of density difference and added as a source term in the momentum equation. The Lagrangian-Eulerian Model is adopted in order to accurately predict the evaporation phenomena. Results showing the impact of droplet characteristics, ambient temperature and relative humidity on the exit velocity, temperature, density, and relative humidity are presented.

MATHEMATICAL MODEL

In FLUENT, the discrete phase model follows a Lagrangian - Eulerian approach. The fluid flow is treated in the continuous phase by solving the time-averaged Navier-Stokes equations, while the discrete phase is solved by tracking droplets in the flow field. The discrete phase can exchange mass, momentum and energy with the continuous phase and this is modeled by the inclusion of source terms from the discrete phase into the continuous phase equations. The downdraft model consists of equations of heat and mass transfer, turbulence, and species. The sets of equations in the continuous phase solve equations of conservation of mass, momentum, energy and species transport whereas equations for the discrete phase calculate the trajectory of water droplets. Both phases are highly interactive, and the two way coupling procedure is solved in FLUENT as follows:

1. Initialize flow field by solving the continuous phase
2. Calculate the discrete phase particle trajectory and obtain source terms
3. Calculate the continuous phase with the inclusion of source terms adopted from the discrete phase
4. Calculate discrete phase in the modified flow field
5. Repeat steps 2-4 until a converged solution is achieved

Continuous Phase Model

In solving the continuous phase, equations of conservation of mass, momentum, energy and species are solved. At the start of the simulation, the flow field is initialized by solving the equations while excluding the interphase exchange coefficients. After the discrete phase equations are solved and the source terms are obtained, the continuous phase equations are solved with the inclusion of the source terms. The equations for conservation of mass, momentum, energy and species respectively are:

$$\frac{\partial \rho}{\partial t} + \nabla \cdot (\rho \mathbf{v}) = S_m \tag{1}$$

$$\frac{\partial (\rho \mathbf{v})}{\partial t} + \nabla \cdot (\rho \mathbf{v}\mathbf{v}) = \nabla \cdot (\bar{\bar{\tau}}) + \rho \mathbf{g} - \nabla p + \mathbf{F}_m \tag{2}$$

$$\frac{\partial (\rho E)}{\partial t} + \nabla \cdot (\rho \mathbf{v} E) = -p \nabla \mathbf{v} + \nabla \cdot (k \nabla T) - \nabla \cdot \left(\sum_j h_j \mathbf{J}_j \right) + \Phi + S_E \tag{3}$$

$$\frac{\partial (\rho m_j)}{\partial t} + \nabla \cdot (\rho \mathbf{v} m_j) = -\nabla \cdot \mathbf{J}_j + S_m \tag{4}$$

Discrete Phase Model

In the discrete phase model, the trajectory of droplets is calculated by integrating the force balance acting on the particles:

$$\frac{\partial \mathbf{v}_p}{\partial t} = F_D\left(\mathbf{v} - \mathbf{v}_p\right) + \mathbf{g}\frac{\left(\rho_p - \rho\right)}{\rho_p} + \mathbf{F} \tag{5}$$

NUMERICAL SIMULATION

Computational Domain and Boundary Conditions

In this study, a two-dimensional axisymmetric model of an Energy Tower is constructed with a height of 20 m and width of 8 m. The computational domain, shown in Figure 1, is composed of 64,000 elements with near wall refinment. A mesh independent test revealed no significant change in results with further grid refinement.
A pressure-inlet, pressure-outlet boundary condition was imposed on the domain in order to predict the quantity of flow that can potentially be entrained from the environment. The walls are assumed to be fully insulated.

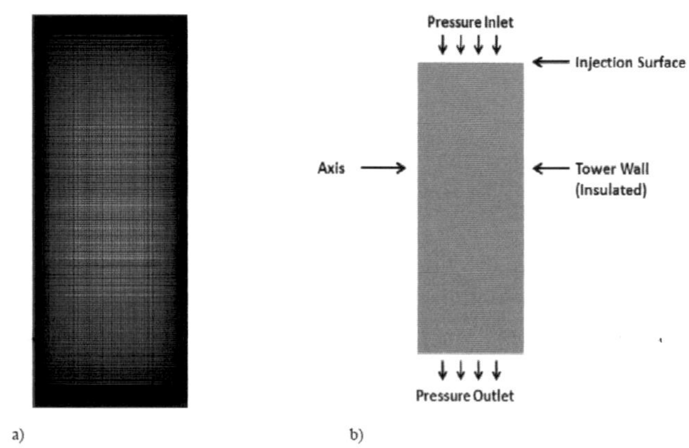

Figure 1 (a) Computational Domain with Grid Regions (b) Boundary Conditions

Numerical Schemes

The finite-volume based numerical solution is obtained by using the ANSYS FLUENT commercial software package. The SIMPLE algorithm is selected to resolve the pressure-velocity coupling of the Navier-Stokes equations, and a second order scheme is chosen to compute the face pressure. Convective terms are evaluated using a Second Order Upwind Scheme. A semi-implicit trapezoidal scheme is adopted for the discretization of the Discrete Phase model equations.

Model Validation

The model was validated against the published measured data from a small scale PDEC test facility at the Conphoebus Institute in Sicily and reported in (Kang and Strand 2013). Dimensions of the tower are 4.1m x 4.4m x 10.7 m with a 1.7m x 3.7m wind tower at the top. Sensors were placed at five different locations at the outlet to measure temperature and relative humidity. T_M and RH_M in Table 1 represent the average measured Temperatures and Relative Humidity respectively, while T_P and RH_P represent the respective predicted temperatures and relative humidity using the CFD model. Seven cases were considered with each case having a different ambient temperature, relative humidity and wind speed.

Table 1. Comparison between Measured and Predicted Data

Case Number	T_M (°C)	T_P (°C)	% Error	RH_M (%)	RH_P (%)	% Error
1	27.00	28.60	5.93	75.9	74.56	1.76
2	26.00	26.16	0.62	79.4	77.07	2.93
3	26.56	27.26	2.64	76.2	72.48	4.88
4	26.90	25.65	4.65	77.4	76.04	1.75
5	25.90	25.69	0.81	72.7	71.31	1.91
6	28.78	28.24	1.88	59.9	56.72	5.30
7	28.26	27.74	1.84	62.5	62.2	0.48

Results of the CFD model and experimental data are in good agreement with maximum errors of 5.93% and 5.30% in predicting temperature and relative humidity, respectively.

RESULTS AND DISCUSSION

Computational analysis was performed on a 20m x 8m tower. Three different scenarios were considered as shown in Table 2.

Table 2. Test Series Data

Case	Base Case "BC"	Vary Ambient Temperature "T25"	Vary Outdoor Relative Humidity "RH50"
$T_{ambient}$ (K)	309.11	298.15	309.11
Density of Mixture (kg/m^3)	1.138	1.182	1.129
Relative Humidity (%)	13	13	50
Species Mass Fraction (kg/kg)	0.004755254	0.00253028	0.018696265
Droplet Temperature (K)	290	290	290
Droplet Diameter (μm)	50	50	50
Water Mass Flow Rate (kg/s)	0.1	0.1	0.1

Impact of Water Mass Flow Rate

The exit conditions of temperature, relative humidity, density and velocity were explored by stepping up the water injection flow rate until saturation conditions are achieved for each of the three cases. Figure 2 shows the impact of increasing the injection rate on the exit temperature, density, velocity and relative humidity. For a hot dry condition (Base Case), the largest temperature drop achieved is 18.62°C. Due to initial low moisture content, a high amount of water can be absorbed until saturation is achieved. For the case where the ambient temperature is 25°C, a temperature drop of 13.99°C is predicted. A lower amount of water is required to reach saturation. For a hot humid condition, the lowest drop of 9.10°C is achieved because the cooling process is hindered by the high moisture content in air.

Initially without spray, the velocity at exit from the domain is 0.3m/s. As Water is injected, a sharp increase in velocity is witnessed and as the quantity of water injected increases, the terminal velocity increases as well. Even if saturation is achieved, an increase in water mass flow rate still increase the exit velocity due to the effect of the weight of unevaporated droplets that enhances the downdraft. Similarly the density of the mixture increases as the air becomes cooler and stabilizes at saturation.

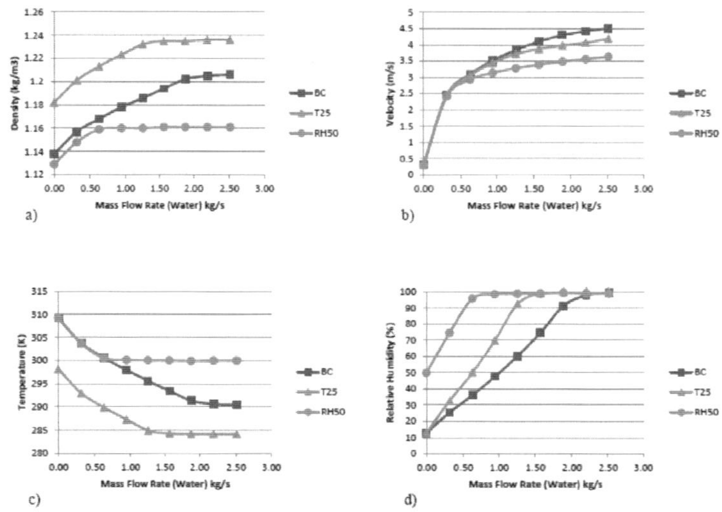

Figure 2 Impact of Injection rate on exit (a) Density (b) Velocity (c) Temperature and (d) Relative Humidity

In order to demonstrate the amount of air that is entrained into the domain, the ratio R_q of Produced Mass Flow Rate over the Initial Mass Flow Rate is computed. Figure 3 shows that injecting only 0.3kg/s of water increases the mass flow rate in the tower by about 8 times. A ratio of 15.6 can be achieved for a hot dry condition, 14.1 for a cold dry condition and 12.2 for a hot humid condition.

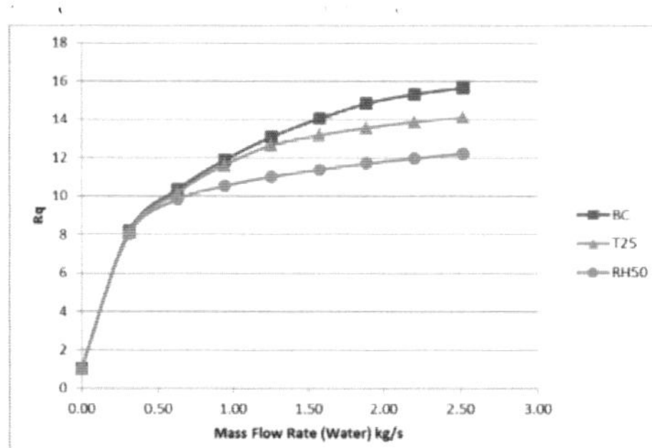

Figure 3 Ratio of Produced Flow Rate vs. Water Mass Flow Rate

Validation in 3D

Results of the 2D simulation are validated against a 3D model in order to verify that results in 2D remain effective in 3D. A 20 m cylindrical tower with diameter of 8 m was studied whereby the model in 2D is reproduced in 3D. Figure 4 shows the average exit conditions of the field field between 2D and 3D simulations on the base case scenario. Results show overlapping trends between 2D and 3D, and therefore results remain valid in 3D simulations

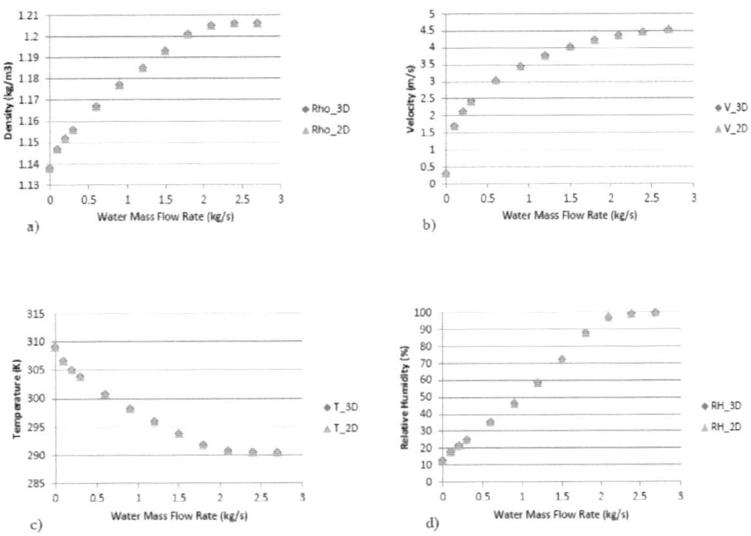

Figure 4 Average exit conditions vs. Injection Rate between 2D and 3D simulations

Impact of Droplet Diameter

Figure 5 shows the impact of increasing droplet diameter in steps of 50μm up to 300μm for an injection flow rate of 0.3 kg/s. As the diameter increases, the amount of flow rate that can be produced decreases, as more energy will be required to evaporate a single droplet.

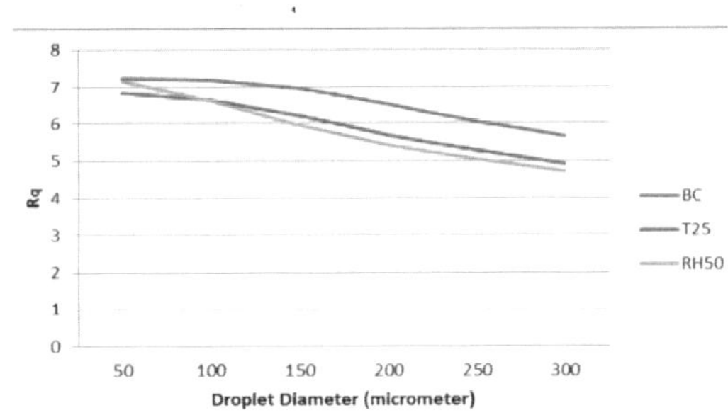

Figure 5 Ratio of Produced Flow Rate vs. Droplet Diameter

CLOSING REMARKS

A two-dimensional numerical study of a solar-wind energy tower was presented. A Lagrangian-Eulerian multiphase technique was adopted for simulating droplet transport and evaporation. The model was first validated with published experimental data and results obtained demonstrated that the model is capable of accurately predicting the evaporation phenomenon that takes place in the tower. The model was then used to predict the effect of spray droplet diameter, spray flow rate, and ambient conditions on the performance of a 20m high wind-energy tower. Results indicated that the entrained mass flow rate in the tower increases with increasing the mass flow rate of injected water and decreases with increasing droplet diameter.

ACKNOWLEDGEMENTS

The financial support provided by the Munib R. and Angela Masri Institute of Energy and Natural Resources at the

American University of Beirut through award number 103027 is gratefully acknowledged.

NOMENCLATURE

ρ	Density (kg/m^3)	g	Gravitational Acceleration (m/s²)
v	Velocity (m/s)	E	Internal energy (J/kg)
p	Pressure (Pa)	k	Effective Thermal conductivity $(W/m\ K)$
T	Temperature (K)	h	Enthalpy of Species $(kg/m^2\ s)$
S_m	Mass Source Term (kg/m³)	Φ	Rayleigh Dissipation Function (kg/s³.m)
F_m	Momentum Source Term (N/m³)	m	Local Mass Fraction
S_e	Energy Source Term (W/m^3)	J	Diffusion of Mass
$\bar{\bar{\tau}}$	Stress tensor (N/m^2)	F_D	Drag Force

Subscripts

p	Particle
j	Species

REFERENCES

UNEP. 2009. *Buildings and Climate Change*. Paris: Sustainable Consumption & Production Branch.

Carlson, P. 1974. Power Generation through Controlled Convection (Aeroelectric Power Generation). *U.S. Patent 3,894,393 A*.

Zaslavsky, D., Guetta, R., Hitron, R., Krivchenko, G., Burt, M., and M. Poreh. 2011. Sharav Sluices Ltd Files Patent Application for Renewable Resource Hydro/Aero-power Generation Plant and Method of Generating Hydro/Aero-power. *U.S. Patent 6,510,687 B1*.

Subramaniam, S. 2013. Lagrangian-Eulerian Methods for Multiphase Flows. *Progress in Energy and Combustion Science* 39:215-245.

Montazeri, H., Blocken, B., and J.L.M. Hensen. 2015. Evaporative Cooling by Water Spray Systems: CFD Simulation, Experimental Validation and Sensitivity Analysis. *Building and Environment* 83:129-141.

Sureshkumar, R., Kale, S.R., and P.L. Dhar. 2008. Heat and Mass Transfer Process between a Water Spray and Ambient Air- I. Experimental Data. *Applied Thermal Engineering* 28:349-360.

Kang, D., and R.K. Strand. 2013. Modeling of Simultaneous Heat and Mass Transfer within Passive Down-Draft Evaporative Cooling (PDEC) Towers with Spray in FLUENT. *Energy and Buildings* 62:196-209.

Cook, M.J., Robinson, D., Lomas, K.J., Bowman., N.T., and H. Eppel. 2000. Passive Downdraught Evaporative Cooling: II. Airflow Modelling. *Indoor and Built Environment* 9:325-334.

Kalantar, V. 2009. Numerical Simulation of Cooling Performance of Wind Tower (Baud-Geer) in Hot and Arid Region. *Renewable Energy* 34:246-254.

Guo, P., Li, J., Wang, Y., and Y. Wang. 2015. Numerical Study on the Performance of a Solar Chimney Power Plant. *Energy Conversion and Management* 105:197-205.

Zhou, X., Yang, J., Xiao, B., Hou, G., and Y. Wu. 2009. Numerical Investigation of a Compressible Flow Through a Solar Chimney. *Heat Transfer Engineering* 30(8):670-676.

Omer, E., Guetta, R., Ioslovich, P., Gutman, P., and M. Borshchevsky. 2008. Optimal Design of an Energy Tower Power Plant. *IEEE Transactions On Energy Conversion* 23:215-225.

Abhinava, G., Swarnkar, N., Behera, S., and G. Edison. 2013. Creation of Artificial Downdraft for Wind Power Plant. *International Conference on Energy Efficient Technologies for Sustainability, Nagercoil, India,* 571-576.

Hassid, S., Merksamer, I., and R. Guetta. 2012. Energy Towers- The Effect of Droplet Coalescence on Power and the Environment. *Solar Energy* 86:1443-1453.

Bauer, M., and I. Gasser. 2012. Modeling, Asymptotic Analysis, and Simulation of an Energy Tower. *SIAM Journal on Applied Mathematics* 72(1):362-381.

Comparative Study Using OpenFOAM© and Fluent Solvers of Smoke Propagation inside Occupied Spaces

Jad Hawi **Iyad Fayssal** **Marwan Darwish** **Fadl Moukalled**

ABSTRACT

Smoke composed of toxic gases is considered the most dangerous aspect in case of fire in occupied buildings; thus, it is essential to design an effective ventilation system capable of creating a suitable environment for people to evacuate. The decision criteria of a proper ventilation system for a certain engineering application requires prior understanding of the transport physical phenomena associated with smoke propagation inside closed buildings and its thermal and hydrodynamic interactions with surrounding environment. The aim of this study is to build and implement a 3D based computational code in an open source platform (OpenFOAM©), providing fire protection engineers with an effective, free, and customizable open source tool to simulate and analyze smoke propagation in occupied spaces. Results are validated by comparison to similar ones obtained using the Fluent commercial CFD solver. The finite volume method is used to numerically mimic the existence of fire via a modified energy and smoke concentration equations. A source term resembling the time-dependent power of fire is implemented in a pre-existing modified energy equation. Further, to account for the transport of smoke, a concentration equation governing the temporal, convection, diffusion, and mass generation variations of smoke is numerically implemented and coupled to the OpenFOAM© solver. Similarly, the Fluent solver is coupled to a developed C code to account for the existence of fire. To validate the implemented model in OpenFOAM©, simulations are conducted on a test case (1mx1mx1m) topologically decomposed into 244,776 cells. The heat release and smoke generation rate due to fire are triggered via an integrated source term in each of the energy and species equations using the t-squared method in the growth region. Temporal and spatial variations of velocity magnitude, temperature, smoke concentration, visibility, and smoke and heat exposure are presented and analyzed in terms of contours and spatial profiles. Results predicted via the developed open source code and the well-known Fluent solver are in good agreement.

INTRODUCTION

Fire is a chemical reaction that occurs when a carbon based material mixes with air and ignites as it comes in contact with a hot enough surface. Worldwide, nearly 15000 deaths from fires and burns and 70 billion USD of damages are sustained yearly with most victims of fires die from smoke or toxic gases and not from burns (Hall, 2001). Hence, smoke is the most dangerous aspect of fires.

Smoke in fires is made up of toxic gases that irritate the eyes and lung, reduce visibility, and decrease mental acuity making it difficult for occupants to escape the building. For this reason, providing safe conditions that gives good visibility of exit signs and small amount of smoke inhalation is recently becoming more and more important in all building codes according to NFPA (2014). To apply these codes, companies are relying on commercial Computational Fluid Dynamics (CFD) packages such as ANSYS Fluent and Pyrosim. However, many designers avoid using such programs due mainly to their high cost.

The current work deals with the analysis of fire propagation in large buildings using an open source CFD code known as "OpenFOAM©", which is a free CFD toolbox with many built-in features. OpenFOAM© adopts a pressure-based finite volume formulation with its structure allowing for a complete freedom to extend and modify its initial functionality.

FIRE DESIGN USING CFD

The hot combustion products caused by a fire rise due to variations in pressure and entrain with them the nearby cold air to form a plume. This intensifying plume reaches the ceiling and forms two streams of smoke that flow horizontally in opposite directions. This process was first studied by Yang et al. in 1977 using the computer code UNDSAFE-I and then in the late 1980s a similar study done by the Fire Research Station, UK using of JASMINE software. FLOW-3D which was developed in the 1990s by the Atomic Energy Authority, UK was used to reconstruct the case of Kings Cross Fire in the London Underground station as stated by Cox et al. (1989) and Simcox et al. (1992). Kandola and Morris (1994) carried out a numerical study using the AEA CFD-FLOW-3D package in which they focused on hazards of smoke accompanying fire as it affects people and discussed fire safety procedures and equipment to be used. Sinai et al. (1995) reported on the AEA attempt to validate fire predictions using different CFD codes. Most papers tackling this subject have either conducted numerical simulations using CFD packages and/or experimental analysis on full scale models, to investigate fires in tunnels and car parking area. Den Boer et al. (2003) performed full scale fire tests in a tunnel in Netherland and compared it with predictions generated using the PHOENICS CFD software; where results showed an acceptable accuracy of the software. In addition, Tabarra et al. (1996) used scale model tests on tunnels to validate the CFD results, while Deng et al. (1996) used CFD to model fire on trains in underground stations.

METHODOLOGY

The objective of this project is to develop a solver in OpenFOAM© able to simulate the propagation of smoke resulting from fire. The solver should be capable, regardless of geometry and boundary and initial conditions, of producing accurate and robust smoke propagation results. Computations involve predicting very complex fields of velocity, temperature, smoke concentration, visibility, and time to lose consciousness due to heat and smoke. Such numerical tool will be helpful for checking and modifying designs of ventilation and smoke extraction systems to guarantee maintaining a lower "smoke-free" layer for a specified period of time for occupants to safely exit and fire fighters to contain and extinguish the fire. The correctness and accuracy of results generated using the OpenFOAM© based solver were checked by direct comparison with similar predictions obtained using the ANSYS software, which is the most trusted CFD tool for predicting smoke propagation by engineers.

THEORETICAL FORMULATION

The equations governing the transport phenomena of interest in this work represent the conservation of mass, momentum, energy, and smoke concentration. In addition, turbulence effects are modeled using the two-equation standard $k - \varepsilon$ turbulence model. The forms of these conservation equations are presented next (Launder et al. 1974).

Continuity Equation. Since the flow is considered to be incompressible, density is independent of time and the continuity equation is reduced to:

$$\frac{\partial \bar{u}_j}{\partial x_j} = 0 \quad (1)$$

Momentum Equation. The momentum equation for incompressible flow with turbulence accounted for within the Boussinesq hypothesis is expressed as:

$$\frac{\partial \bar{u}_i}{\partial t} + \frac{\partial}{\partial x_j}(\bar{u}_j \bar{u}_i) = -\frac{1}{\rho}\nabla p + v_{eff}\nabla^2 \bar{u}_i \bar{u}_j + F_i \quad (2)$$

Energy Equation. The final form of the energy equation used in this study is given by

$$\frac{\partial}{\partial t}(\bar{T}) + \frac{\partial}{\partial x_j}(\bar{T}\bar{u}_j) = \frac{\partial}{\partial x_i}\left[\left(\frac{\vartheta}{Pr} + \frac{\vartheta_t}{Pr_t}\right)\frac{\partial \bar{T}}{\partial x_i}\right] + S_t \quad (3)$$

Smoke Concentration Equation. Smoke is tracked by solving its conservation equation given by

$$\frac{\partial}{\partial t}(\bar{C}) + \frac{\partial}{\partial x_j}(\bar{C}\bar{u}_j) = \frac{\partial}{\partial x_i}\left[\left(D_i + \frac{\mu_t}{Sc_t}\right)\frac{\partial \bar{C}}{\partial x_i}\right] + S_t \quad (4)$$

As discussed by Moghtaderi et al. (1997), in fire regions, flow will be strongly buoyancy driven; and not taking turbulence into account, simulations will show relatively uniform thermal layers rather than strong stratification. Thus, turbulence will be accounted by using k- ε model since it is widely used and best validated.

Heat Source Term In fire simulations, growth and fully developed stages should be identified to determine the initial heat release rate for the flaming fire.

In the growth stage, the heat release rate (HRR) in the growth stage is described by the t^2-method (Hertzberg, 2003) as

$$\dot{Q} = \alpha t^2 \quad (5)$$

Table 1. Values for α (Karlsson et al., 2000)

Growth Rate	α in kW/s²
Slow	0.003
Medium	0.012
Fast	0.047
Ultra-Fast	0.188

According to Boverket (2010) and British Standard Institution (2001), the fire growth rates in schools and offices occupancy is assumed to be medium but fast for Karlsson and Quintiere (2000) as shown in Table 1. This depends on the material burned in the simulations. After determining the type of fire growth, the next step is to specify the fire characteristics in the fully developed regime. Thus, we should identify the maximum value of heat release rate, which is limited either by the presence of ventilation or the presence of combustible material. Therefore, peak HRR for both ventilation and fuel-controlled fires should be calculated and the lower value between them should be selected as described by Karlsson and Quintiere (2000).

The peak heat release rate for the fuel-controlled fire is estimated according to

$$\dot{Q} = \dot{Q}'' \times A_f \quad (8)$$

For offices and schools, the European standard (2002) indicates that the heat release rate per unit area is equal 250 kW/m². The peak HRR for ventilation control fire is given according to Karlsson and Quintiere (Karlsson et al., 2000)

$$\dot{Q}_V = 1500 A_0 \sqrt{H_0} \quad (9)$$

where A_0 is the summation of all the openings in the studied space and H_0 is the summation of A_0 times their height over A_0

Smoke Source Term. The smoke production rate is calculated according to the equation established by the Fire Protection Handbook (Karlsson et al., 2000)

$$Smokemass = \frac{\dot{Q}}{\Delta H_{eff}} \quad (10)$$

Visibility. Klote and Milke (2002) describe the visibility according to:

$$S = \frac{K}{\alpha_m \times m_p} \quad (11)$$

Table 2. Proportionality Constant for Different Situations (Klote, 2002)

Situation	Proportionality Constant K
Illuminated Signs	8
Reflecting Signs	3
Building Components in Reflected Light	3

Table 3. Specific Extinction Coefficient for Smoldering and Flaming Combustion (Klote, 2002)

Mode of Combustion	Specific Extinction Coefficient α_m (m^2/kg)
Smoldering Combustion	4301
Flaming Combustion	7578

Exposure to Smoke. The relationship between the concentration of smoke in air and the time to lose consciousness is presented by Purser (2000), in SFPE handbook of Fire Protection Engineering for an average person with high level of activity as follows:

$$t = \frac{30}{8.2925\times10^{-4}\times(10^{4}\times C)^{1.036}} \qquad (12)$$

Exposure to Heat. For the exposure of human to heat, Purser (2000) proposes the following equation that relates temperature to the time of consciousness is:

$$t = e^{5.1849-0.0273\times(T-273)} \qquad (13)$$

COMPUTATIONAL APPROACH

Case Geometry

The test case considered in this study is a room (1x1x1 m^3) with a smoke source of 0.2x0.2x0.2 m^3 located at the floor in the middle of the space as shown in Figure 1. Three points having respectively the heights of 0.25m, 0.5 m and 0.75 m from the floor surface are selected to carry on the analysis of results and discussions.

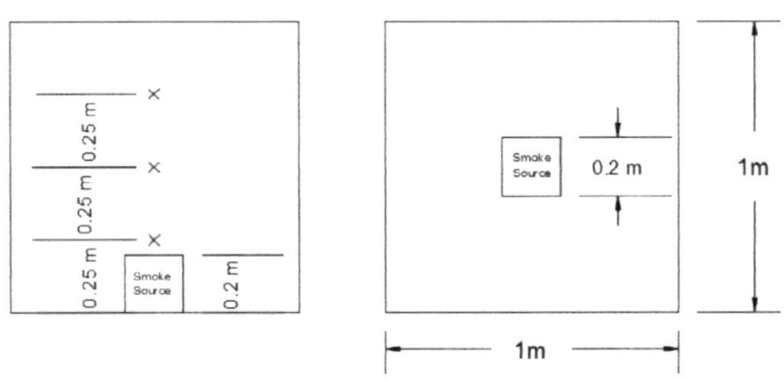

Figure 1. Case Geometry

Numerical Scheme

The finite volume discretization method is adopted in the simulations. First and second order numerical behavior for each subcategory implemented is summarized in Table 4.

Table 4. Numerical Behavior for each Scheme used in the Test Case

Subcategory	1st Order Numerical Scheme	2nd Order Numerical Scheme
Transient	Crank Nicolson 0.5	Crank Nicolson 0.5
Gradient	Gauss linear	Gauss linear
Laplacian	Gauss linear limited 0.5	Gauss linear limited 0.5
Divergence	Gauss upwind	Gauss linear

Simulations Description

Three simulations were performed in both software (OpenFOAM & ANSYS) as shown in Table 5. In all cases, the worst case scenario is applied with no smoke removal systems or sprinklers.

Table 5. Different Simualtion Aproach

	Numerical Scheme Order	Fire Growth Coefficient α
Simulation 1 (Base Case)	2nd Order	47
Simulation 2	1st Order	47
Simulation 3	2nd Order	12

RESULTS AND DISCUSSIONS

Simulation 1: Base Case

After simulating the base case in both OpenFOAM© and ANSYS, a set of comparative plots for velocity (m/s), temperature (K), concentration (kg_{smoke}/kg_{air}), visibility (m) and time to lose consciousness due to smoke and heat exposure (min) are presented for three different heights of 0.25 m, 0.5 m, 0.75 m, located at the plume centerline. Growth and fully developed regions are the two stages studied in the fire simulation; where the time to reach fully developed one is 14.589 seconds.

As shown in Figure 2, the results of the studied parameters on the plume centerline generated by the smoke source using both programs are in close proximity. The comparison of temperature plots gives similar trends with a maximum difference of 2.78% during the growth period and 3% in the fully developed regime. A slight difference is noticed at the height of 0.25m where the slope of the results acquired from ANSYS is higher than OpenFOAM©. However, at the heights of 0.5m and 0.75m, the difference is mainly in the fully developed regime, where temperature values obtained by ANSYS are higher than the ones generated by OpenFOAM© (not shown due to space constraint).

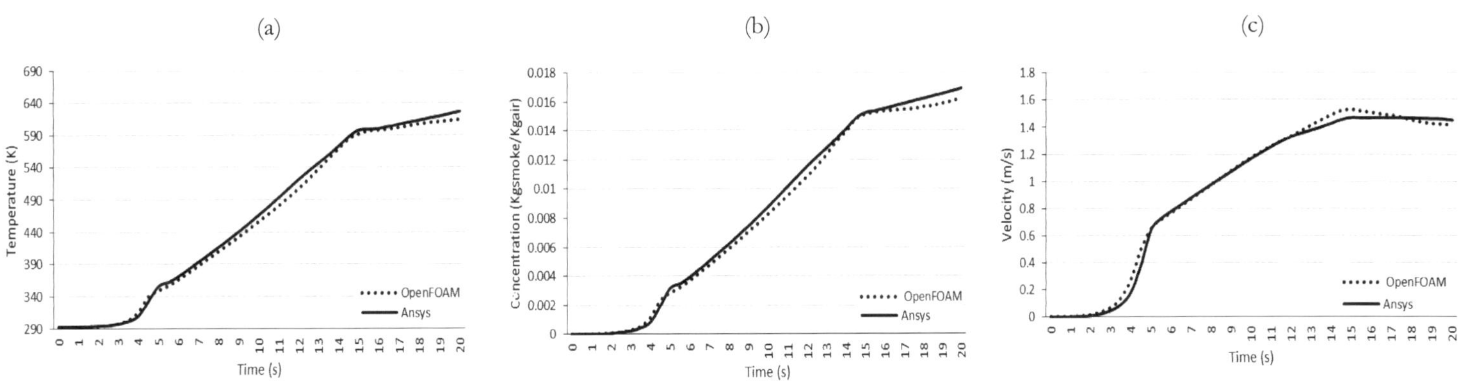

Figure 2. At height 0.25m, the variation of: (a) Temperature, (b) Concentration, and (c) Velocity for simulation 1

Regarding the velocity variation, the outcome of both software simulations are roughly the same with a parallel trend at a height of 0.25m near the smoke source.are in good agreement. On the other hand, at higher heights of 0.5 m and 0.75 m, the calculated velocities using OpenFOAM© are higher than the ones predicted by ANSYS while following the same trend.

In summary, results for temperature, smoke concentration and velocity obtained with both packages using a second order scheme were close. It is important to mention that the growth phase is divided into two main periods, each possessing a different behavior. The first period extends over 5 s showing an exponential increase in the temperature and concentration field at the closest point to the fire location (located at 5 cm from top of the fire source). In this phase, the process is diffusion dominant, where the velocity field is not yet well established in the computational domain. In the second period, both temperature and concentration, being majorly controlled by the convective process, continuously increases, however, at a lower rate as compared to the rate of increase during the first phase period.

As for the effects of fire, as related to visibility and time to lose consciousness due to heat and smoke exposure, the increase in smoke production leads to a reduction in the visibility and to a higher rate in losing consciousness of the occupants in the area. These effects have higher values near the source and they decrease gradually with height. The comparison of the plots for both OpenFOAM© and ANSYS shows that there is minor difference between the values and maintain the same profile at all heights.

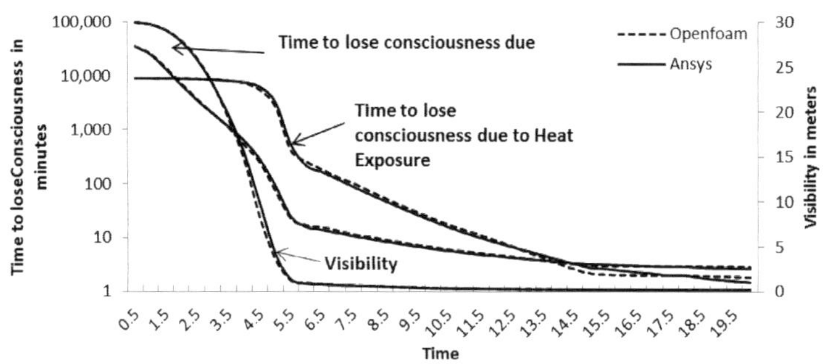

Figure 3. Effects of fire variation at height 0.5m for simulation 1

Simulation 2

For this case, plots for temperature, concentration and velocity at a height of 0.25m obtained from both packages are presented in Figure 3. Temperature and concentration have similar comparative profiles with the highest percentage difference equal to 4.13% and 12.55% respectively. Data from ANSYS has higher slope than OpenFOAM© in the fully developed region. Regarding velocity, the values predicted by OpenFOAM© are higher than the corresponding ones generated by ANSYS Fluent while maintaining approximately the same profile variation. By adopting the 1st order for the divergence scheme, the error increases of the numerical discretization resulting in less accurate values than the ones obtained by high order schemes.

(a) (b) (c)

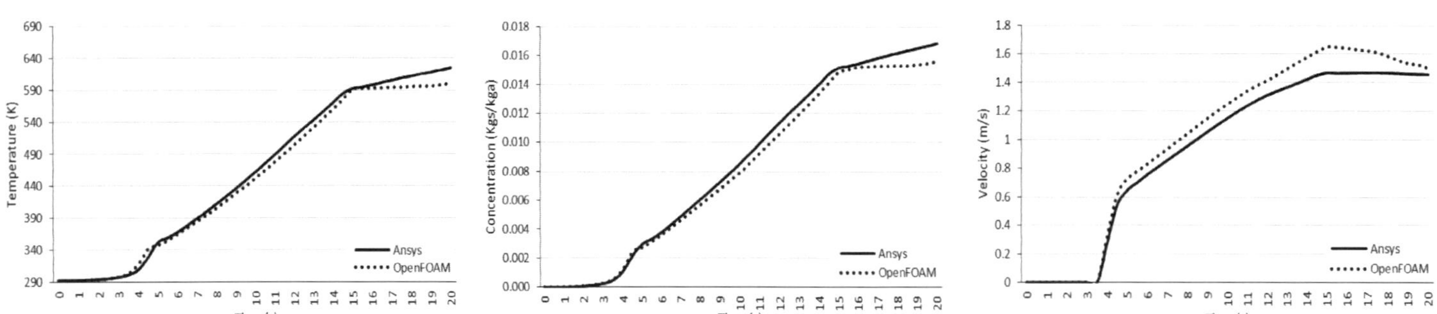

Figure 3. At height 0.25m, the variation of: (a) Temperature, (b) Concentration, and (c) Velocity for simulation 2

Simulation 3

This case investigates the effects of the heat release rate on fire. For this purpose, the fire growth coefficient is reduced to 12 kW/s² to simulate a medium type fire. The first 20 seconds of the growth period are simulated using both software packages. Results show similar trends with the maximum relative differences in temperature and concentration values being 2 % and 8.9 %, respectively. This indicates that even with different types of fire, simulations using OpenFOAM© are expected to give appropriate results. The comparison results obtained in simulations 1 & 3 reveals as expected that both temperature and concentration reach higher values in case 1. In addition, the growth region in the medium fire takes more time to reach the fully developed one. For this reason, fire engineers should carefully assess the studied building to identify the different types of burned materials and the different scenarios expected to occur.

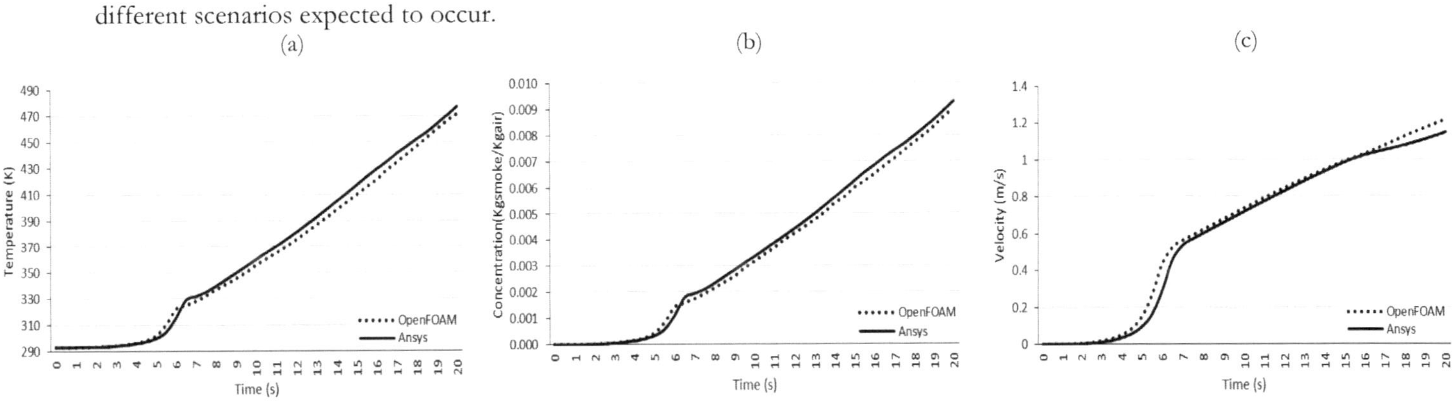

Figure 4. At height 0.25m, the variation of: (a) Temperature, (b) Concentration, and (c) Velocity for simulation 3

CONCLUSION

A special solver "SmokeFoam" capable of simulating smoke inside buildings was developed using the open source platform "OpenFOAM©". By means of this code, one can generate, after implementing the appropriate inputs, values of concentration and temperature of the smoke. These values can be used to calculate the visibility and time to lose consciousness from smoke and heat in order to produce a safe design of the studied building. The results obtained were compared to a commercial and reliable software called ANSYS Fluent to validate its accuracy. The difference in results between the two software is minor and acceptable; this indicates that the OpenFOAM© code can be used in new designs and complex structures conducted by engineers to simulate smoke in case of fire. In addition, further improvement may be added to the code in order to expand its capabilities and enhance its efficiency.

ACKNOWLEDGMENTS

The financial support by the University Research Board (URB) of the American University of Beirut is gratefully acknowledged.

NOMENCLATURE

g = gravity acceleration, [m/s^2]
τ = stress tensor, [Pa]
υ = kinematic viscosity, [m^2/s]
β = thermal expansion coefficient, [1/K]
T = temperature, [K]
$\dot{Q}$ = heat release rate in kW
α = fire growth coefficient in kW/s^2
t = time, [s]

A = horizontal area, [m^2]
ΔH = heat of combustion, [kJ/kg]
S = Visibilty through smoke, [m]
K = proportionality constant from Table 2
α_m = specific extinction, [m^2/kg]
m = mass concentration, [kg/m^3]
C = concentration, [kg_{smoke}/kg_{air}]

Subscripts

t = turbulent

eff = effective

0	=	reference	*f*	=	fuel
"	=	per unit area	*v*	=	ventilation control fire

REFERENCES

B. Karlsson and J. G. Quintiere, Enclosure fire dynamics. Boca Raton, Florida: CRC Press, 2000, pp. 17-18, 39-43, 100, 121, 125, 129, 228-238, 244-253.

B.S. Kandola and M. Morris, "Smoke hazard assessment using computational fluid dynamics(CFD) modeling", Institution of Chemical Engineers Symposium Series, No. 134, pp. 171-194 (1994).

British Standards Institution, BS 7974:2001, Application of fire safety engineering principles to the design of building - Code of practice. London: British Standards Institution, 2001, pp. 10, 15, 19-20.

Cox, G. Chitty, R and Kumar, S. (1989). Fire Modeling and the King's Cross Fire Investigation. *Fire Safety J.* , Vol 15, pp. 103-106.

D.J. den Boer, J.J. Mertens, N.J. van Oerle and J.F. Wijnia, "Validation of CFD calculations of full scale medium sized fires in a two lane road tunnel", 11th International Symposium on Aerodynamics and Ventilation of Vehicle Tunnels, Vol. 2, pp. 657-667 (2003).

European Standard, "Eurocode 1: Actions on structures - Part 1-2: General actions - Actions on structures exposed to fires," EN 1991-1-2, 2002.

Hall, J.R., Jr., "Brief History of Home Smoke Alarms(Abridged)," Proceedings of the Research and Practice: Bridging the Gap Fire Suppression and Detection Research Application Symposium. Orlando, FL. February 7-9, 2001, National Fire Protection Research Foundation , Quincy, MA, 2001, pp.258-281.

J. Schabacker et al. (n.d.). *CFD Study of Temperature and Smoke Distribution in a Railway Tunnel with Natural Ventilation System.* Switzerland: HBI Haerter AG.

Klote, J.H. and Milke, J.A. (2002). Principles of Smoke Management. American Society of Heating, Refrigerating and Air-Conditioning Engineers & Society of Fire Protection Engineers, Atlanta.

Launder, B.E. and spalding, D.B., (1974), "The numerical computation of turbulent flows", Computer Methods in Applied Mechanics and Engineering, 3(2): 269-289.

Moghtaderi, B., Novozhilov, V., Fletcher, D., and Kent, J., A new correlation of bench-scale piloted ignition data of wood. *Fire Safety Journal* 1997, *29*, 41-59.

M. Tabarra, B. Kenrick and R.D. Matthews, "CFD validation of natural smoke movement in a model tunnel", American Society of Mechanical Engineers, Fluids Engineering Division (Publication) FED, Vol. 238, No. 3, pp. 543-546 (1996).

M.P. Deng, P.C. Miclea and D. McKinney, "CFD modeling considerations for train fires in underground subway stations", American Society of Mechanical Engineers, Fluids Engineering Division (Publication) FED, Vol. 238, No. 3, pp. 547-555 (1996).

National Board of Housing, *Building and Planning General Council 2011 – Guidance on analytical design of fire protection of buildings – Refferral. Karlskrona*: National Board of Housing, 2010.

National Fire Protection Association. (2014). Retrieved September 10, 2014, from *National Fire Protection Association website*: http://www.nfpa.org/press-room/reporters-guide-to-fire-and-nfpa/consequences-of-fire.

Novozhilov, V. (2001). Computational Fluid Dynamics Modeling of Compartment Fire, Prog. Energy Combust. Sci., Vol. 27, pp. 611-666.

Society of Fire Protection Engineers, *The SFPE engineering guide to performance-based fire protection analysis and design,* 1st ed. Quincy, Massachusetts: National Fire Protection Association, 2000, pp. 14-18, 23-24, 27-29, 31-36, 41, 58.

Simcox, S., Wilkes, N.S. and Jones, I.P. (1992). Computer Simulation of the Flows of Hot Gases from the Fire at King's Cross Underground Station, *Fire Safety J.*,Vol. 18, pp 49-73.

T. Hertzberg, B. Sundström, and P. van Hees, "Design fires for enclosures - A first attempt to create design fires based on Euroclasses for linings," SP Fire Technology, *SP Swedish National Testing and Research Institute*, Borås, SP Report 2003:02, 2003, pp. 10-11.

Y.L. Sinai, M.P. Owens and P. Smith, "Advances in CFD assessment of fire and smoke movement", Proceedings of the International Conference on Offshore Mechanics and Arctic Engineering - OMAE, Vol. 2, pp. 389-395 (1995).

Yang, K.T. and Chang, J. C. (1977). UNDSAFE-I. A Computer Code for Buoyant Flow in an Enclosure, University of Notre Dame Technical Report TR 79002-77-1.

AUTHOR INDEX

American Society of Heating, Refrigerating and Air-Conditioning Engineers (ASHRAE)
1791 Tullie Circle
N.E. Atlanta, GA 30329
USA

ISBN 978-1-5108-3708-9

9 781510 837089

CURRAN ASSOCIATES INC.
proceedings.com